Lecture Notes in Computer Science

Lecture Notes in Artificial Intelligence 14650

Founding Editor

Jörg Siekmann

Series Editors

The series Lecture Notes in Artificial Intelligence (LNAI) was established in 1988 as a topical subseries of LNCS devoted to artificial intelligence.

The series publishes state-of-the-art research results at a high level. As with the LNCS mother series, the mission of the series is to serve the international R & D community by providing an invaluable service, mainly focused on the publication of conference and workshop proceedings and postproceedings.

De-Nian Yang · Xing Xie · Vincent S. Tseng ·
Jian Pei · Jen-Wei Huang · Jerry Chun-Wei Lin
Editors

Advances in Knowledge Discovery and Data Mining

28th Pacific-Asia Conference on Knowledge Discovery and Data Mining, PAKDD 2024
Taipei, Taiwan, May 7–10, 2024
Proceedings, Part VI

Editors
De-Nian Yang
Academia Sinica
Taipei, Taiwan

Xing Xie
Microsoft Research Asia
Beijing, China

Vincent S. Tseng
National Yang Ming Chiao Tung University
Hsinchu, Taiwan

Jian Pei
Duke University
Durham, NC, USA

Jen-Wei Huang
National Cheng Kung University
Tainan, Taiwan

Jerry Chun-Wei Lin
Silesian University of Technology
Gliwice, Poland

ISSN 0302-9743 ISSN 1611-3349 (electronic)
Lecture Notes in Artificial Intelligence
ISBN 978-981-97-2265-5 ISBN 978-981-97-2266-2 (eBook)
https://doi.org/10.1007/978-981-97-2266-2

LNCS Sublibrary: SL7 – Artificial Intelligence

This Springer imprint is published by the registered company Springer Nature Singapore Pte Ltd.
The registered company address is: 152 Beach Road, #21-01/04 Gateway East, Singapore 189721, Singapore

Paper in this product is recyclable.

General Chairs' Preface

On behalf of the Organizing Committee, we were delighted to welcome attendees to the 28th Pacific-Asia Conference on Knowledge Discovery and Data Mining (PAKDD 2024). Since its inception in 1997, PAKDD has long established itself as one of the leading international conferences on data mining and knowledge discovery. PAKDD provides an international forum for researchers and industry practitioners to share their new ideas, original research results, and practical development experiences across all areas of Knowledge Discovery and Data Mining (KDD). This year, after its two previous editions in Taipei (2002) and Tainan (2014), PAKDD was held in Taiwan for the third time in the fascinating city of Taipei, during May 7–10, 2024. Moreover, PAKDD 2024 was held as a fully physical conference since the COVID-19 pandemic was contained.

We extend our sincere gratitude to the researchers who submitted their work to the PAKDD 2024 main conference, high-quality tutorials, and workshops on cutting-edge topics. The conference program was further enriched with seven high-quality tutorials and five workshops on cutting-edge topics. We would like to deliver our sincere thanks for their efforts in research, as well as in preparing high-quality presentations. We also express our appreciation to all the collaborators and sponsors for their trust and cooperation. We were honored to have three distinguished keynote speakers joining the conference: Ed H. Chi (Google DeepMind), Vipin Kumar (University of Minnesota), and Huan Liu (Arizona State University), each with high reputations in their respective areas. We enjoyed their participation and talks, which made the conference one of the best academic platforms for knowledge discovery and data mining. We would like to express our sincere gratitude for the contributions of the Steering Committee members, Organizing Committee members, Program Committee members, and anonymous reviewers, led by Program Committee Chairs De-Nian Yang and Xing Xie. It is through their untiring efforts that the conference had an excellent technical program. We are also thankful to the other Organizing Committee members: Workshop Chairs, Chuan-Kang Ting and Xiaoli Li; Tutorial Chairs, Jiun-Long Huang and Philippe Fournier-Viger; Publicity Chairs, Mi-Yen Yeh and Rage Uday Kiran; Industrial Chairs, Kun-Ta Chuang, Wei-Chao Chen and Richie Tsai; Proceedings Chairs, Jen-Wei Huang and Jerry Chun-Wei Lin; Registration Chairs, Chih-Ya Shen and Hong-Han Shuai; Web and Content Chairs, Cheng-Te Li and Shan-Hung Wu; Local Arrangement Chairs, Yi-Ling Chen, Kuan-Ting Lai, Yi-Ting Chen, and Ya-Wen Teng. We feel indebted to the PAKDD Steering Committee for their constant guidance and sponsorship of manuscripts. We are also grateful to the hosting organizations, National Yang Ming Chiao Tung University and Academia Sinica, and all our sponsors for continuously providing institutional and financial support to PAKDD 2024.

May 2024

Vincent S. Tseng
Jian Pei

PC Chairs' Preface

It is our great pleasure to present the 28th Pacific-Asia Conference on Knowledge Discovery and Data Mining (PAKDD 2024) as Program Committee Chairs. PAKDD is one of the longest-established and leading international conferences in the areas of data mining and knowledge discovery. It provides an international forum for researchers and industry practitioners to share their new ideas, original research results, and practical development experiences in all KDD-related areas, including data mining, data warehousing, machine learning, artificial intelligence, databases, statistics, knowledge engineering, big data technologies, and foundations.

This year, PAKDD received a record number of 720 submissions, among which 86 submissions were rejected at a preliminary stage due to policy violations. There were 595 Program Committee members and 101 Senior Program Committee members involved in the double-blind reviewing process. For submissions entering the double-blind review process, each one received at least three quality reviews from PC members. Furthermore, each valid submission received one meta-review from the assigned SPC member, who also led the discussion with the PC members. The PC Co-chairs then considered the recommendations and meta-reviews from SPC members and looked into each submission as well as its reviews and PC discussions to make the final decision.

As a result of the highly competitive selection process, 175 submissions were accepted and recommended to be published, with 133 oral-presentation papers and 42 poster-presentation papers. We would like to thank all SPC and PC members whose diligence produced a high-quality program for PAKDD 2024. The conference program also featured three keynote speeches from distinguished data mining researchers, eight invited industrial talks, five cutting-edge workshops, and seven comprehensive tutorials.

We wish to sincerely thank all SPC members, PC members, and external reviewers for their invaluable efforts in ensuring a timely, fair, and highly effective paper review and selection procedure. We hope that readers of the proceedings will find the PAKDD 2024 technical program both interesting and rewarding.

May 2024

De-Nian Yang
Xing Xie

Organization

Organizing Committee

Honorary Chairs

Philip S. Yu	University of Illinois at Chicago, USA
Ming-Syan Chen	National Taiwan University, Taiwan

General Chairs

Vincent S. Tseng	National Yang Ming Chiao Tung University, Taiwan
Jian Pei	Duke University, USA

Program Committee Chairs

De-Nian Yang	Academia Sinica, Taiwan
Xing Xie	Microsoft Research Asia, China

Workshop Chairs

Chuan-Kang Ting	National Tsing Hua University, Taiwan
Xiaoli Li	A*STAR, Singapore

Tutorial Chairs

Jiun-Long Huang	National Yang Ming Chiao Tung University, Taiwan
Philippe Fournier-Viger	Shenzhen University, China

Publicity Chairs

Mi-Yen Yeh	Academia Sinica, Taiwan
Rage Uday Kiran	University of Aizu, Japan

Industrial Chairs

Kun-Ta Chuang	National Cheng Kung University, Taiwan
Wei-Chao Chen	Inventec Corp./Skywatch Innovation, Taiwan
Richie Tsai	Taiwan AI Academy, Taiwan

Proceedings Chairs

Jen-Wei Huang	National Cheng Kung University, Taiwan
Jerry Chun-Wei Lin	Silesian University of Technology, Poland

Registration Chairs

Chih-Ya Shen	National Tsing Hua University, Taiwan
Hong-Han Shuai	National Yang Ming Chiao Tung University, Taiwan

Web and Content Chairs

Shan-Hung Wu	National Tsing Hua University, Taiwan
Cheng-Te Li	National Cheng Kung University, Taiwan

Local Arrangement Chairs

Yi-Ling Chen	National Taiwan University of Science and Technology, Taiwan
Kuan-Ting Lai	National Taipei University of Technology, Taiwan
Yi-Ting Chen	National Yang Ming Chiao Tung University, Taiwan
Ya-Wen Teng	Academia Sinica, Taiwan

Steering Committee

Chair

Longbing Cao	Macquarie University, Australia

Vice Chair

Gill Dobbie	University of Auckland, New Zealand

Treasurer

Longbing Cao	Macquarie University, Australia

Members

Ramesh Agrawal	Jawaharlal Nehru University, India
Gill Dobbie	University of Auckland, New Zealand
João Gama	University of Porto, Portugal
Zhiguo Gong	University of Macau, Macau SAR
Hisashi Kashima	Kyoto University, Japan
Hady W. Lauw	Singapore Management University, Singapore
Jae-Gil Lee	KAIST, Korea
Dinh Phung	Monash University, Australia
Kyuseok Shim	Seoul National University, Korea
Geoff Webb	Monash University, Australia
Raymond Chi-Wing Wong	Hong Kong University of Science and Technology, Hong Kong SAR
Min-Ling Zhang	Southeast University, China

Life Members

Longbing Cao	Macquarie University, Australia
Ming-Syan Chen	National Taiwan University, Taiwan
David Cheung	University of Hong Kong, China
Joshua Z. Huang	Chinese Academy of Sciences, China
Masaru Kitsuregawa	Tokyo University, Japan
Rao Kotagiri	University of Melbourne, Australia
Ee-Peng Lim	Singapore Management University, Singapore
Huan Liu	Arizona State University, USA
Hiroshi Motoda	AFOSR/AOARD and Osaka University, Japan
Jian Pei	Duke University, USA
P. Krishna Reddy	IIIT Hyderabad, India
Jaideep Srivastava	University of Minnesota, USA
Thanaruk Theeramunkong	Thammasat University, Thailand
Tu-Bao Ho	JAIST, Japan
Vincent S. Tseng	National Yang Ming Chiao Tung University, Taiwan
Takashi Washio	Osaka University, Japan
Kyu-Young Whang	KAIST, Korea
Graham Williams	Australian National University, Australia
Chengqi Zhang	University of Technology Sydney, Australia

Ning Zhong	Maebashi Institute of Technology, Japan
Zhi-Hua Zhou	Nanjing University, China

Past Members

Arbee L. P. Chen	Asia University, Taiwan
Hongjun Lu	Hong Kong University of Science and Technology, Hong Kong SAR
Takao Terano	Tokyo Institute of Technology, Japan

Senior Program Committee

Aijun An	York University, Canada
Aris Anagnostopoulos	Sapienza Università di Roma, Italy
Ting Bai	Beijing University of Posts and Telecommunications, China
Elisa Bertino	Purdue University, USA
Arnab Bhattacharya	IIT Kanpur, India
Albert Bifet	Université Paris-Saclay, France
Ludovico Boratto	Università degli Studi di Cagliari, Italy
Ricardo Campello	University of Southern Denmark, Denmark
Longbing Cao	University of Technology Sydney, Australia
Tru Cao	UTHealth, USA
Tanmoy Chakraborty	IIT Delhi, India
Jeffrey Chan	RMIT University, Australia
Pin-Yu Chen	IBM T. J. Watson Research Center, USA
Bin Cui	Peking University, China
Anirban Dasgupta	IIT Gandhinagar, India
Wei Ding	University of Massachusetts Boston, USA
Eibe Frank	University of Waikato, New Zealand
Chen Gong	Nanjing University of Science and Technology, China
Jingrui He	UIUC, USA
Tzung-Pei Hong	National University of Kaohsiung, Taiwan
Qinghua Hu	Tianjin University, China
Hong Huang	Huazhong University of Science and Technology, China
Jen-Wei Huang	National Cheng Kung University, Taiwan
Tsuyoshi Ide	IBM T. J. Watson Research Center, USA
Xiaowei Jia	University of Pittsburgh, USA
Zhe Jiang	University of Florida, USA

Toshihiro Kamishima	National Institute of Advanced Industrial Science and Technology, Japan
Murat Kantarcioglu	University of Texas at Dallas, USA
Hung-Yu Kao	National Cheng Kung University, Taiwan
Kamalakar Karlapalem	IIIT Hyderabad, India
Anuj Karpatne	Virginia Tech, USA
Hisashi Kashima	Kyoto University, Japan
Sang-Wook Kim	Hanyang University, Korea
Yun Sing Koh	University of Auckland, New Zealand
Hady Lauw	Singapore Management University, Singapore
Byung Suk Lee	University of Vermont, USA
Jae-Gil Lee	KAIST, Korea
Wang-Chien Lee	Pennsylvania State University, USA
Chaozhuo Li	Microsoft Research Asia, China
Gang Li	Deakin University, Australia
Jiuyong Li	University of South Australia, Australia
Jundong Li	University of Virginia, USA
Ming Li	Nanjing University, China
Sheng Li	University of Virginia, USA
Ying Li	AwanTunai, Singapore
Yu-Feng Li	Nanjing University, China
Hao Liao	Shenzhen University, China
Ee-peng Lim	Singapore Management University, Singapore
Jerry Chun-Wei Lin	Silesian University of Technology, Poland
Shou-De Lin	National Taiwan University, Taiwan
Hongyan Liu	Tsinghua University, China
Wei Liu	University of Technology Sydney, Australia
Chang-Tien Lu	Virginia Tech, USA
Yuan Luo	Northwestern University, USA
Wagner Meira Jr.	UFMG, Brazil
Alexandros Ntoulas	University of Athens, Greece
Satoshi Oyama	Nagoya City University, Japan
Guansong Pang	Singapore Management University, Singapore
Panagiotis Papapetrou	Stockholm University, Sweden
Wen-Chih Peng	National Yang Ming Chiao Tung University, Taiwan
Dzung Phan	IBM T. J. Watson Research Center, USA
Uday Rage	University of Aizu, Japan
Rajeev Raman	University of Leicester, UK
P. Krishna Reddy	IIIT Hyderabad, India
Thomas Seidl	LMU München, Germany
Neil Shah	Snap Inc., USA

Yingxia Shao	Beijing University of Posts and Telecommunications, China
Victor S. Sheng	Texas Tech University, USA
Kyuseok Shim	Seoul National University, Korea
Arlei Silva	Rice University, USA
Jaideep Srivastava	University of Minnesota, USA
Masashi Sugiyama	RIKEN/University of Tokyo, Japan
Ju Sun	University of Minnesota, USA
Jiliang Tang	Michigan State University, USA
Hanghang Tong	UIUC, USA
Ranga Raju Vatsavai	North Carolina State University, USA
Hao Wang	Nanyang Technological University, Singapore
Hao Wang	Xidian University, China
Jianyong Wang	Tsinghua University, China
Tim Weninger	University of Notre Dame, USA
Raymond Chi-Wing Wong	Hong Kong University of Science and Technology, Hong Kong SAR
Jia Wu	Macquarie University, Australia
Xindong Wu	Hefei University of Technology, China
Xintao Wu	University of Arkansas, USA
Yiqun Xie	University of Maryland, USA
Yue Xu	Queensland University of Technology, Australia
Lina Yao	University of New South Wales, Australia
Han-Jia Ye	Nanjing University, China
Mi-Yen Yeh	Academia Sinica, Taiwan
Hongzhi Yin	University of Queensland, Australia
Min-Ling Zhang	Southeast University, China
Ping Zhang	Ohio State University, USA
Zhao Zhang	Hefei University of Technology, China
Zhongfei Zhang	Binghamton University, USA
Xiangyu Zhao	City University of Hong Kong, Hong Kong SAR
Yanchang Zhao	CSIRO, Australia
Jiayu Zhou	Michigan State University, USA
Xiao Zhou	Renmin University of China, China
Xiaofang Zhou	Hong Kong University of Science and Technology, Hong Kong SAR
Feida Zhu	Singapore Management University, Singapore
Fuzhen Zhuang	Beihang University, China

Program Committee

Zubin Abraham	Robert Bosch, USA
Pedro Henriques Abreu	CISUC, Portugal
Muhammad Abulaish	South Asian University, India
Bijaya Adhikari	University of Iowa, USA
Karan Aggarwal	Amazon, USA
Chowdhury Farhan Ahmed	University of Dhaka, Bangladesh
Ulrich Aïvodji	ÉTS Montréal, Canada
Esra Akbas	Georgia State University, USA
Shafiq Alam	Massey University Auckland, New Zealand
Giuseppe Albi	Università degli Studi di Pavia, Italy
David Anastasiu	Santa Clara University, USA
Xiang Ao	Chinese Academy of Sciences, China
Elena-Simona Apostol	Uppsala University, Sweden
Sunil Aryal	Deakin University, Australia
Jees Augustine	Microsoft, USA
Konstantin Avrachenkov	Inria, France
Goonmeet Bajaj	Ohio State University, USA
Jean Paul Barddal	PUCPR, Brazil
Srikanta Bedathur	IIT Delhi, India
Sadok Ben Yahia	University of Southern Denmark, Denmark
Alessandro Berti	Università di Pisa, Italy
Siddhartha Bhattacharyya	University of Illinois at Chicago, USA
Ranran Bian	University of Sydney, Australia
Song Bian	Chinese University of Hong Kong, Hong Kong SAR
Giovanni Maria Biancofiore	Politecnico di Bari, Italy
Fernando Bobillo	University of Zaragoza, Spain
Adrian M. P. Brasoveanu	Modul Technology GmbH, Austria
Krisztian Buza	Budapest University of Technology and Economics, Hungary
Luca Cagliero	Politecnico di Torino, Italy
Jean-Paul Calbimonte	University of Applied Sciences and Arts Western Switzerland, Switzerland
K. Selçuk Candan	Arizona State University, USA
Fuyuan Cao	Shanxi University, China
Huiping Cao	New Mexico State University, USA
Jian Cao	Shanghai Jiao Tong University, China
Yan Cao	University of Texas at Dallas, USA
Yang Cao	Hokkaido University, Japan
Yuanjiang Cao	Macquarie University, Australia

Robert Churchill	Georgetown University, USA
Chaoran Cui	Shandong University of Finance and Economics, China
Alfredo Cuzzocrea	Università della Calabria, Italy
Bi-Ru Dai	National Taiwan University of Science and Technology, Taiwan
Honghua Dai	Zhengzhou University, China
Claudia d'Amato	University of Bari, Italy
Chuangyin Dang	City University of Hong Kong, China
Mrinal Das	IIT Palakkad, India
Debanjan Datta	Virginia Tech, USA
Cyril de Runz	Université de Tours, France
Jeremiah Deng	University of Otago, New Zealand
Ke Deng	RMIT University, Australia
Zhaohong Deng	Jiangnan University, China
Anne Denton	North Dakota State University, USA
Shridhar Devamane	KLE Institute of Technology, India
Djellel Difallah	New York University, USA
Ling Ding	Tianjin University, China
Shifei Ding	China University of Mining and Technology, China
Yao-Xiang Ding	Zhejiang University, China
Yifan Ding	University of Notre Dame, USA
Ying Ding	University of Texas at Austin, USA
Lamine Diop	EPITA, France
Nemanja Djuric	Aurora Innovation, USA
Gillian Dobbie	University of Auckland, New Zealand
Josep Domingo-Ferrer	Universitat Rovira i Virgili, Spain
Bo Dong	Amazon, USA
Yushun Dong	University of Virginia, USA
Bo Du	Wuhan University, China
Silin Du	Tsinghua University, China
Jiuding Duan	Allianz Global Investors, Japan
Lei Duan	Sichuan University, China
Walid Durani	LMU München, Germany
Sourav Dutta	Huawei Research Centre, Ireland
Mohamad El-Hajj	MacEwan University, Canada
Ya Ju Fan	Lawrence Livermore National Laboratory, USA
Zipei Fan	Jilin University, China
Majid Farhadloo	University of Minnesota, USA
Fabio Fassetti	Università della Calabria, Italy
Zhiquan Feng	National Cheng Kung University, Taiwan

Len Feremans	Universiteit Antwerpen, Belgium
Edouard Fouché	Karlsruher Institut für Technologie, Germany
Dongqi Fu	UIUC, USA
Yanjie Fu	University of Central Florida, USA
Ken-ichi Fukui	Osaka University, Japan
Matjaž Gams	Jožef Stefan Institute, Slovenia
Amir Gandomi	University of Technology Sydney, Australia
Aryya Gangopadhyay	UMBC, USA
Dashan Gao	Hong Kong University of Science and Technology, China
Wei Gao	Nanjing University, China
Yifeng Gao	University of Texas Rio Grande Valley, USA
Yunjun Gao	Zhejiang University, China
Paolo Garza	Politecnico di Torino, Italy
Chang Ge	University of Minnesota, USA
Xin Geng	Southeast University, China
Flavio Giobergia	Politecnico di Torino, Italy
Rosalba Giugno	Università degli Studi di Verona, Italy
Aris Gkoulalas-Divanis	Merative, USA
Djordje Gligorijevic	Temple University, USA
Daniela Godoy	UNICEN, Argentina
Heitor Gomes	Victoria University of Wellington, New Zealand
Maciej Grzenda	Warsaw University of Technology, Poland
Lei Gu	Nanjing University of Posts and Telecommunications, China
Yong Guan	Iowa State University, USA
Riccardo Guidotti	Università di Pisa, Italy
Ekta Gujral	University of California Riverside, USA
Guimu Guo	Rowan University, USA
Ting Guo	University of Technology Sydney, Australia
Xingzhi Guo	Stony Brook University, USA
Ch. Md. Rakin Haider	Purdue University, USA
Benjamin Halstead	University of Auckland, New Zealand
Jinkun Han	Georgia State University, USA
Lu Han	Nanjing University, China
Yufei Han	Inria, France
Daisuke Hatano	RIKEN, Japan
Kohei Hatano	Kyushu University/RIKEN AIP, Japan
Shogo Hayashi	BizReach, Japan
Erhu He	University of Pittsburgh, USA
Guoliang He	Wuhan University, China
Pengfei He	Michigan State University, USA

Ming Jin	Monash University, Australia
Ruoming Jin	Kent State University, USA
Wei Jin	University of North Texas, USA
Mingxuan Ju	University of Notre Dame, USA
Wei Ju	Peking University, China
Vana Kalogeraki	Athens University of Economics and Business, Greece
Bo Kang	Ghent University, Belgium
Jian Kang	University of Rochester, USA
Ashwin Viswanathan Kannan	Amazon, USA
Tomi Kauppinen	Aalto University School of Science, Finland
Jungeun Kim	Kongju National University, Korea
Kyoung-Sook Kim	National Institute of Advanced Industrial Science and Technology, Japan
Primož Kocbek	University of Maribor, Slovenia
Aritra Konar	Katholieke Universiteit Leuven, Belgium
Youyong Kong	Southeast University, China
Olivera Kotevska	Oak Ridge National Laboratory, USA
P. Radha Krishna	NIT Warangal, India
Adit Krishnan	UIUC, USA
Gokul Krishnan	IIT Madras, India
Peer Kröger	CAU, Germany
Marzena Kryszkiewicz	Warsaw University of Technology, Poland
Chuan-Wei Kuo	National Yang Ming Chiao Tung University, Taiwan
Kuan-Ting Lai	National Taipei University of Technology, Taiwan
Long Lan	NUDT, China
Duc-Trong Le	Vietnam National University, Vietnam
Tuan Le	New Mexico State University, USA
Chul-Ho Lee	Texas State University, USA
Ickjai Lee	James Cook University, Australia
Ki Yong Lee	Sookmyung Women's University, Korea
Ki-Hoon Lee	Kwangwoon University, Korea
Roy Ka-Wei Lee	Singapore University of Technology and Design, Singapore
Yue-Shi Lee	Ming Chuan University, Taiwan
Dino Lenco	INRAE, France
Carson Leung	University of Manitoba, Canada
Boyu Li	University of Technology Sydney, Australia
Chaojie Li	University of New South Wales, Australia
Cheng-Te Li	National Cheng Kung University, Taiwan
Chongshou Li	Southwest Jiaotong University, China

Fengxin Li	Renmin University of China, China
Guozhong Li	King Abdullah University of Science and Technology, Saudi Arabia
Huaxiong Li	Nanjing University, China
Jianxin Li	Beihang University, China
Lei Li	Hong Kong University of Science and Technology (Guangzhou), China
Peipei Li	Hefei University of Technology, China
Qian Li	Curtin University, Australia
Rong-Hua Li	Beijing Institute of Technology, China
Shao-Yuan Li	Nanjing University of Aeronautics and Astronautics, China
Shuai Li	Cambridge University, UK
Shuang Li	Beijing Institute of Technology, China
Tianrui Li	Southwest Jiaotong University, China
Wengen Li	Tongji University, China
Wentao Li	Hong Kong University of Science and Technology (Guangzhou), China
Xin-Ye Li	Bytedance, China
Xiucheng Li	Harbin Institute of Technology, China
Xuelong Li	Northwestern Polytechnical University, China
Yidong Li	Beijing Jiaotong University, China
Yinxiao Li	Meta Platforms, USA
Yuefeng Li	Queensland University of Technology, Australia
Yun Li	Nanjing University of Posts and Telecommunications, China
Panagiotis Liakos	University of Athens, Greece
Xiang Lian	Kent State University, USA
Shen Liang	Université Paris Cité, France
Qing Liao	Harbin Institute of Technology (Shenzhen), China
Sungsu Lim	Chungnam National University, Korea
Dandan Lin	Shenzhen Institute of Computing Sciences, China
Yijun Lin	University of Minnesota, USA
Ying-Jia Lin	National Cheng Kung University, Taiwan
Baodi Liu	China University of Petroleum (East China), China
Chien-Liang Liu	National Yang Ming Chiao Tung University, Taiwan
Guiquan Liu	University of Science and Technology of China, China
Jin Liu	Shanghai Maritime University, China
Jinfei Liu	Emory University, USA
Kunpeng Liu	Portland State University, USA

Maxwell McNeil	University at Albany SUNY, USA
Massimo Melucci	Università degli Studi di Padova, Italy
Alex Memory	Johns Hopkins University, USA
Ernestina Menasalvas	Universidad Politécnica de Madrid, Spain
Xupeng Miao	Carnegie Mellon University, USA
Matej Miheli	University of Zagreb, Croatia
Fan Min	Southwest Petroleum University, China
Jun-Ki Min	Korea University of Technology and Education, Korea
Tsunenori Mine	Kyushu University, Japan
Nguyen Le Minh	JAIST, Japan
Shuichi Miyazawa	Graduate University for Advanced Studies, Japan
Songsong Mo	Nanyang Technological University, Singapore
Jacob Montiel	Amazon, USA
Yang-Sae Moon	Kangwon National University, Korea
Sebastian Moreno	Universidad Adolfo Ibáñez, Chile
Daisuke Moriwaki	CyberAgent, Inc., Japan
Tsuyoshi Murata	Tokyo Institute of Technology, Japan
Charini Nanayakkara	Australian National University, Australia
Mirco Nanni	Consiglio Nazionale delle Ricerche, Italy
Wilfred Ng	Hong Kong University of Science and Technology, Hong Kong SAR
Cam-Tu Nguyen	Nanjing University, China
Canh Hao Nguyen	Kyoto University, Japan
Hoang Long Nguyen	Meharry Medical College, USA
Shiwen Ni	Chinese Academy of Sciences, China
Jian-Yun Nie	Université de Montréal, Canada
Tadashi Nomoto	National Institute of Japanese Literature, Japan
Tim Oates	UMBC, USA
Eduardo Ogasawara	CEFET-RJ, Brazil
Kouzou Ohara	Aoyama Gakuin University, Japan
Kok-Leong Ong	RMIT University, Australia
Riccardo Ortale	Consiglio Nazionale delle Ricerche, Italy
Arindam Pal	CSIRO, Australia
Eliana Pastor	Politecnico di Torino, Italy
Dhaval Patel	IBM T. J. Watson Research Center, USA
Martin Pavlovski	Yahoo Inc., USA
Le Peng	University of Minnesota, USA
Nhan Pham	IBM T. J. Watson Research Center, USA
Thai-Hoang Pham	Ohio State University, USA
Chengzhi Piao	Hong Kong Baptist University, Hong Kong SAR
Marc Plantevit	EPITA, France

Mario Prado-Romero	Gran Sasso Science Institute, Italy
Bardh Prenkaj	Sapienza Università di Roma, Italy
Jianzhong Qi	University of Melbourne, Australia
Buyue Qian	Xi'an Jiaotong University, China
Huajie Qian	Columbia University, USA
Hezhe Qiao	Singapore Management University, Singapore
Biao Qin	Renmin University of China, China
Zengchang Qin	Beihang University, China
Tho Quan	Ho Chi Minh City University of Technology, Vietnam
Miloš Radovanović	University of Novi Sad, Serbia
Thilina Ranbaduge	Australian National University, Australia
Chotirat Ratanamahatana	Chulalongkorn University, Thailand
Chandra Reddy	IBM T. J. Watson Research Center, USA
Ryan Rossi	Adobe Research, USA
Morteza Saberi	University of Technology Sydney, Australia
Akira Sakai	Fujitsu Laboratories Ltd., Japan
David Sánchez	Universitat Rovira i Virgili, Spain
Maria Luisa Sapino	Università degli Studi di Torino, Italy
Hernan Sarmiento	UChile & IMFD, Chile
Badrul Sarwar	CloudAEye, USA
Nader Shakibay Senobari	University of California Riverside, USA
Nasrin Shabani	Macquarie University, Australia
Ankit Sharma	University of Minnesota, USA
Chandra N. Shekar	RGUKT RK Valley, India
Chih-Ya Shen	National Tsing Hua University, Taiwan
Wei Shen	Nankai University, China
Yu Shen	Peking University, China
Zhi-Yu Shen	Nanjing University, China
Chuan Shi	Beijing University of Posts and Telecommunications, China
Yue Shi	Meta Platforms, USA
Zhenwei Shi	Beihang University, China
Motoki Shiga	Tohoku University, Japan
Kijung Shin	KAIST, Korea
Kai Shu	Illinois Institute of Technology, USA
Hong-Han Shuai	National Yang Ming Chiao Tung University, Taiwan
Zeren Shui	University of Minnesota, USA
Satyaki Sikdar	Indiana University, USA
Dan Simovici	University of Massachusetts Boston, USA
Apoorva Singh	IIT Patna, India

Bikash Chandra Singh	Islamic University, Bangladesh
Stavros Sintos	University of Illinois at Chicago, USA
Krishnamoorthy Sivakumar	Washington State University, USA
Andrzej Skowron	University of Warsaw, Poland
Andy Song	RMIT University, Australia
Dongjin Song	University of Connecticut, USA
Arnaud Soulet	Université de Tours, France
Ja-Hwung Su	National University of Kaohsiung, Taiwan
Victor Suciu	University of Wisconsin, USA
Liang Sun	Alibaba Group, USA
Xin Sun	Technische Universität München, Germany
Yuqing Sun	Shandong University, China
Hirofumi Suzuki	Fujitsu Laboratories Ltd., Japan
Anika Tabassum	Oak Ridge National Laboratory, USA
Yasuo Tabei	RIKEN, Japan
Chih-Hua Tai	National Taipei University, Taiwan
Hiroshi Takahashi	NTT, Japan
Atsuhiro Takasu	National Institute of Informatics, Japan
Yanchao Tan	Fuzhou University, China
Chang Tang	China University of Geosciences, China
Lu-An Tang	NEC Laboratories America, USA
Qiang Tang	Luxembourg Institute of Science and Technology, Luxembourg
Yiming Tang	Hefei University of Technology, China
Ying-Peng Tang	Nanjing University of Aeronautics and Astronautics, China
Xiaohui (Daniel) Tao	University of Southern Queensland, Australia
Vahid Taslimitehrani	PhysioSigns Inc., USA
Maguelonne Teisseire	INRAE, France
Ya-Wen Teng	Academia Sinica, Taiwan
Masahiro Terabe	Chugai Pharmaceutical Co. Ltd., Japan
Kia Teymourian	University of Texas at Austin, USA
Qing Tian	Nanjing University of Information Science and Technology, China
Yijun Tian	University of Notre Dame, USA
Maksim Tkachenko	Singapore Management University, Singapore
Yongxin Tong	Beihang University, China
Vicenç Torra	University of Umeå, Sweden
Nhu-Thuat Tran	Singapore Management University, Singapore
Yash Travadi	University of Minnesota, USA
Quoc-Tuan Truong	Amazon, USA

Wei-Yao Wang	National Yang Ming Chiao Tung University, Taiwan
Wendy Hui Wang	Stevens Institute of Technology, USA
Xiao Wang	Beihang University, China
Xiaoyang Wang	University of New South Wales, Australia
Xin Wang	University of Calgary, Canada
Xinyuan Wang	George Mason University, USA
Yanhao Wang	East China Normal University, China
Yuanlong Wang	Ohio State University, USA
Yuping Wang	Xidian University, China
Yuxiang Wang	Hangzhou Dianzi University, China
Hua Wei	Arizona State University, USA
Zhewei Wei	Renmin University of China, China
Yimin Wen	Guilin University of Electronic Technology, China
Brendon Woodford	University of Otago, New Zealand
Cheng-Wei Wu	National Ilan University, Taiwan
Fan Wu	Central South University, China
Fangzhao Wu	Microsoft Research Asia, China
Jiansheng Wu	Nanjing University of Posts and Telecommunications, China
Jin-Hui Wu	Nanjing University, China
Jun Wu	UIUC, USA
Ou Wu	Tianjin University, China
Shan-Hung Wu	National Tsing Hua University, Taiwan
Shu Wu	Chinese Academy of Sciences, China
Wensheng Wu	University of Southern California, USA
Yun-Ang Wu	National Taiwan University, Taiwan
Wenjie Xi	George Mason University, USA
Lingyun Xiang	Changsha University of Science and Technology, China
Ruliang Xiao	Fujian Normal University, China
Yanghua Xiao	Fudan University, China
Sihong Xie	Lehigh University, USA
Zheng Xie	Nanjing University, China
Bo Xiong	Universität Stuttgart, Germany
Haoyi Xiong	Baidu, Inc., China
Bo Xu	Donghua University, China
Bo Xu	Dalian University of Technology, China
Guandong Xu	University of Technology Sydney, Australia
Hongzuo Xu	NUDT, China
Ji Xu	Guizhou University, China

Aoying Zhou	East China Normal University, China
Bing Zhou	Sam Houston State University, USA
Nianjun Zhou	IBM T. J. Watson Research Center, USA
Qinghai Zhou	UIUC, USA
Xiangmin Zhou	RMIT University, Australia
Xiaoping Zhou	Beijing University of Civil Engineering and Architecture, China
Xun Zhou	University of Iowa, USA
Jonathan Zhu	Wheaton College, USA
Ronghang Zhu	University of Georgia, China
Xingquan Zhu	Florida Atlantic University, USA
Ye Zhu	Deakin University, Australia
Yihang Zhu	University of Leicester, UK
Yuanyuan Zhu	Wuhan University, China
Ziwei Zhu	George Mason University, USA

External Reviewers

Zihan Li	University of Massachusetts Boston, USA
Ting Yu	Zhejiang Lab, China

Sponsoring Organizations

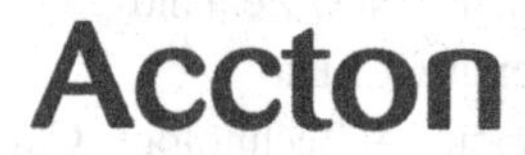

Accton

ACSI

Appier

Chunghwa Telecom Co., Ltd

DOIT, Taipei

ISCOM

Metaage

NSTC

Pegatron

Quanta Computer

TWS

Wavenet Co., Ltd

Contents – Part VI

Time-Series and Streaming Data

Scientific Data

FR^{3}LS: A Forecasting Model with Robust and Reduced Redundancy Latent Series

Abdallah Aaraba[1(✉)], Shengrui Wang[1], and Jean-Marc Patenaude[2]

[1] University of Sherbrooke, Sherbrooke, QC, Canada
{abdallah.aaraba,shengrui.wang}@usherbrooke.ca
[2] Laplace Insights, Sherbrooke, QC, Canada
jeanmarc@laplaceinsights.com

Abstract. While some methods are confined to linear embeddings and others exhibit limited robustness, high-dimensional time series factorization techniques employ scalable matrix factorization for forecasting in latent space. This paper introduces a novel factorization method that employs a non-contrastive approach, guiding an autoencoder-like architecture to extract robust latent series while minimizing redundant information within the embeddings. The resulting learned representations are utilized by a temporal forecasting model, generating forecasts within the latent space, which are subsequently decoded back to the original space through the decoder. Extensive experiments demonstrate that our model achieves state-of-te-art performance on numerous commonly used datasets.

Keywords: Time series factorization · Probabilistic forecasting · Non-contrastive learning

1 Introduction

Modern time series forecasting, involving correlated multivariate time series over an extended period, encounters challenges with conventional methods like autoregressive models (AR, ARIMA) [12], especially when handling large datasets with hundreds of thousands of time series due to scalability issues. Deep learning, exemplified by LSTM [8] and Temporal Convolution Networks (TCN) [1], addresses this by training on the entire dataset, utilizing shared model parameters. However, these deep learning methods inherently struggle with capturing inter-series interactions and correlations observed in diverse domains [24].

A promising research direction explores factorizing time series relationships into a low-rank matrix, yielding a concise latent time series representation [22,27,32]. Temporal Regularized Matrix Factorization (TRMF) [32] achieves this by representing each time series with a linear combination of a few latent series and applying linear temporal regularization to ensure temporal dependencies. Forecasted values in the latent space are transformed back using matrix multiplication. DeepGLO [27] extends TRMF with nonlinear regularization, incorporating iterative training between linear matrix factorization and fitting a latent

D.-N. Yang et al. (Eds.): PAKDD 2024, LNAI 14650, pp. 3–15, 2024.
https://doi.org/10.1007/978-981-97-2266-2_1

space Temporal Convolutional Network (TCN). Another advancement, named Temporal Latent AutoEncoder (TLAE) [22], incorporates a nonlinear framework through an autoencoder to extract the latent time series space. TLAE forecasts within the embedding space, and then transforms forecast to the original space through a decoder. All these factorization methods lack robustness, overlooking random noise and distortions in data that may lead to issues like overfitting.

To address the aforementioned limitations and extend the existing line of factorization research, we introduce the FR^{3}LS forecasting model. Illustrated in Fig. 1, FR^{3}LS nonlinearly projects high-dimensional time series into a latent space with manageable dimensions, facilitating future value forecasting using latent representations. The final predictions are derived by decoding latent forecasts generated by the middle layer forecasting model. This latent prediction approach serves to regularize the embedding space (latent space), capturing temporal dependencies between embeddings. Furthermore, latent representations undergo additional regularization through a non-contrastive objective with only positive samples. This means that the model is trained to produce embeddings robust to distortions applied to input subseries (i.e., augmented context views), while minimizing redundancy between components of the vector embedding. Thus, we present an all-in-one model capable of learning robust representations while maintaining a connection between forecasts in latent and original spaces.

2 Related Work

We focus on recent deep learning approaches beyond traditional methods. Further details on classical methods (e.g., AR and ARIMA) can be found in [2,12,20]. Deep learning methods, encompassing RNNs [23,26], CNNs [1], GNNs (Graph NNs) [4], and Transformers [36], have gained acclaim for their effectiveness in time series forecasting, surpassing classical models like ARIMA and VAR (Vector AR). For instance, TCN [1] uses dilated convolutions to enhance efficiency and predictive performance over traditional RNNs. Models like LSTnet [15] combine CNNs and RNNs to capture short-term local dependencies and long-term trends. Additionally, LogTrans [16] and Informer [36] address self-attention efficiency and excel in forecasting tasks with extended sequences. Furthermore, GNNs such as StemGNN [4], offer competitive results in multivariate time series forecasting by exploring the spectral domain of the data.

Several deep neural network (DNN) models have been proposed for multivariate forecast distributions [6,24,25,31]. A low-rank Gaussian copula model was proposed in [25] using a multi-task univariate LSTM. In [31], a deep factor generative model using a linear combination of RNN latent global factors plus parametric noise was introduced. Normalizing flows for probabilistic forecasting with a multivariate RNN as well as a normalizing flow approach was used in [24]. VRNN was proposed in [6] as a model that uses a variational AE (VAE) in every hidden state of a RNN across the input series. However, such methods suffer from one of the following shortcomings: limited flexibility in modeling distributions in high-dimensional settings [25], only linear combinations of global series and noise distributions are modeled [31], invertible flow needs equal latent and input dimensions [24], and multistep prediction propagation through the whole model is required [6]. Additionally, scaling these methods for high-dimensional multivariate series presents a significant challenge.

3 Problem Setup

Consider a high-dimensional multivariate time series dataset $\mathbb{Y} \in \mathbb{R}^{T \times N}$, represented by $\mathbb{Y}_{1:T} = (y_1, y_2, \ldots, y_T)^T$, where each time point y_t is a vector of dimensionality N (i.e., $y_t \in \mathbb{R}^N$). The objective is to forecast the next τ values $\mathbb{Y}_{T+1:T+\tau} = (y_{T+1}, y_{T+2}, \ldots, y_{T+\tau})^T$ based on the original time series within the training time-range $Y_{1:T}$. The challenging yet intriguing task is to develop a model capable of capturing the conditional probability distribution in the high-dimensional space:

$$p(y_{T+1}, \ldots, y_{T+\tau}|y_{1:T}) = \prod_{i=1}^{\tau} p(y_{T+i}|y_{1:T+i-1}). \tag{1}$$

4 Model Architecture

Following TLAE [22], we introduce an autoencoder-like structure for extracting latent series with temporal regularization in the latent space. The complete FR^{3}LS architecture is illustrated in Fig. 1. Starting with the input $Y \in \mathbb{R}^{w \times N}$, our model is trained to extract meaningful latent series X. A forecasting model is then employed to predict the next values in the latent series. Finally, a decoder is applied to the latent forecasts, producing forecasts in the original space.

The encoder $\mathcal{E}_{\theta_{\mathcal{E}}}(.)$ comprises three components: an Input Projection Layer (IPL), a Timestamp Noising (TN) module, and a Feed Forward neural network (FF), inspired by the work in [33]. The Input Projection Layer consists of a fully connected layer that maps each vector $y_t \in \mathbb{R}^N$ at a timestamp to an intermediate latent vector $z_t \in \mathbb{R}^{d_z}$, with $d_z \in \mathbb{N}^*$. The Timestamp Noising module introduces small noise to selected entries of $Z = (z_1, \ldots, z_w)^T$ at randomly chosen timestamps, generating distorted outputs $\tilde{Z}^{(1)}$ and $\tilde{Z}^{(2)}$. These distortions are applied to Z (intermediary subsequence) instead of directly on raw values Y or latent ones X for enhanced model learning stability. The Feed Forward neural network then projects the intermediate latent vectors $\tilde{Z}^{(j)}, j \in \{1, 2\}$ into the two augmented views $\tilde{X}^{(1)}$ and $\tilde{X}^{(2)} \in \mathbb{R}^{w \times d}$. Subsequently, we apply the Timestamp Mean module to produce the latent series given as input to the forecasting model: $X = (x_1, x_2, \ldots, x_w)^T \in \mathbb{R}^{w \times d}$, where $x_t = \frac{\tilde{x}_t^{(1)} + \tilde{x}_t^{(2)}}{2}$ and $d << N$ is the dimensionality of the latent space. It is worth noting that we opted for two context views for model simplification and computational ease, but users can choose a different number if desired

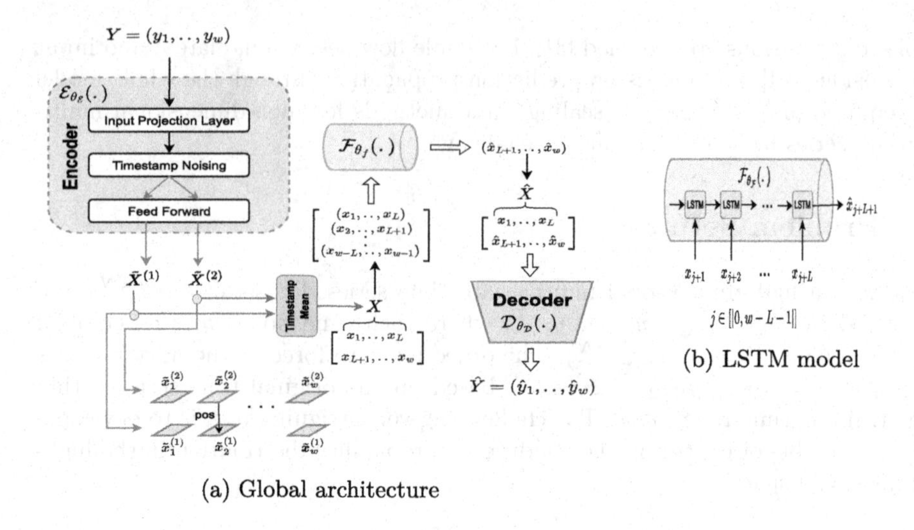

Fig. 1. Model architecture overview.

4.1 Temporal Contextual Consistency

Constructing positive pairs is essential in non-contrastive learning. Following the recommendations of [33], we adopt the temporal contextual consistency paradigm, treating representations at the same timestamp in two augmented contexts as positive pairs to avoid the generation of false positives. We create a context by applying timestamp (light) noising to the intermediary subsequence Z. This approach leverages the fact that timestamp light noising does not alter the magnitude of the time series. Moreover, it encourages the model to learn robust representations at different timestamps capable of reconstructing themselves in distinct contexts.

Timestamp Noising generates an augmented context view for an input series by randomly introducing noise to some of its timestamps. Specifically, it adds small noise to the latent vectors $z_t \in \mathbb{R}^{d_z}$ obtained immediately after the application of the Input Projection Layer, defined as $\tilde{z}_t = z_t + b_t\epsilon_t$ along the time axis. Here, $b_t \in \{0, 1\}$ is a random variable drawn from a Bernoulli distribution with a probability of $p = 0.5$ (i.e., $b_t \sim \mathcal{B}(0.5)$), $\epsilon_t \sim \mathcal{N}(0, 1)$, and both random variables b_t and z_t are independently sampled in every forward pass.

4.2 Non-contrastive Representations Learning

As self-supervised learning (SSL) has demonstrated significant advancements in enabling models to acquire meaningful representations across various domains, Ts2Vec [33] incorporated the concept of contrastive learning to learn time series representations. However, this learning approach is susceptible to selecting false negatives in series with homogeneous distributions, where $p(y_t) \simeq p(y_{t+l})$ for some small integer l. To address this issue, we propose a *non-contrastive* strategy, aiming to encourage the model to learn exclusively from positive samples. This strategy has shown substantial potential in previous works, such as [5,10,34].

As depicted in Fig. 1a, after generating the augmented views $\tilde{X}^{(1)}, \tilde{X}^{(2)}$ by applying the timestamp noising module followed by the feed forward module, we treat representations at the same timestamp from the two views as positive samples. The *Barlow Twins* loss ($\mathcal{L}_{\mathcal{BT}}$) [34] serves as the loss function describing the non-contrastive error of the model, defined by:

$$\mathcal{L}_{\mathcal{BT}} \triangleq \sum_{i=1}^{d}(1 - \mathcal{C}_{ii})^2 + \lambda_{NC} \sum_{i=1}^{d} \sum_{\substack{j=1 \\ i \neq j}}^{d} \mathcal{C}_{ij}^2, \tag{2}$$

where $\lambda_{NC} > 0$ and $\mathcal{C}$ is the cross-correlation matrix computed between the two augmented views along the timestamps (batch) dimension:

$$\mathcal{C}_{ij} \triangleq \frac{\sum_t \tilde{x}_{t,i}^{(1)} \tilde{x}_{t,j}^{(2)}}{\sqrt{\sum_t (\tilde{x}_{t,i}^{(1)})^2} \sqrt{\sum_t (\tilde{x}_{t,j}^{(2)})^2}}. \tag{3}$$

This loss function encourages the cross-correlation matrix between embedded outputs to be as close to the identity matrix as possible. Specifically, we aim to equate the diagonal elements to 1, promoting invariance to distortions applied, and the off-diagonal elements to 0, thereby decorrelating different vector components of the embedding and reducing redundant information in the embeddings.

4.3 Deterministic Forecasting

When the latent representation X effectively captures the information in Y, tasks such as forecasting in the original space can be efficiently performed within the much smaller latent space. To this end, we introduce a layer between the encoder and decoder to extract the temporal structure of the latent representations while enforcing forecasting abilities. The central idea is illustrated in Fig. 1b: a forecasting model $\mathcal{F}_{\theta_{\mathcal{F}}}(.)$, such as LSTM [11], is employed in the middle layer to capture the long-range dependencies of the embeddings.

The latent matrix $X = (x_1, \ldots, x_w)$ is divided into two subseries: $X_{1:L} = (x_1, \ldots, x_L)$ and $X_{L+1:w} = (x_{L+1}, \ldots, x_w)$. During the training phase, the forecasting model is utilized to estimate the second subsequence $\hat{X}_{L+1:w}$. Subsequences of length $L < w$ denoted by the set $\{(x_{j+1}, \ldots, x_{j+L}) | j \in [|0, w-L-1|]\}$ ($[|a, b|]$ denotes a closed integer interval) serve as inputs to the forecasting model, producing latent forecasts $\hat{X}_{L+1:w} = \left(\hat{x}_{j+L+1} = \mathcal{F}_{\theta_{\mathcal{F}}}((x_{j+1}, ..., x_{j+L}))\right)_{j=0}^{w-L-1}$. The forecasting model is then trained using the deterministic loss $\mathcal{L}\mathcal{F}_{\mathcal{D}}$ with a ℓ_q norm, described as:

$$\mathcal{L}_{\mathcal{F}_{\mathcal{D}}} \triangleq \frac{1}{d(w-L)} \sum_{j=0}^{w-l-1} \| x_{j+L+1} - \mathcal{F}_{\theta_{\mathcal{F}}}((x_{j+1}, ..., x_{j+L})) \|_{\ell_q}^q. \tag{4}$$

4.4 Probabilistic Forecasting

In high-dimensional settings, probabilistic modeling of forecasts conditioned on observed ones, i.e., $p(y_{T+1}, \dots, y_{T+\tau}|y_{1:T})$, poses a significant challenge. Previous research has predominantly focused on either modeling each individual time series independently or considering the joint distribution as Gaussian. However, the former approach neglects inter-series interactions, while the latter suffers from a quadratic increase in the number of learned parameters with the data dimension.

Once again, in the context of probabilistic modeling, we advocate for the non-linear encoding of input data into a significantly lower-dimensional space [22]. Assuming that the encoder function $\mathcal{E}_{\theta_\mathcal{E}}$ is sufficiently trained to be considered a one-to-one function, we can then associate the probability of the latent series X with that of the original series Y ($p(x) = p(y)$). Subsequently, we could incorporate a fairly simple probabilistic structure, such as a Gaussian distribution, in the latent space and still be able to model complex distributions of multivariate data through the decoder mapping:

$$p(x_{i+1}|x_{1:i}) = \mathcal{N}(x_{i+1}; \mu_i, 1), \quad i \in [|L, w|]. \tag{5}$$

Here, we identify the conditional distribution as a multivariate Gaussian, with the identity matrix as the covariance matrix to guide the embeddings in capturing different orthogonal patterns in the data. The mean μ_i is computed using the function $\mathcal{F}_{\theta_\mathcal{F}}$ as $\mu_i = \mathcal{F}_{\theta_\mathcal{F}}(x_1, ..., x_i)$. We employ the reparameterization trick [14] to generate latent forecasts needed for backpropagation through input data. In other words, we estimate the value of x_{i+1} using $\hat{x}_{i+1} = \mu_i + 1\epsilon = \mathcal{F}_{\theta_\mathcal{F}}(x_1, ..., x_i) + 1\epsilon = \mathcal{R} \circ \mathcal{F}_{\theta_\mathcal{F}}(x_1, ..., x_i)$, where $\epsilon \sim \mathcal{N}(0, 1)$, and $\mathcal{R}(.)$ describes the reparameterization trick function, such that $\mathcal{R}(x) = x + 1\epsilon$. Similar to the deterministic setting, the forecasting loss function is defined as $\mathcal{L}_{\mathcal{F}_\mathcal{P}}$ using a ℓ_q norm as:

$$\mathcal{L}_{\mathcal{F}_\mathcal{P}} \triangleq \frac{1}{d(w-L)} \sum_{j=0}^{w-l-1} \|\hat{x}_{j+L+1} - x_{j+L+1}\|_{\ell_q}^q. \tag{6}$$

4.5 End-to-End Training

After producing elements $\hat{X}_{L+1:w}$, the decoder takes the matrix $\hat{X} = (X_{1:L}; \hat{X}_{L+1:w}) \in \mathbb{R}^{w \times d}$ as input and generates the matrix $\hat{Y} = (\hat{y}_1, \hat{y}_2, \dots, \hat{y}_w) \in \mathbb{R}^{w \times N}$ as $\hat{Y} = \mathcal{D}_{\theta_\mathcal{D}}(\hat{X})$. Consequently, the output $\hat{Y}$ comprises two components: the first consists of elements $\hat{y}_i$, with $i \in [|1, L|]$, decoded directly from the encoder output without passing through the middle layer, defined as $\hat{y}_i = \mathcal{D}_{\theta_\mathcal{D}} \circ \mathcal{E}_{\theta_\mathcal{E}}(y_i)$, whereas the forecasting model is involved in decoding the second part $\hat{y}_i = \mathcal{D}_{\theta_\mathcal{D}} \circ \mathcal{R}^* \circ \mathcal{F}_{\theta_\mathcal{F}} \circ \mathcal{E}_{\theta_\mathcal{E}}((y_{i-L+1}, \dots, y_i))$, with $i \in [|L+1, w|]$, and $\mathcal{R}^*(.) \triangleq Identity(.)$ in the point estimate problem or $\mathcal{R}^*(.) \triangleq \mathcal{R}(.)$ otherwise.

Minimizing the error $\mathcal{L}_{\mathcal{AE}} \triangleq \frac{1}{Nw}\|\hat{Y} - Y\|_{\ell_p}^p$ could then be thought of as enabling latent representations to have predictive abilities while also being capable of faithfully reconstructing data. The objective function for a batch Y is defined as:

$$\mathcal{L}_Y(\theta_\mathcal{E}, \theta_\mathcal{F}, \theta_\mathcal{D}) \triangleq \lambda_{AE}\mathcal{L}_{A\mathcal{E}} + \lambda_F \mathcal{L}_\mathcal{F} + \lambda_{BT}\mathcal{L}_{BT}, \tag{7}$$

where $\lambda_{AE}, \lambda_F, \lambda_{BT} \in \mathbb{R}^+$ are positive constants, and $\mathcal{L}_\mathcal{F} \triangleq \mathcal{L}_{\mathcal{F}_\mathcal{D}}$ for the deterministic case or $\mathcal{L}_\mathcal{F} \triangleq \mathcal{L}_{\mathcal{F}_\mathcal{P}}$ otherwise.

Once the model is trained, several steps ahead forecasting is performed using rolling windows. Given the past input data $(y_{T-L+1}, \ldots, y_T)$, the trained model constructs the latent prediction $\hat{x}_{T+1} = \mathcal{R}^* \circ \mathcal{F}_{\theta_\mathcal{F}}\big((x_{T-L+1}, \ldots, x_T)\big)$. Subsequently, the predicted point $\hat{y}_{T+1}$ is decoded from $\hat{x}_{T+1}$. This operation can be repeated τ times to predict τ future points of the input time series $\mathbb{Y}$, where we produce the latent prediction $\hat{x}_{T+2}$ by providing the subsequence $(x_{T-L}, \ldots, \hat{x}_{T+1})$ as input to the forecasting model.

5 Experiments

Deterministic Forecasting Experimental Setup: For point estimation, we conduct a comparative analysis with state-of-the-art multivariate and univariate forecasting methods, following the approach in [27] and [32]. Our evaluation employs three popular datasets: *electricity* [30]: hourly consumption of 370 houses, *traffic* [7]: hourly traffic on 963 car lanes in San Francisco, and *wiki* [18]: daily web traffic of about 115k Wikipedia articles. We conduct rolling forecasting with 24 time points per window, reserving the last 7 windows for testing in both the traffic and electricity datasets, and 14 points per window with the last 4 windows for testing in the wiki dataset. Evaluation metrics include mean absolute percent error (MAPE), symmetric MAPE (SMAPE), and weighted average percentage error (WAPE) as in [27].

The model architecture and optimization setup align with TLAE, featuring a bottleneck feed-forward network with RELU nonlinearity functions on all layers except the last ones of both the encoder and decoder modules. The dimensions of the layers vary according to the dataset. In the latent space, a 4-layer LSTM network is employed, with 32 hidden units for traffic and wiki datasets, and 64 for electricity, following the recommendations of [22]. The ℓ_1 loss is used in the $\mathcal{L}_{AE}$ loss, and the ℓ_2 loss is used in the $\mathcal{L}_{\mathcal{F}_\mathcal{D}}$ loss. Regularization parameters λ_{AE}, λ_F, λ_{BT} are all set to 1, and λ_{NC} is set to 0.005 as suggested in [34]. Additional setup and training details are provided in Tables 1a and 1b as well as Sub-sect. 5.3.

Probabilistic Forecasting Experimental Setup: for the analysis of probabilistic estimation, we introduce two additional datasets: ***solar***: hourly photovoltaic production data from 137 stations used in [15], and ***taxi***: New York taxi rides taken every 30 min from 1214 locations [28]. Our evaluation compares the performance of our model against state-of-the-art probabilistic multivariate methods introduced in [22,25,31], as well as univariate forecasting methods [16,23,26], all utilizing the same data setup. It's essential to note that the data

Table 1. Statistics and network architectures of datasets.

(a) Statistics of datasets used in both deterministic and probabilistic forecasting experiments.

Dataset	Time Steps T	Dimension N	Predicted Steps τ	Rolling Window k	Frequency
Traffic	10392	963	24	7	hourly
Electricity (large)	25920	370	24	7	hourly
Electricity (small)	5833	370	24	7	hourly
Wiki (large)	635	115084	14	4	daily
Wiki (small)	792	2000	30	5	daily
Solar	7009	137	24	7	hourly
Taxi	1488	1214	24	56	30-min

(b) Network architecture per dataset. Encoder dims = number neurons/layer of the encoder.

Dataset	Encoder dims	LSTM layers	LSTM hidden dim	sequence length L
Traffic	[256, 128, 64]	4	32	32
Electricity (large)	[256, 128, 64]	4	64	32
Electricity (small)	[128, 64]	4	64	32
Wiki (large)	[256, 128, 64]	4	32	16
Wiki (small)	[128, 64]	4	32	64
Solar	[256, 128, 64]	4	32	32
Taxi	[256, 128, 64]	4	32	112

processing and splits utilized in this analysis differ from those of point forecasting. We maintain an identical network architecture as in our previous experimental setup and use the same values for regularization parameters, as well as ℓ_2 loss for $\mathcal{L}_{\mathcal{F}_\mathcal{P}}$.

To assess the quality of our probabilistic estimates, we employ two distinct error metrics: the first metric is the Continuous Ranked Probability Score across Summed time series (CRPS-Sum) [9,19,22,25], which measures the overall fit of the joint distribution pattern. The second metric is the mean square error (MSE), which assesses the fit of the joint distribution central tendency. Together, these two metrics provide a comprehensive evaluation of the precision of our predictive distribution fit.

5.1 Experimental Results

Table 2a presents a comparison of various deterministic prediction approaches. Results for all models, except the FR^{3}LS model, were originally reported in [22] under the same experimental setup. Here we do not compare our model with classic methods such as VAR, ARIMA etc., as it has already been shown that they obtain performance inferior to TLAE, TRMF and DeepAR methods ([22, 26,32]). Global models leverage global features for multivariate forecasting, while local models employ univariate models to predict individual series separately.

In Table 2b, we display the error scores comparison for probabilistic algorithms. Most results are drawn from Table 2 of [22], with our FR^{3}LS results provided at the end. Conventional statistical multivariate techniques, such as VAR and GARCH ([2,17]), and Vec-LSTM methods, which use a single global LSTM to process and predict all series simultaneously, are included. Additionally, GP methods encompass DNN Gaussian process techniques proposed in [25], with GP-Copula being the primary approach. Further details can be found in [25].

As seen in Table 2a, our method outperforms other global factorization methods in 8 out of 9 dataset-metric combinations. Compared to TLAE, our method achieves an average gain of up to 15% in traffic performance and 13.4% in electricity. Furthermore, compared to other methods, we observe a gain of up to 50% in performance in both the traffic and electricity datasets. For probabilistic forecasting, as shown in Table 2b, our proposed model demonstrates superior performance in the majority of dataset-metric combinations (7 out of 10), with significant gains observed in the Solar, Traffic, and Taxi datasets.

The improvement of the results seen in Tables 2a and 2b over our direct competitor model, TLAE, in both deterministic and probabilistic settings, can be attributed to the enhancement of the latent representations learned by the model. Indeed, enabling the model to be robust against distortions applied to the embeddings enhances its stability in capturing the underlying latent series with predictive power for the time series at hand. Moreover, constraining the model to have embeddings with decorrelated vector components, that is, minimizing the $\mathcal{C}_{ij}^2$ terms in (2), effectively reduces the redundancy of the latent representations. This, in turn, aids the model in focusing on learning latent series that are well-distributed in the latent space.

It is important to note that we utilized LSTM as both the forecasting model and a standard feed-forward autoencoder architecture. We did not employ more advanced models such as TCNs and N-Beats, which could potentially lead to further enhancements. Moreover, in the deterministic prediction case, our model did not use additional local modeling or exogenous features, in contrast to local and combined methods, yet achieved superior performance on 2 out of 3 datasets across all metrics. Finally, we emphasize that our model does not require any further retraining during the testing phase.

5.2 Visualization of Latent and Original Series Forecasts

Figure 2a illustrates the dynamics of trained latent variables and their predictions on the traffic dataset. The blue curve represents the original latent series, while the orange curve depicts their mean predictions. The light-shaded gray area signifies the 90% prediction interval. For each of the 168 predicted timestamps (7×24), 1000 prediction samples were generated. The figure demonstrates that latent variables possess the ability to capture global trends in individual time series. Despite having unique local properties, these latent variables exhibit similar global repeating patterns.

Additionally, Fig. 2b showcases a selection of real-time series variables from the traffic dataset alongside their corresponding predictions. The original time series, Y, is depicted in blue, while the predicted time series, $\hat{Y}$, is shown in orange. The light-shaded gray area represents the 90% prediction interval. The predictive power of the latent representations in FR^{3}LS enables the model to accurately capture the overall pattern of the original time series. Furthermore, the model performs well in predicting the local variability associated with individual time series.

5.3 Further Experimental Setup Details

To train the model, we use the Adam optimizer [13] with a learning rate of 0.0001, commonly recommended for stability. Our observations indicate that higher learning rates often lead to unstable performance. To prevent exploding gradients and stabilize model training, we employ gradient clipping. This

Table 2. Comparison of different forecasting algorithms.

(a) Deterministic algorithms comparison in terms of WAPE/MAPE/SMAPE metrics. Best global factorisation results are indicated in bold, best overall performance with *.

Model	Algorithm	Datasets		
		Traffic	Electricity	Wiki
Global factorisation	**FR³LS** (proposed method)	**0.102*/0.116*/0.090***	**0.071*/0.127*/0.105***	**0.290**/0.463 /**0.380**
	TLAE [22]	0.117/0.137/0.108	0.080/0.152/0.120	0.334/**0.447**/0.434
	DeepGLO-TCN-MF [27]	0.226/0.284/0.247	0.106/0.525/0.188	0.433/1.59/0.686
	TRMF [32]	0.159/0.226/0.181	0.104/0.280/0.151	0.309/0.847/0.451
	SVD+TCN	0.329/0.687/0.340	0.219/0.437/0.238	0.639/2.000/0.893
Local & combined	DeepGLO-combined [27]	0.148/0.168/0.142	0.082/0.341/0.121	0.237/0.441/0.395
	LSTM [11]	0.270/0.357/0.263	0.109/0.264/0.154	0.789/0.686/0.493
	DeepAR [26]	0.140/0.201/0.114	0.086/0.259/0.141	0.429/2.980/0.424
	TCN (no LeveldInit) [1]	0.204/0.284/0.236	0.147/0.476/0.156	0.511/0.884/0.509
	TCN (LeveldInit) [1]	0.157/0.201/0.156	0.092/0.237/0.126	0.212*/0.316*/0.296*
	Prophet [29]	0.303/0.559/0.403	0.197/0.393/0.221	-

(b) Probabilistic comparison in terms of CRPS-Sum/MSE metrics. Lower scores indicate better results. A '–' denotes a method failed (e.g., due to the lack of scalability).

Algorithm	Solar	Electricity-small	Traffic	Taxi	Wiki-small
VAR	0.524/7.0e3	0.031/1.2e7	0.144/5.1e–3	0.292/–	3.400/–
GARCH	0.869/3.5e3	0.278/1.2e6	0.368/3.3e–3	–/–	–/–
Vec-LSTM-ind	0.470/9.9e2	0.731/2.6e7	0.110/6.5e–4	0.429/5.2e1	0.801/5.2e7
Vec-LSTM-ind-scaling	0.391/9.3e2	0.025/2.1e5	0.087/6.3e–4	0.506/7.3e1	0.113/7.2e7
Vec-LSTM-fullrank	0.956/3.8e3	0.999/2.7e7	–/–	–/–	–/–
Vec-LSTM-fullrank-scaling	0.920 /3.8e3	0.747/3.2e7	–/–	–/–	–/–
Vec-LSTM-lowrank-Copula	0.319/2.9e3	0.064/5.5e6	0.103/1.5e–3	0.4326/5.1e1	0.241/3.8e7
LSTM-GP [25]	0.828/3.7e3	0.947/2.7e7	2.198/5.1e–1	0.425/5.9e1	0.933/5.4e7
LSTM-GP-scaling [25]	0.368/1.1e3	**0.022**/1.8e5	0.079/5.2e–4	0.183/2.7e1	1.483/5.5e7
LSTM-GP-Copula [25]	0.337/9.8e2	0.024/2.4e5	0.078/6.9e–4	0.208/3.1e1	**0.086**/4.0e7
VRNN [6]	0.133/7.3e2	0.051/2.7e5	0.181/8.7e–4	0.139/3.0e1	0.396/4.5e7
TLAE [22]	0.124/6.8e2	0.040/2.0e5	0.069/4.4e–4	0.130/2.6e1	0.241/**3.8e7**
FR³LS (proposed method)	**0.091/3.5e2**	0.038/**1.4e5**	**0.056/3.7e–4**	**0.123/2.5e1**	0.244/3.9e7

technique limits the magnitude of gradients during backpropagation [21]. Furthermore, we implement an adaptive learning rate scheduling strategy to control the training optimization process's convergence while enhancing stability [3].

We adhere to the recommendation of TLAE [22] for setting the subsequence lengths L and w as $w = 2 \times L$. Additionally, when selecting input data for training, a potential approach is to employ sliding windows that overlap entirely, except for one time point, between two subsequences. For instance, we can use two batches, $Y_{t:t+w}$ and $Y_{t+1:t+w+1}$, at times t and $t+1$, respectively, where w represents the input subsequence length. However, to expedite the training process, we choose the nonoverlapping regions as follows: 12, 24, 12, 12, 12, 1, and 12 for the traffic, electricity (large), electricity (small), solar, taxi, wiki (large), and wiki (small) datasets, respectively. In other words, smaller nonoverlapping window sizes were used for smaller datasets.

The source code, along with reproducibility instructions for the model and experiments, is publicly available at github.com/Abdallah-Aaraba/FR3LS.

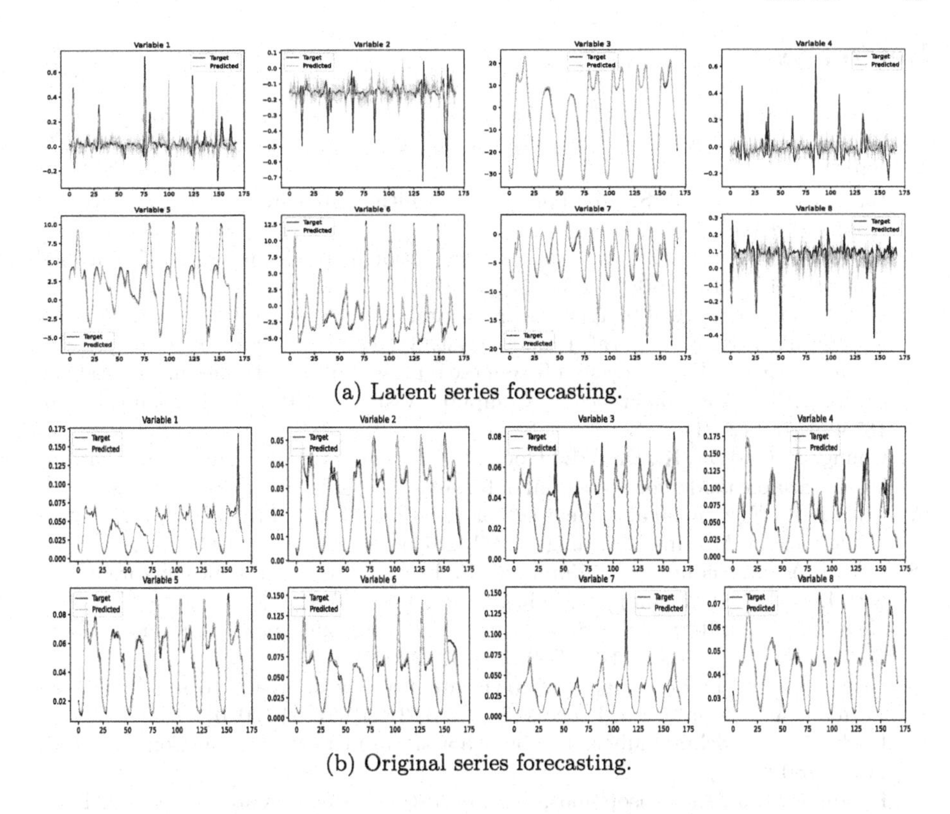

(a) Latent series forecasting.

(b) Original series forecasting.

Fig. 2. Latent and original series forecasting visualization.

6 Conclusion

This paper introduces an efficient approach for high-dimensional multivariate time series forecasting, advancing the current state-of-the-art in global factorization methods. The method achieves this by combining a flexible nonlinear autoencoder mapping, regularized through a non-contrastive self-supervised learning approach, along with a forecasting model capturing latent temporal dynamics. Furthermore, the proposed approach enables end-to-end training and demonstrates its capability to generate complex predictive distributions by modeling the distribution in the latent space through a nonlinear decoder. Our experiments showcase the superior performance of this method when compared to other state-of-the-art techniques across various commonly used time series datasets. Future research directions may involve exploring alternative temporal models and considering a Transformer-based approach [35] to mitigate the accumulation of forecast errors in sequential predictions and reduce training time.

Acknowledgement. Funded by Deutsche Forschungsgemeinschaft (DFG, German Research Foundation) under Germany's Excellence Strategy – EXC 2075 – 390740016.

References

1. Bai, S., Kolter, J.Z., Koltun, V.: An empirical evaluation of generic convolutional and recurrent networks for sequence modeling. arXiv preprint arXiv:1803.01271 (2018)
2. Bauwens, L., Laurent, S., Rombouts, J.V.: Multivariate garch models: a survey. J. Appl. Economet. **21**(1), 79–109 (2006)
3. Bottou, L., Curtis, F.E., Nocedal, J.: Optimization methods for large-scale machine learning. SIAM Rev. **60**(2), 223–311 (2018)
4. Cao, D., et al.: Spectral temporal graph neural network for multivariate time-series forecasting. Adv. Neural. Inf. Process. Syst. **33**, 17766–17778 (2020)
5. Chen, X., He, K.: Exploring simple siamese representation learning. In: Proceedings of the IEEE/CVF Conference on Computer Vision and Pattern Recognition, pp. 15750–15758 (2021)
6. Chung, J., Kastner, K., Dinh, L., Goel, K., Courville, A.C., Bengio, Y.: A recurrent latent variable model for sequential data. Adv. Neural Inf. Process. Syst. **28** (2015)
7. Cuturi, M.: Fast global alignment kernels. In: Proceedings of the 28th International Conference on Machine Learning (ICML-2011), pp. 929–936 (2011)
8. Gers, F.A., Schmidhuber, J., Cummins, F.: Learning to forget: continual prediction with LSTM. Neural Comput. **12**(10), 2451–2471 (2000)
9. Gneiting, T., Raftery, A.E.: Strictly proper scoring rules, prediction, and estimation. J. Am. Stat. Assoc. **102**(477), 359–378 (2007)
10. Grill, J.B., et al.: Bootstrap your own latent-a new approach to self-supervised learning. Adv. Neural. Inf. Process. Syst. **33**, 21271–21284 (2020)
11. Hochreiter, S., Schmidhuber, J.: Long short-term memory. Neural Comput. **9**(8), 1735–1780 (1997)
12. Hyndman, R.J., Athanasopoulos, G.: Forecasting: principles and practice. OTexts (2018)
13. Kingma, D.P., Ba, J.: Adam: a method for stochastic optimization. arXiv preprint arXiv:1412.6980 (2014)
14. Kingma, D.P., Welling, M.: Stochastic gradient vb and the variational auto-encoder. In: Second International Conference on Learning Representations, ICLR, vol. 19, p. 121 (2014)
15. Lai, G., Chang, W.C., Yang, Y., Liu, H.: Modeling long-and short-term temporal patterns with deep neural networks. In: The 41st International ACM SIGIR Conference on Research & Development in Information Retrieval, pp. 95–104 (2018)
16. Li, S., et al.: Enhancing the locality and breaking the memory bottleneck of transformer on time series forecasting. Adv. Neural Inf. Process. Syst. **32** (2019)
17. Lütkepohl, H.: New Introduction to Multiple Time Series Analysis. Springer, Heidelberg (2005). https://doi.org/10.1007/978-3-540-27752-1
18. Maggie, O.A., Vitaly, K., Will, C.: Web traffic time series forecasting (2017). https://kaggle.com/competitions/web-traffic-time-series-forecasting
19. Matheson, J.E., Winkler, R.L.: Scoring rules for continuous probability distributions. Manag. Sci. **22**(10), 1087–1096 (1976)
20. McKenzie, E.: General exponential smoothing and the equivalent arma process. J. Forecast. **3**(3), 333–344 (1984)
21. Mikolov, T., et al.: Statistical language models based on neural networks. In: Present. Google Mountain View, 2nd April **80**(26) (2012)
22. Nguyen, N., Quanz, B.: Temporal latent auto-encoder: a method for probabilistic multivariate time series forecasting. In: Proceedings of the AAAI Conference on Artificial Intelligence, vol. 35, pp. 9117–9125 (2021)

23. Rangapuram, S.S., Seeger, M.W., Gasthaus, J., Stella, L., Wang, Y., Januschowski, T.: Deep state space models for time series forecasting. Adv. Neural Inf. Process. Syst. **31**, 1–10 (2018)
24. Rasul, K., Sheikh, A.S., Schuster, I., Bergmann, U., Vollgraf, R.: Multivariate probabilistic time series forecasting via conditioned normalizing flows. arXiv preprint arXiv:2002.06103 (2020)
25. Salinas, D., Bohlke-Schneider,. M., Callot, L., Medico, R., Gasthaus, J.: High-dimensional multivariate forecasting with low-rank gaussian copula processes. Adv. Neural Inf. Process. Syst. **32**, 1–11 (2019)
26. Salinas, D., Flunkert, V., Gasthaus, J., Januschowski, T.: DeepAR: probabilistic forecasting with autoregressive recurrent networks. Int. J. Forecast. **36**(3), 1181–1191 (2020)
27. Sen, R., Yu, H.F., Dhillon, I.S.: Think globally, act locally: a deep neural network approach to high-dimensional time series forecasting. Adv. Neural Inf. Process. Syst. **32**, 1–10 (2019)
28. Taxi, N.: New york city taxi and limousine commission (tlc) trip record data (2015). https://www1nyc.gov/site/tlc/about/tlc-trip-record-data
29. Taylor, S.J., Letham, B.: Forecasting at scale. Am. Stat. **72**(1), 37–45 (2018)
30. Trindade, A.: Electricityloaddiagrams20112014 data set. Center for Machine Learning and Intelligent Systems (2015)
31. Wang, Y., Smola, A., Maddix, D., Gasthaus, J., Foster, D., Januschowski, T.: Deep factors for forecasting. In: International Conference on Machine Learning, pp. 6607–6617. PMLR (2019)
32. Yu, H.F., Rao, N., Dhillon, I.S.: Temporal regularized matrix factorization for high-dimensional time series prediction. Adv. Neural Inf. Process. Syst. **29** (2016)
33. Yue, Z., Wang, Y., Duan, J., Yang, T., Huang, C., Tong, Y., Xu, B.: Ts2vec: towards universal representation of time series. In: Proceedings of the AAAI Conference on Artificial Intelligence, vol. 36, pp. 8980–8987 (2022)
34. Zbontar, J., Jing, L., Misra, I., LeCun, Y., Deny, S.: Barlow twins: self-supervised learning via redundancy reduction. In: International Conference on Machine Learning, pp. 12310–12320. PMLR (2021)
35. Zerveas, G., Jayaraman, S., Patel, D., Bhamidipaty, A., Eickhoff, C.: A transformer-based framework for multivariate time series representation learning. In: Proceedings of the 27th ACM SIGKDD Conference on Knowledge Discovery & Data Mining, pp. 2114–2124 (2021)
36. Zhou, H., et al.: Informer: beyond efficient transformer for long sequence time-series forecasting. In: Proceedings of the AAAI Conference on Artificial Intelligence, vol. 35, pp. 11106–11115 (2021)

Knowledge-Infused Optimization for Parameter Selection in Numerical Simulations

Julia Meißner[1(✉)], Dominik Göddeke[1], and Melanie Herschel[2]

[1] Institute of Applied Analysis and Numerical Simulation, University of Stuttgart, Stuttgart, Germany
{julia.meissner,dominik.goeddeke}@ians.uni-stuttgart.de

[2] Institute for Parallel and Distributed Systems, University of Stuttgart, Stuttgart, Germany
melanie.herschel@ipvs.uni-stuttgart.de

Abstract. Many engineering applications rely on simulations based on partial differential equations. Different numerical schemes to approximate solutions exist. These schemes typically require setting parameters to appropriately model the problem at hand. We study the problem of parameter selection for applications that rely on simulations, where standard methods like grid search are computationally prohibitive. Our solution supports engineers in setting parameters based on knowledge gained through analyzing metadata acquired while partially executing specific simulations. Selecting these so-called farming runs of simulations is guided by an optimization algorithm that leverages the acquired knowledge. Experiments demonstrate that our solution outperforms state-of-the-art approaches and generalizes to a wide range of application settings.

Keywords: Data Science · Parameter Selection · Optimization

1 Introduction

Many application problems are modeled mathematically by partial differential equations (PDEs), i.e., equations describing the interaction and evolution of quantities and their derivatives. An example of a prominent model problem is the diffusion equation $\partial_t u + \Delta u = f$. It appears as a submodel in, e.g., fluid dynamics, structural mechanics, social sciences, or semiconductor design. In general, no solution formulae exist for PDEs. Hence, engineers employ *numerical schemes* to express a problem in computer-tractable form and to compute (approximate) solutions. Such schemes comprise multiple interacting *subproblems*, e.g., discretization, linearization, preconditioning, and iterative solving, usually involving multiple parameters to be set. Choosing a scheme and appropriate parameters for a particular problem also depends on requirements, e.g., maximizing accuracy in safety-critical, or minimizing runtime in time-critical scenarios. The common practice of manually selecting and parameterizing schemes based on expert knowledge is tedious, time-consuming, and error-prone.

D.-N. Yang et al. (Eds.): PAKDD 2024, LNAI 14650, pp. 16–28, 2024.
https://doi.org/10.1007/978-981-97-2266-2_2

Appropriately selecting parameters is a well-studied problem in machine learning (ML) [9,16]. Indeed, hyperparameter tuning is analogous to our setting, except that we focus on PDE-based simulation models rather than ML models. Common techniques for hyperparameter optimization rely on grid search, random search [1], gradient-based optimization, Bayesian optimization (BO) [13,14], or hyperband optimization [9]. For ML models, hyperparameter optimization techniques, albeit effective, are potentially wasteful, e.g., hyperband tuning builds upon successive halving, where the worse performing half of tunings are not only discarded, but the knowledge about their behavior is lost [9].

Contributions. For simulation models that regularly run on HPC clusters for long periods of time, we cannot, in general, afford computation-hungry parameter optimization strategies. To still support engineers in this task, we focus on *minimizing the overhead* of acquiring sufficient metadata to optimize a *wide range of possible (future) applications.* We make the following contributions: (1) we identify relevant metadata that serve to predict the performance of simulation runs (Sect. 3); (2) we present methods to efficiently capture the described metadata and integrate the knowledge gained into a domain-specific optimization algorithm (Sect. 4); (3) we experimentally validate that the proposed solution is efficient and effective in a wide range of applications (Sect. 5). Further information and details about our code are available on Anonymous GitHub.[1]

2 Preliminaries on Numerical Simulations

Our parameter optimization approach supports pipelines that define schemes in numerical simulations. This section provides the background for a model problem and a computational scheme underlying our subsequent discussion.

Model Problem. We consider anisotropic Poisson problems. They are prototypical for many elliptic PDEs and can be found, as a submodel, in many applications. On a domain $\Omega =]0,1[^2$, we have $-\alpha\frac{\partial^2}{\partial_x^2}u - \beta\frac{\partial^2}{\partial_y^2}u = f$ in Ω, $u = 0$ on $\partial\Omega$, with model parameters α, $\beta \in \mathbb{R}$. For $\alpha = \beta = 1$, we have $-\Delta u = f$. Intuitively, α and β determine the degree of anisotropy in the system. For $\alpha = \beta$, we have a homogeneous material in which the modeled stationary diffusion process takes place, while for $\beta \neq \alpha$, the material has preferred directions, e.g., slate.

Scheme. Our scheme consists of several steps, i.e., discretization, use of a numerical solver, and preconditioning to boost solver performance.

Through *discretization*, we cover Ω with an equidistant grid of spacing $h > 0$, and apply second-order centered finite differences. This results in a linear system $Av = b$ for the discrete unknown values v_i that approximate the unknown continuous solution u in the grid points. The size of the system matrix depends on h^{-1}. We choose powers of $\frac{1}{2}$ for h, so that A quickly becomes large. The increase in dimensionality reduces the discretization error. Also, A is symmetric, positive definite, and sparse. We choose b as Au for $u(x,y) = (x(1-x)y(1-y))^2$.

[1] anonymous.4open.science/r/simulation-optimizer-E691.

Due to the properties of A, the conjugate gradient method (CG) [5] is a favorable *solver* for the linear system. CG is an iterative method: in each iteration, the dimension of a basis of the solution space is increased by one through a clever projection, and an approximate solution is formed in that basis. For our model problem, the larger the quotient of α to β, the more iterations are needed.

To boost solver performance, we apply preconditioning, such that CG effectively solves the system $PAv = Pb$. For $P \approx A^{-1}$, the system becomes trivial, but computing A^{-1} is infeasible. We use the *dual threshold incomplete LU preconditioner (ILUT)* [12], which has been shown to be highly efficient for the model problem we consider. It approximates the lower and upper triangular factors $\tilde{L} \approx L$ and $\tilde{U} \approx U$ of the LU factorization $A = LU$. Then, in each iteration of the CG algorithm, two auxiliary triangular systems have to be solved. A *dropping strategy* determines which entries are used for the incomplete LU factorization, and which entries are ignored. This dropping strategy depends on two parameters, the drop-tolerance τ and the fill-in p (aka fill-factor). First, the drop-tolerance τ is multiplied by the average magnitude of the current row i of A, resulting in the relative drop-tolerance τ_i. During the factorization, all elements that are smaller than this relative drop-tolerance τ_i are replaced by zero. Then, only the p largest elements in the rows of L and U are kept in addition to the diagonal element. A small drop-tolerance τ and a big fill-in p result in a high amount of nonzero entries in the preconditioner, and high costs to compute it a priori and to apply it in each CG iteration. As τ and p influence $\tilde{L}$ and $\tilde{U}$ in a highly nonlinear fashion, it is in general impossible to determine these *scheme parameters* a priori. Bad choices may lead to too dense and thus too expensive factors, even though the CG solver would then converge rapidly. It may also happen that the preconditioned system PA is no longer symmetric positive definite, causing CG to fail. An optimized selection of the parameters is therefore essential and typically involves careful balancing of all related costs and the improvement of the convergence speed.

3 Identification and Analysis of Relevant Metadata

We use the above sample setting to discuss our parameter optimization strategy for simulation applications. The optimization is based on knowledge gained by analyzing metadata captured during simulation runs. To identify which metadata could be used to predict the behavior of scheme runs, we first log metadata of runs in a brute-force manner, systematically sweeping through the parameter ranges and performing a run for each sampled combination. This "exhaustive farming" for metadata (cf. Sect. 5) requires running a simulation for 90 parameter combinations, each with up to 1000 iterations. For each run, we log the model parameters α and β, the scheme parameters drop-tolerance τ and fill-factor, and the residual norms in every iteration. We also track performance metrics related to possible requirements (e.g., runtime, required memory). To detect solver failure, we prescribe a maximum number of iterations, $\theta_{it} = 1000$. In successful configurations, CG terminates upon reaching an absolute tolerance $\theta_{tol} = 10^{-14}$. Note that Sect. 4 describes strategies for reducing the computational cost.

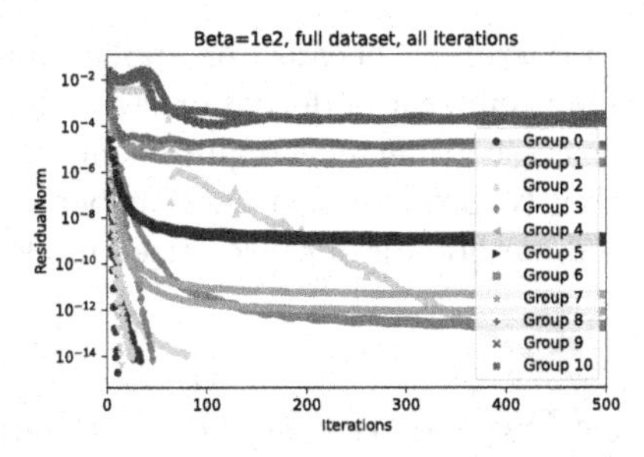

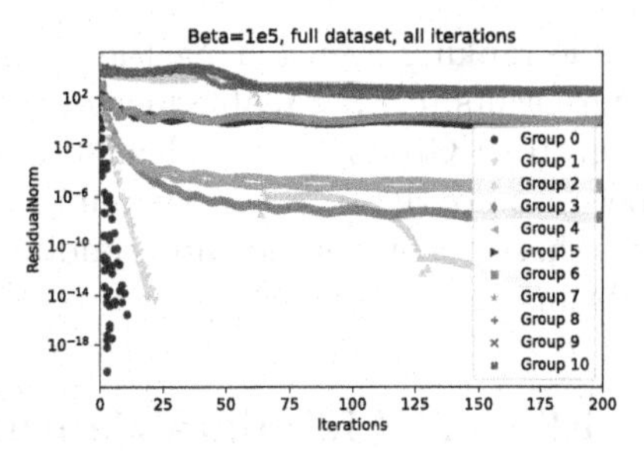

Fig. 1. $\beta = 1e2$ and $\beta = 1e5$, trajectories based on whole dataset, automatic k

By analyzing the metadata captured by this "exhaustive" metadata farming, we acquire deeper knowledge of the behavior of the simulation with different parameters. This goes beyond knowledge useful to identify which parameter combinations are good or bad in regards to simulation performance (i.e., residual norm over iterations, the main indicator tracked by iterative solvers to decide on termination due to the relation to solution accuracy), as it also allows us to identify reasonable parameter settings for different application requirements or restrictions, e.g., on memory usage. Previous work [7] has shown that different parameter combinations exhibit three basic performance behaviors, fast or slow converging or not converging at all. In terms of performance analysis, relevant knowledge may be acquired through clustering of simulation run trajectories that record the change of residual norm with increasing iterations. Essentially, good trajectories exhibit a fast reduction of the residual norm, while undesirable trajectories showcase a stagnating residual norm as iterations increase.

For our exhaustive metadata farming runs, we validate that for certain ratios of α to β, the metadata (1) form distinctive clusters and (2) different clusters correspond to different parameterization qualities. The algorithm and analysis devised and employed for this validation are illustrated based on Fig. 1. It shows trajectories and clusters identified by applying k-means clustering and ranked by their performance, when setting $\alpha = 1$ while varying β (remember, only their ratio is relevant). We automatically set the parameter k of k-means based on the *S_Dbw* Score [4]. Other clustering algorithms were tested as well, k-means performed best overall. We base the cluster ranking on the average residual norm within a cluster in the last iteration. For clusters that have reached the norm tolerance, the ranking algorithm takes the final iteration into account.

The clusters obtained for representative β values by applying the above algorithm are colored in Fig. 1. The order of cluster labels reflects the ranking, Group 0 (dark red) being the best. Throughout all experiments using exhaustive farming, we make the following important observations: (1) The best trajectories (i.e., quickly converging and reaching the tolerance threshold) are commonly assigned the best cluster. (2) Several trajectories exhibit a stagnating residual norm, at different levels. Comparing for example the grey (Group 3) and the purple (Group 7) trajectories, the grey trajectories are to be preferred (lower residual norm). Hence, our ranking is effective. It is important that we have a reliable ranking even among the stagnant trajectories, because of application requirements (e.g., available memory) that may render a preferred cluster (e.g., Group 0 or Group 3) inapplicable due to excessive memory requirement. (3) There is an outlier-trajectory exhibiting sudden "drops" for each β. This behavior is characteristic of the unpreconditioned CG method for our discretized model problem.

4 Efficient Metadata Capture for Parameter Optimization

Based on our exhaustive farming runs, we identified a clustering algorithm and a ranking function that successfully reveal relevant information about model behavior. While we may now leverage knowledge of the ranked clusters for parameter optimization, the brute-force metadata farming of the previous section that leads to these clusters is computationally prohibitive overall. We propose two strategies to reduce computational cost: First we present how farming runs can be terminated early, i.e., after very few iterations, while still allowing proper clustering of their (projected) trajectories. Subsequently, we discuss an algorithm that selectively probes the parameter space instead of performing the full sweep of the brute-force approach, thereby improving efficiency without a significant loss in quality of the final ranked clusters.

4.1 Early Termination of Farming Runs

The first strategy solves the model problem (for each parameter combination) for just a few iterations. This relies on the general mathematical property of Krylov methods (including the CG method we use as representative) that a good preconditioner shows its effect already after few iterations. Consequently, we can accurately predict the effect of the preconditioner on the overall trajectory already after few iterations (validated in Sect. 5). This allows us to alter our farming algorithm to collect data only for the first n iterations of a solver. Based on the data collected about the few iterations, we apply the modified clustering algorithm summarized in Algorithm 1. Essentially, it determines the optimal clustering not only for the best k using S_Dbw, but for the best combination of reasonable numbers of clusters k and iterations considered $l \leq n$ with respect to S_Dbw. The corresponding clusters are then ranked as previously described.

Algorithm 1: ETClustering

```
Input: #of farmed iterations n and
       associated data D_n; max.
       number of clusters m
Output: Optimal clusters BestC
MinScore ← ∞; BestC ← ⊥ ;
for l = 3, 4, ...n do
    for k = 2, 3..., m do
        Clusters = k-means(D_l, k)
        Score = S_Dbw(Clusters)
        if Score < LowestScore then
            LowestScore = Score ;
            BestC ← Clusters;
return BestC;
```

4.2 PROBE: Probing Specific Parameter Combinations

The partial solving of early termination remains computationally expensive when running it for every parameter combination. Here, we discuss how to select specific parameter combinations to gather metadata, without compromising the effectiveness of parameter optimization.

We define the PROBE algorithm that, for a given model problem, adaptively probes the parameter space, based on the metadata obtained through parameter combinations considered so far. PROBE exploits mathematical properties underlying the parameter space, illustrated in Fig. 2. It visualizes the parameter combinations for fill-factor and drop-tolerance as points on a two-dimensional discretized parameter space for a selected β. Point colors match the clusters of the ranked trajectories obtained with the respective parameter combinations. Clearly, the points of a same cluster are not randomly distributed, they form contiguous areas. Mathematically, the behavior of two preconditioner configurations sharing one of the solver parameters can be similar, up to some parameter boundary that entails a change in performance. Our algorithm efficiently identifies these boundaries by exploiting the knowledge about this characteristic distribution of clusters in the parameter space.

PROBE proceeds as illustrated in Fig. 3. As input, PROBE requires a (small) set of initial points, i.e., parameter combinations for which ETClustering has been run. This set needs to include the extremal "corner" points of the parameter space, optionally more. The finite set of parameter combinations to consider is defined by fixing the distance between two parameter values. The top-left state of the parameter space shown in Fig. 3 exemplifies the result of initially probing 9 parameter combinations, different colors again illustrating different performance clusters. Observe that the bottom and right "border" points belong to the same cluster, whereas all other points exhibit significantly different performance and thus fall in different clusters. Intuitively, the algorithm identifies the contiguous areas of parameter combinations known, at a given point of processing, to belong to the same cluster (e.g., "coloring" areas in the middle of Fig. 3). To select the next points, PROBE exploits the knowledge that the boundary between clusters must lie in the "non-colored" zones in-between differently colored clusters of the parameter space. Consider for instance the middle point labeled as 1 in Iteration 1 as starting point. Following the vertical and horizontal axis (i.e., changing one parameter while fixing another), we encounter different clusters in every direction. This leads to four new points to be probed next, selected equidistantly between point 1 and the next cluster boundary in each direction. Applying this to each point available in Iteration 1, we obtain the new points illustrated in the rightmost image of

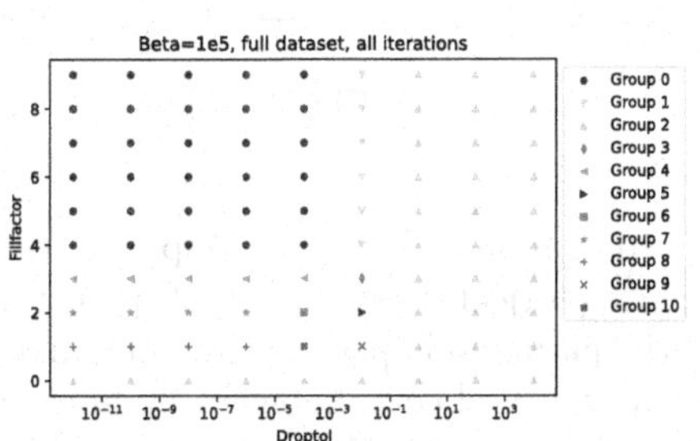

Fig. 2. $\beta = 1e5$, parameter space colored based on whole dataset, automatic k

Iteration 1. Iteration 2 (second line in Fig. 3) starts by running the simulations for the new points (with early termination), then applying ETClustering for all available metadata. Performing clustering independently from previous iterations is crucial for the algorithm to self-correct as additional metadata become available through new probes. The remainder of Iteration 2 applies the same procedure as described for Iteration 1, i.e., identify areas and determine next points in between covered areas. At the end of Iteration 2, a large portion of the whole space is covered through colored clusters, visually showing how the probing strategy reduces the search space for next probes at every iteration. PROBE terminates when the whole parameter grid is colored.

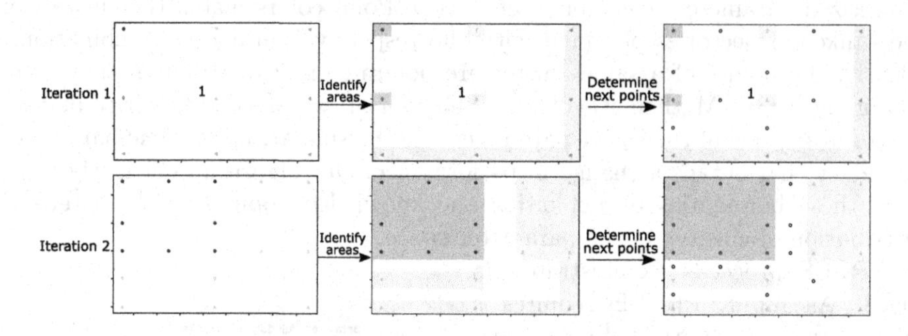

Fig. 3. Illustration of the PROBE algorithm

In summary, PROBE groups parameter combinations yielding similar performance (trajectories of different quality), only probing few points at the "borders" of the parameter space where performance shifts happen. In addition to reducing computational cost, knowledge about the performance of these specific points is also most useful for identifying suitable parameter combinations for applications with varying requirements. Indeed, due to inherent mathematical properties, these points exhibit the largest variation in properties like required storage.

5 Experimental Evaluation

We focus our report of our experimental evaluation on the following important aspects: (1) Quality of the parameter optimization through ranked clusters when employing PROBE (see Sect. 4) compared to the exhaustive farming introduced in Sect. 3; (2) runtime improvement of PROBE compared to (a) exhaustive farming and (b) parameter optimization techniques not leveraging domain knowledge to "skip" probes; (3) possible reuse of metadata acquired through PROBE for (a) similar problems and (b) different application requirements; and (4) generalization of our method to other model problems and computational schemes.

We implement the scheme outlined in Sect. 2 through the linear system stemming from the anisotropic Poisson problem, using a discretization that leads to 16129 unknowns. We implement CG in C++ and employ an existing ILUT

Table 1. Parameters variations

β	1, 1e1, ..., 1e10
Solver tolerance θ_{tol}	1e–14
Max iterations	1000
Drop-tolerance τ	1e–12, 1e–10,..., 1e4
Fill-factor	0,1,...,9

Table 2. Quality evaluation of PROBE

β	1e0	1e1	1e2	1e3	1e4	1e5	1e6	1e7	1e8	1e9	1e10
μ_{et}	3.5	3.5	3.7	1.6	1.9	1.9	2.0	1.9	2.0	1.9	1.9
$\mu_{\overline{et}}$	1.1	1.1	2.0	0.4	0.2	0.4	0.4	0.5	0.5	0.4	0.5
μ_{PROBE}	1.6	0.2	0.0	0.0	0.0	0.0	0.0	0.0	0.0	0.0	0.0

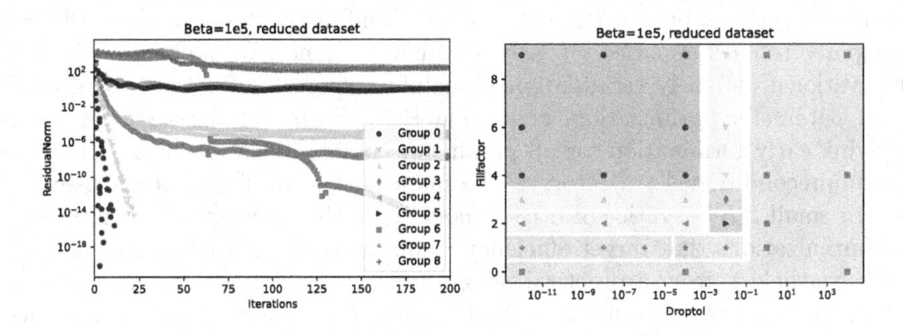

Fig. 4. Clusters based on PROBE for $\beta = 1\text{e}5$

implementation [3]. Our Python implementation of PROBE uses s_dbw score for Python [8], coupled with the computation method of [11] based on [4], producing the best results based on preliminary experiments.

Using this implementation, we run experiments based on the parameter variations summarized in Table 1, i.e., for all combinations of varying β (thus the quotient of $\alpha = 1$ to β), drop-tolerance, and fill-factor. The solver tolerance and maximum number of iterations are fixed as described earlier.

Beyond standard metrics to quantify runtime and cluster quality, the evaluation of the quality of the ranked clusters also needs to be compared to the quality of "gold standard" ranked clusters obtained through exhaustive farming. We measure ranked cluster similarity using μ, which takes the element-wise absolute values of the difference between two cluster arrays, sums them up, and divides them by the total amount of considered combinations. The larger μ, the worse the match, and vice versa.

5.1 Quality of Parameter Optimization Using PROBE

To evaluate PROBE, we initialize it with a start set of 9 equidistantly distributed points, efficiently processed using ETClustering with $n = 9$, $m = 11$ and data D_n captured using early termination. Figure 4 shows representative results for $\beta = 1\text{e}5$.

Visually comparing both the trajectories and parameter space of Fig. 4 with those in Fig. 1 and Fig. 2 (same β but brute-force approach), we see only minor differences in either clusters formed (e.g., purple Group 7 in Fig. 4 comprises two points for PROBE, whereas exhaustive farming separates them) or their ranking (e.g., the "unpreconditioned" trajectory is ranked as third-best for exhaustive farming, whereas its rank drops to 6 for PROBE). This high similarity translates to a score $\mu = 1.9$ when comparing to exhaustive farming. We recognize that this score is mostly affected by the unpreconditioned case that is inherently hard to correctly predict when using early termination: when ignoring it, μ drops to 0.4.

Table 2 summarizes the different ranked cluster qualities we obtain for all tested β when probing all parameter combinations with early termination (μ_{et}) but ignoring the unpreconditioned case (μ_{et}^{-}) and lastly when applying PROBE (compared to clusters based on early termination alone). We make the following observations: (1) early termination is mainly responsible for the loss of quality of parameter optimization, given that PROBE obtains the same results as applying early termination for all parameter combinations for all $\beta \geq 1e2$. (2) The unpreconditioned trajectory is the main reason for slightly high μ-scores. (3) For small β (e.g., ratios of α to β modeling rather homogeneous materials), our optimizations that target efficiency compromise the most in terms of quality. Nevertheless, from a practical perspective, by looking at all trajectories in detail, we see that the quickly converging trajectories are consistently ranked in the top-3, considered high quality in practice, even for larger μ. Note that the high quality of the clustering also reflects in the very low S_{Dbw} reached by PROBE, ranging between 0.0002 an 0.004 for all experiments.

5.2 Efficiency Evaluation

Efficiency of PROBE Compared to Exhaustive Farming. For PROBE, we measure the runtime $\mathsf{CaptureD_n}$ of capturing relevant metadata D_n for specific parameter combinations with early termination, and the time ProbeSelecting needed for clustering with ETClustering and identifying contiguous areas and next points to probe (cf. Sect. 4). Figure 5 summarizes the results for all tested values of β.

We see that PROBE significantly outperforms the baseline for small β values, whereas the performance is only marginally better or comparable for larger β values. Runtime of PROBE is dominated by the ProbeSelecting phase and remains stable on this series of experiments that uses rather small system matrices. In this setting, PROBE offers most gain for lower values of β or when the same metadata probed can be reused multiple times (c.f. Sect. 5.3). The real benefit of PROBE in terms of runtime for parameter optimization unfolds on larger matrices, found e.g., for the steady-state thermal problem discussed in Sect. 5.4. Its dimension of 82,654 increases the time for matrix-vector multiplications and leads to a computation time of 2805 s for the exhaustive approach. PROBE terminates after 105.5 s (85.0 s for $\mathsf{CaptureD_n}$, 20.5 s for ProbeSelecting). The above observations also hold on all other experiments in Sect. 5.4.

Efficiency of PROBE w.r.t. Alternative Probe Selection Strategies. Early termination also applies to alternative probing strategies. We compare the strategy of PROBE, informed by mathematical knowledge, to standard techniques, i.e., random search and Baysean optimization (BO), all coupled with

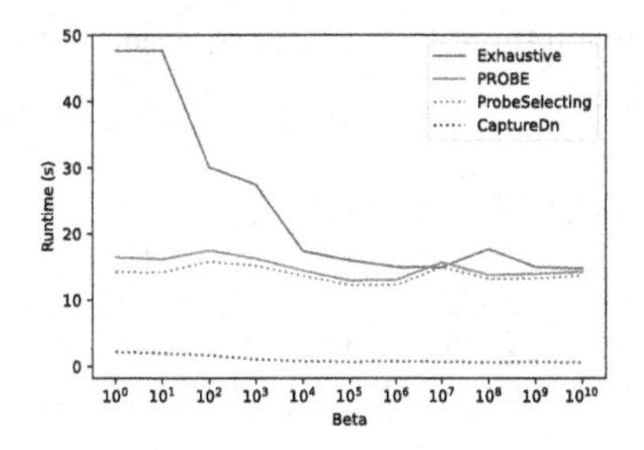

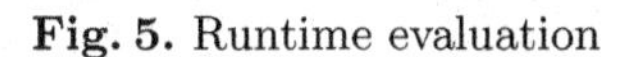

Fig. 5. Runtime evaluation

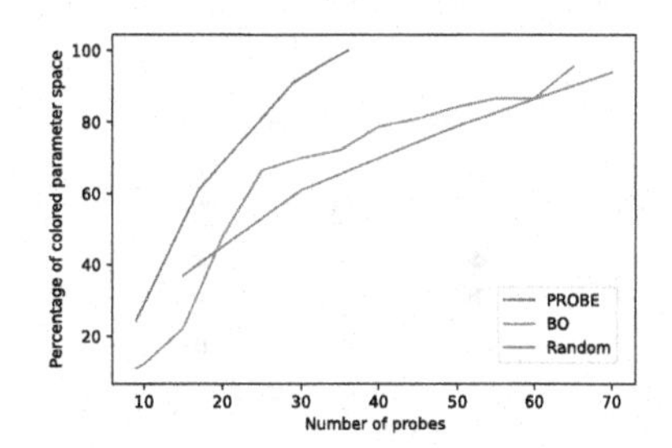

Fig. 6. Comparative evaluation ($\beta = 1e6$)

early termination. The optimization goal to be reached to effectively maximize reuse of farming runs throughout applications is full coloration of the parameter space.

Our implementation of random search selects a random set of parameter combinations that includes the extremal points of the parameter space, on which we then run ETClustering . We repeat this random selection five times and report averaged results over all runs. For BO, our implementation relies on [6]. Given the goal of full coloration, our BO implementation uses an entropy-based acquisition function that reduces the uncertainty over the whole parameter space. We pass the same initial points to both BO and PROBE.

Figure 6 plots our detailed analysis of the percentage of colored parameter space for PROBE, random search, and BO for a representative β. Clearly, PROBE outperforms the other two algorithms when it comes to quickly (i.e., with least number of probes) coloring the whole parameter space.

5.3 Reuse of Metadata Acquired Through PROBE

Reuse for Similar Problems. This series of experiments validates that the information gained through selective metadata farming using PROBE can be reused for similar problems without having to run additional farming runs. In particular, we study the reusability of farming runs for parameter optimization on new instances of a model class. For the anisotropic Poisson model problem, this corresponds to optimally configuring the ILUT preconditioner for yet unseen values of the anisotropy parameter β, based on farming runs for β values.

We compare clusterings across all the different β values that we obtained in experiments described previously. This comparison leads to the following observations: (1) The clusters with $\beta = 1e0$ and $\beta = 1e1$ significantly differ from the clusters for higher values of β, leading to different optimal parameter settings for the lower and upper β-ranges. (2) As β increases beyond $1e2$, the best ranked cluster remains in the upper left corner of the parameter space, but expands its area, merging previous clusters into a single performance class, and (3) beyond $\beta = 1e6$, the clusters remain mostly stable. These observations hold for exhaus-

Table 3. Scenarios

Scenario	Accuracy	Memory	Runtime
S1	●	○	◑
S2	◕	◑	◔
S3	◕	◑	◑
S4	◕	◕	◑
S5	◕	●	◑

Table 4. Evaluation for varying requirements

$\beta = 1e6$	PROBE	BO-20	BO-30	BO-36	BO-40	BO-50
S1	4	9	5	5	4	3
S2	1	x	x	x	x	x
S3	1	5	3	3	1	1
S4	4	9	5	5	4	3
S5	1	9	3	3	3	3

tive farming (with all iterations), early-termination for all parameter combinations, and PROBE. This allows us to conclude that optimal parameter combinations for smaller β-neighbors commonly lead to a good parameter selection for previously unseen β, without requiring any additional farming runs.

Reuse for Varying Simulation Requirements. Recall from Sect. 1 that different applications lead to different simulation requirements, e.g., on accuracy, used memory, or runtime. This series of experiments validates our claim that full coloration using PROBE allows us to cover the parameter space such that probed parameter combinations are relevant, and can thus be reused, for a wide range of simulation requirements.

Table 3 summarizes five scenarios with different simulation requirements. For each scenario, we want to find optimal parameter settings, relying solely on readily farmed (executed) simulations, i.e., without any new farming runs. This means that reuse is achieved by a posteriori filtering farmed information. Red symbols indicate a hard constraint (e.g., a threshold on available memory) from moderate limitation (◑) to extreme limitation (●), whereas **black** symbols indicate optimization of the performance, e.g., minimize runtime. For multiple optimization criteria, the filling of the symbols translates the priority of the given criteria. For instance, S2 describes a scenario where only best clusters (e.g., top-3) in terms of accuracy are retained (◕), so candidate parameter combinations are limited to those satisfying this constraint. These candidates are then ranked first by their memory usage (◑), then by their runtime (◔).

For each scenario, we determine the ground truth, i.e., the optimal parameter combination given the requirements, by probing all combinations using early termination and processing these following the requirements. We then compare the quality of parameter combinations obtainable for the five scenarios using either the probing strategy of PROBE or BO to the ground truth. Given the definition of PROBE, it guarantees full coloration upon termination, unlike BO. We therefore run five variants of BO differing in the number of probed combinations (four variants fixing the number of probes between 20 to 50 in increments of 10 and one variant where the number of BO-probes matches the number of probes needed by PROBE). We run the experiments for all β. Applying a scenario S_i to the results of PROBE and the BO variants by filtering and ranking the performed probes, we can identify a best parameter combination (ranked first), denoted p_{best}. We then check which rank p_{best} has in the ground-truth. The lower the rank of p_{best} in the ground truth, the better.

Table 4 shows representative results for $\beta = 1e6$. Across all experiments, we generally observe the following: (1) BO very rarely outperforms PROBE (only 6 cases across all β and scenarios) and this only when allowed more probes (e.g., BO-50 on S1). These cases correlate with a high coloration also reached by BO (cf. Fig. 6). (2) When BO outperforms PROBE in terms of the rank achieved, we observe that the results of both algorithms are highly similar (e.g., in the worst case, the best combination from PROBE is only 0.003 s slower than the best parameter combination returned by BO). (3) BO is prone to returning solutions as p_{best} that are not even part of the ground truth (e.g., x-marks for S2, where BO's p_{best} is not part of the ground-truth top-3 clusters). (4)In the vast majority of experiments, BO is worse than PROBE, and many times, so significantly worse that the results of BO are not usable from an application perspective.

We conclude that PROBE, optimized for fast and full coloration, yields the best results when it comes to reusing farming runs for similar problems on different requirements.

5.4 Generalization to Other Model Problems and Schemes

To validate the generalization of our approach beyond the model problem and computational scheme introduced in Sect. 2, we consider further model problems from the field of solid and fluid mechanics (e.g., a convection-diffusion-reaction equation [15] and a steady-state thermal problem [2]). Some of these problems require changing the computational scheme, i.e., by switching the iterative solver to the BiCGStab method [15] or GMRES and combining them either with the ILUT or SpiLU preconditioner [10].

For all examined problems, the quality of the clusters and ranking were comparable to the results of our model problem even though in one case the assumption of connected clusters was not met. Furthermore, we observe that the time saving increase with the matrix sizes.

6 Conclusion and Outlook

This paper studies the problem of efficient and effective parameter selection for engineering applications that rely on simulations. Our solution supports engineers in setting parameters based on knowledge gained through analyzing metadata acquired while partially executing specific simulations. Selecting these so-called farming runs of simulations is guided by an optimization algorithm that leverages the acquired knowledge. Experiments demonstrate that our solution outperforms state-of-the-art approaches and generalizes to varying application requirements, other simulation problems, and computational schemes.

Acknowledgement. Funded by Deutsche Forschungsgemeinschaft (DFG, German Research Foundation) under Germany's Excellence Strategy – EXC 2075 – 390740016.

References

1. Bergstra, J., Bengio, Y.: Random search for hyper-parameter optimization. J. Mach. Learn. Res. **13**, 281–305 (2012)
2. Davis, T.A., Hu, Y.: The university of Florida sparse matrix collection. ACM Trans. Math. Softw. (TOMS) **38**(1), 1–25 (2011)
3. Guennebaud, G., Jacob, B., et al.: Eigen v3 (2010). http://eigen.tuxfamily.org
4. Halkidi, M., Vazirgiannis, M.: Clustering validity assessment: finding the optimal partitioning of a data set. In: IEEE ICDM (2001)
5. Hestenes, M.R., Stiefel, E.: Methods of conjugate gradients for solving linear systems. J. Res. Nat. Bur. Stand. **49**(6) (1952)
6. Jiménez, J.: pyGPGO Python package (2020). https://github.com/josejimenezluna/pyGPGO
7. Kühnert, J., Göddeke, D., Herschel, M.: Provenance-integrated parameter selection and optimization in numerical simulations. In: USENIX TAPP (2021)
8. Lashkov, A., Rubinsky, S., Eistrikh-Heller, P.: S_dbw 0.4.0 (2019). https://pypi.org/project/s-dbw/
9. Li, L., Jamieson, K., DeSalvo, G., Rostamizadeh, A., Talwalkar, A.: Hyperband: a novel bandit-based approach to hyperparameter optimization. J. Mach. Learn. Res. **18**(1), 1–52 (2017)
10. Li, X.S., Shao, M.: A supernodal approach to incomplete lu factorization with partial pivoting. ACM Trans. Math. Softw. (TOMS) **37**(4), 1–20 (2011)
11. Liu, Y., Li, Z., Xiong, H., Gao, X., Wu, J.: Understanding of internal clustering validation measures. In: IEEE ICDM (2010)
12. Saad, Y.: ILUT: a dual threshold incomplete LU factorization. Numer. Linear Algebra Appl. **1**(4) (1994)
13. Seeger, M.: Gaussian processes for machine learning. Int. J. Neural Syst. **14**(02), 69–106 (2004)
14. Shahriari, B., Swersky, K., Wang, Z., Adams, R.P., Freitas, N.D.: Taking the human out of the loop: a review of bayesian optimization. Proc. IEEE **104**(1), 148–175 (2015)
15. Van der Vorst, H.: Bi-CGSTAB: a fast and smoothly converging variant of Bi-CG for the solution of nonsymmetric linear systems. SIAM J. Sci. Stat. Comput. **13**(2), 631–644 (1992)
16. Yang, L., Shami, A.: On hyperparameter optimization of machine learning algorithms: theory and practice. Neurocomputing **415**, 295–316 (2020)

Material Microstructure Design Using VAE-Regression with a Multimodal Prior

Avadhut Sardeshmukh[1,2(✉)], Sreedhar Reddy[1], B. P. Gautham[1], and Pushpak Bhattacharyya[2]

[1] TCS Research, Tata Consultancy Services Ltd., Pune 411057, India
{avadhut.sardeshmukh,sreedhar.readdy,bp.gautham}@tcs.com
[2] Department of Computer Science and Engineering, Indian Institute of Technology Bombay, Mumbai 400076, India
pb@cse.iitb.ac.in

Abstract. We propose a variational autoencoder (VAE)-based model for building forward and inverse structure-property linkages, a problem of paramount importance in computational materials science. Our model systematically combines VAE with regression, linking the two models through a two-level prior conditioned on the regression variables. The regression loss is optimized jointly with the reconstruction loss of the variational autoencoder, learning microstructure features relevant for property prediction and reconstruction. The resultant model can be used for both forward and inverse prediction i.e., for predicting the properties of a given microstructure as well as for predicting the microstructure required to obtain given properties. Since the inverse problem is ill-posed (one-to-many), we derive the objective function using a multi-modal Gaussian mixture prior enabling the model to infer multiple microstructures for a target set of properties. We show that for forward prediction, our model is as accurate as state-of-the-art forward-only models. Additionally, our method enables direct inverse inference. We show that the microstructures inferred using our model achieve desired properties reasonably accurately, avoiding the need for expensive optimization loops.

Keywords: Materials Infromatics · Inverse Problems · Variational Inference · Microstructure design

1 Introduction

Materials science and engineering involve studying different materials, their processing and the resulting properties that govern the performance of the material in operation. Processes such as heating, tempering and rolling modify the material's internal structure, altering properties such as tensile strength, ductility

Supplementary Information The online version contains supplementary material available at https://doi.org/10.1007/978-981-97-2266-2_3.

D.-N. Yang et al. (Eds.): PAKDD 2024, LNAI 14650, pp. 29–41, 2024.
https://doi.org/10.1007/978-981-97-2266-2_3

and so on. The structure is commonly represented by microscopy images known as the microstructure, which contain information about the micro-constituents (also known as phases), the grains, their geometry (shape, size), and their orientations. These structural features impact the material properties. Modeling the relationships between processing, structure and properties (also known as the P-S-P linkages) is at the core of computational materials science. Materials scientists and engineers are often interested in inverse analysis i.e., predicting the candidate structure for target properties and the processing route required to get the target structure. This involves systematically exploring a large design space consisting of several possible initial compositions, processing steps, parameters of these processing steps and the resulting structures. Also, the problem is often ill-posed since multiple processing routes can lead to the same structure and multiple structures can lead to the same selected target properties. Traditionally, materials scientists have used a combination of experimentation and physics-based numerical simulations for inverse analysis. However, experimental exploration has limitations because of the time and cost involved. On the other hand, physics-based models, which are based on solving underlying differential equations, are only useful for predicting the forward path. That is, predicting structure from composition and processing conditions, and properties from structure. They cannot be used directly for the inverse problem. Instead, they have to be used inside an optimization loop [23]. However, physics-based models are often computationally too expensive to be useful for design space exploration and optimization.

Machine learning can be an alternative to physics-based simulations for forward prediction. For example, deep convolutional neural networks have been used for predicting properties from microstructure images [3,16]. Recently, probabilistic deep generative models have been proposed to learn features from unlabeled (i.e. no properties data) microstructure images and then train a property-prediction model in the low-dimensional feature space using small labeled data [2,18]. This is an advantage since labeled data is more challenging to get. However, these models are not capable of inverse inference themselves. They still have to be used inside an optimization loop. While probabilistic deep generative models can be leveraged for direct inverse inference, they have not been explored much for this purpose. Our work is aimed at addressing this gap.

We propose a probabilistic generative model of structure-property linkage by combining variational autoencoder (VAE) with regression such that the joint model can be used for both forward and inverse inference without requiring the optimization step for the inverse. The VAE and the regression model are joined through the VAE prior by making the prior conditional on the predicted property. After training, latent representations for a target property value can be sampled from the conditional prior and decoded to get microstructures with that property value. Further, since there can be more than one microstructure for a target property value, the commonly used uni-modal Gaussian prior does not model this accurately. So we replace it with a mixture of Gaussians.

Our contributions are - i) a method for forward prediction of properties from structure, ii) a method for direct inverse prediction of candidate structures given the target properties, effectively handling ill-posedness. Using a reference dataset of 3-D microstructures and elasticity properties, we show that our model is as accurate as state-of-the-art methods for forward inference (Table 1) while additionally enabling direct inverse inference. For inverse inference, we show that the microstructures inferred using our model achieve the desired properties reasonably accurately (Fig. 3b). Further, we show that optimal points lie very close to these inferred microstructures and more precise solutions can be quickly found by searching in the neighborhood of these microstructures using a detailed physics based model (Fig. 4b and 4c).

2 Methodology

Variational autoencoders [14] pose the problem of representation learning as probabilistic inference with the underlying generative model $p(x,z) = p(x|z)p(z)$, where x is the input and z is the latent representation. The posterior $p(z|x)$ is to be inferred. An approximate posterior $q(z|x)$ is found by minimizing the Evidence Lower BOound (ELBO):

$$\mathcal{L} = -D_{KL}(q(z|x)\|p(z)) + \mathbb{E}_{q(z|x)}[log\ p(x|z)] \tag{1}$$

The KL-Divergence term enforces the prior $p(z)$ as a regularization while the second term quantifies how well an x is reconstructed.

VAE-Regression. The use of VAE in semi-supervised or supervised regression settings (predicting a scalar or vector of real numbers from an image) has been relatively less explored. A "VAE for regression" model was proposed by Zhao et al. [22] for predicting a subject's age from their structural Magnetic Resonance (MR) images, as follows. Assuming that the latent representation z is also dependent on the quantity c to be predicted, the generative model is: $p(x,z,c) = p(c)p(z|c)p(x|z)$, leading to a two-level prior (see Fig. 1a). The approximate posterior $q(z,c|x)$ is found using variational inference assuming that q factorizes as $q(z,c|x) = q(z|x)q(c|x)$[1]. Note however that the dependence between z and c is indirectly preserved through the prior $p(z|c)$. With that assumption, Zhao et al. [22] derive the modified ELBO as:

$$\mathcal{L} = \underbrace{-D_{KL}\left(q(c|x)\|p(c)\right)}_{\text{Regression loss}} + \underbrace{\mathbb{E}_{q(z|x)}[log\ p(x|z)]}_{\text{Rec loss}} - \underbrace{\mathbb{E}_{q(c|x)}[D_{KL}(q(z|x)\|p(z|c))]}_{\text{Regularization (cond. prior)}} \tag{2}$$

In supervised settings, the first term can be replaced by $log\ q(c|x)$, which is proportional to the mean squared error (i.e. regression loss) when $q(c|x)$ is a

[1] The mean-field assumption [1], commonly used in variational inference derivations (e.g., [15]).

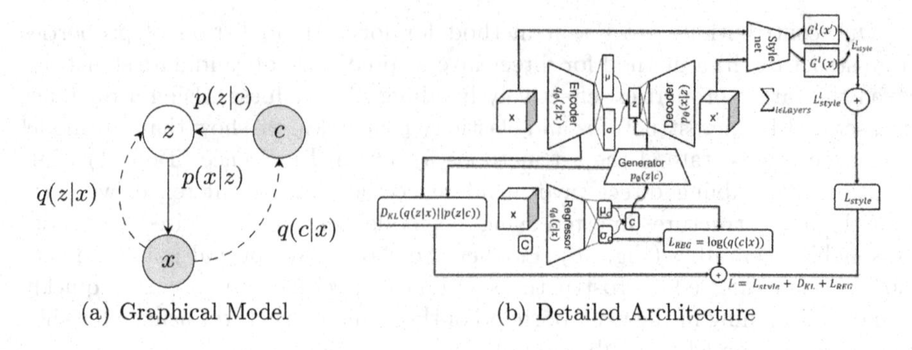

(a) Graphical Model (b) Detailed Architecture

Fig. 1. Architecture: VAE-Regression with Style Loss.

Gaussian. So $q(c|x)$ is parameterized by a "regressor" network. The second term – the reconstruction loss – is the same as in the original ELBO from equation (1), so $q(z|x)$ and $p(x|z)$ are parameterized by "encoder" and "decoder" networks, respectively. The last term is a counterpart of the regularizer from Eq. (1), except the prior is now conditional on c and the KL divergence is now in expectation with respect to $q(c|x)$. This term encourages the encoder posterior $q(z|x)$ and the conditional prior $p(z|c)$ to be similar, aligning the features for reconstruction and property prediction. Since the expectation is with respect to $q(c|x)$, it links the VAE and the regressor. During training, the expectation is estimated Monte-Carlo, using the c predicted by the regressor (i.e., one sample from $q(c|x)$). The distribution $p(z|c)$ is parameterized by a "generator" network. After training, inverse inference can be performed by sampling latent representations for a target c from the generator and decoding them.

We make two modifications to the original formulation from [22] to make it suitable to our use case. First, we replace the reconstruction loss with the style loss, which is based on comparing the statistics of the microstructures. Second, instead of the standard, uni-modal Gaussian prior, we incorporate a Gaussian mixture prior which models the many-to-one structure-property relation better.

VAE-Regression with Style Loss. The reconstruction loss from vanilla VAE objective function typically leads to a pixel-by-pixel comparison between input and reconstruction, which is not suitable for microstructure images [18]. Rather, only a comparison between statistics is suited to quantify the difference between two microstructures. The "style loss", originally proposed for the problem of style-transfer [6], is based on a comparison between statistics. It is the sum of squared differences between Gram matrices of the input and reconstruction as computed from a deep pre-trained network such as VGG19 [20]. Say layer l has C_l feature maps of size $W_l \times H_l$ then the Gram matrix at layer l is $G^l_{ij} = \Sigma_k F^l_{ik} F^l_{jk}$, and the style loss is: $\mathcal{L}_{style}(x, \hat{x}) = \sum_{l=0}^{L} w_l \left[\frac{1}{4C_l^2 M_l^2} \sum_{i,j} (G^l_{ij} - \hat{G}^l_{ij})^2\right]$ Where, $M_l = W_l * H_l$ and F^l is the $C_l \times M_l$ matrix of flattened feature maps. We propose to replace the reconstruction loss from the modified ELBO (Eq. (2)) with the style loss. The architecture of "VAE-regression" after incorporating the style loss is shown in Fig. 1b.

Multi-modal Prior. Since multiple structures can lead to the same properties, there could be more than one likely latent representation z, for a given c. So, the standard Gaussian prior, which a uni-modal distribution is not suitable for modeling $p(z|c)$. Instead, we propose a mixture-of-Gaussians prior $p(z|c) \sim \sum_{k=1}^{K} \pi_k N_k(\mu_k, \sigma_k^2)$ with K components, where π are the probabilities of components. We assume a diagonal co-variance matrix for all mixture components. The generator network outputs K pairs of μ and σ and the K component probabilities. Note that K is a hyperparameter that needs to be tuned. Since the posterior $q(z|x)$ is still a Gaussian, the reparameterization trick from the original VAE works. The only difficulty is in computing the KL-Divergence of the posterior from the conditional prior, that is the third term from Eq. (2). We need to compute the KL-Divergence of a Gaussian from a mixture-of-Gaussians, which is intractable. We use a variational approximation proposed in speech recognition literature [9], which for Gaussian mixtures f and g is: $D_{KL}(f\|g) \approx \sum_k \omega_k log \frac{\sum_i \omega_i e^{-D_{KL}(f_k\|f_i)}}{\sum_j \pi_j e^{-D_{KL}(f_k\|g_j)}}$ where f_k, f_i and g_j denote the component Gaussians of f and g and ω and π are their component weights respectively. Since in our case the first distribution f which is the posterior, is a uni-modal Gaussian, the numerator reduces to 1 (since $D_{KL}(f\|f) = 0$). So the expression is:

$$D_{KL}(f\|g) \approx log \frac{1}{\sum_j \pi_j e^{-D_{KL}(f\|g_j)}} \tag{3}$$

We replace the third term from Eq. (2) with the RHS of Eq. (3) to get our final loss function as follows:

$$\mathcal{L}_{VAE-REG} = -log\,(q(c|x)) + \mathcal{L}_{style} - log \frac{1}{\sum_j \pi_j e^{-D_{KL}(q(z|x)\|p_j(z|c))}} \tag{4}$$

3 Related Work

Due to recent advancements in machine learning, there is a renewed interest among materials scientists to leverage state-of-the art deep learning models for P-S-P linkages [3,16,21]. However, all these works focus on forward prediction using discriminative models. While Cang et al. [2] propose a generative VAE model, new images are generated unconditionally and used as additional training data for a downstream forward property-prediction model. In contrast, our approach enables forward and direct inverse inference combining VAE and regression through a conditional prior.

Recently, deep generative models such as variational autoencoders and Generative Adversarial Networks (GAN) have been explored for inverse inference in structure-property linkage [12,23] and other similar problems such as drug design [7,8]. These methods train a forward regression model from the GAN/VAE latent space to properties. For inverse inference, an optimization loop is setup around the forward model. While the GAN/VAE latent space enables efficient navigation through the space of microstructures, it does not alleviate the need for optimization. As opposed to this, in our approach once the model is trained, inverse inference is same as prediction and does not need optimization. More recently, Mao et al. [17] extended the GAN-based inverse inference method from Yang et al. [23] using mixture density networks (MDN). The GAN is trained on microstructure images and used to create training data for MDN. A set of images are generated from the GAN and the corresponding properties are obtained through FEM simulations on these images. The MDN is then trained to predict the GAN inputs z from the properties c. While this method does not need optimization every time, it can't utilize existing labeled data between microstructures and properties since the MDN needs pairs of z and c.

In machine learning literature, use of VAE in semi-supervised settings has been explored for various objectives such as conditional generation [13], learning disentangled representations [10], multi-modal representation learning [11] and so on. Most works on conditional generation treat the label (c) as one more latent variable, leading to a graphical model like $z \rightarrow x \leftarrow c$, where x is the input and z is the latent representation. However in our problem, c represents the material property which affects the microstructure features relevant for that property. So we treat the label as an auxiliary variable affecting the latent representation, leading to $c \rightarrow z \rightarrow x$. Probably the closest to our work is Characteristic Capturing VAE [10], which focuses on capturing the label characteristics in the latent representation. While our probabilistic generative model is same as theirs, the variational assumptions are slightly different. We assume that the variational posterior factorizes as $q(c, z|x) = q(c|x)q(z|x)$ whereas they assume $q(c, z|x) = q(c|z)q(z|x)$. Our assumption greatly simplifies the mathematical derivation while preserving the essential dependencies indirectly (via KL-divergence).

Some recent works have focused on incorporating stronger priors (including Gaussian mixture) in VAE (e.g., [4,15]) to improve the quality of generations. However, in these formulations, c is typically a discrete latent variable that represents hidden modalities of data, making the joint $p(z, c)$ a Gaussian mixture. Whereas our motivation for a multi-modal prior comes from the many-to-one relationship between structures x and properties c. So c is observed and continuous, and the conditional $p(z|c)$ itself is a Gaussian mixture. This introduces an intractable KL divergence term in our loss function unlike others, which we have dealt with using variational approximation, as explained in Sect. 2.

4 Experimental Results

We now describe the results obtained on a dataset of 3-D microstructures of a high-contrast composite and the associated elastic stiffness property.

Dataset. We use a dataset presented in Fernandez-Zelaia et al. [5] as part of their work on an efficient finite element method for micro-mechanics simulation in high-contrast composites. The authors first synthetically generated a large ensemble of voxelized 3D microstructures with diverse morphological features. This was done by starting with random 3D inputs of size $51 \times 51 \times 51$ in [0,1] and convolution with various Gaussian filters with zero mean and diagonal covariance, and thresholding the results to obtain a binary microstructure. The diagonal of the covariance matrix of the 3D Gaussian filters was of the form $\sigma = [i, j, k], i, j, k \in \{1, 3, 5, 7\}$. Thus there are 64 different Gaussian filters, which were applied to 150 random inputs resulting in ~8900 3D microstructures with a wide variety of morphologies. For example, $\sigma = [1, 1, 1]$ results in small, isotropic grains whereas $\sigma = [7, 7, 7]$ results in very large grains. Other asymmetric choices such as $\sigma = [7, 1, 1]$ or $[1, 1, 7]$ result in anisotropic grains with elongation in the corresponding directions. Please see Fig. 2a and 2b for examples.

The authors further estimate the effective elastic stiffness property of these microstructures using FEM simulations [5]. The crucial assumptions are: The black and white colors correspond to the hard and soft phases, with Young's moduli 120 GPa and 2.4 GPa, respectively, and the Poisson's ratio for both phases is 0.3. Elastic stiffness relates stress σ with strain ϵ. Since these are second-rank tensors, a fourth-rank tensor is required to relate them, i.e., $\sigma_{ij} = C_{ijkl}\epsilon_{kl}$ (Generalized Hooke's law [19]). The effective elastic stiffness parameter mentioned above is the element $\langle 1, 1, 1, 1 \rangle$ of the stiffness tensor C, and is denoted as C_{11} for short.

Training. We split the data into 60% training, 20% validation, and 20% test data. The validation data is used only to decide when to stop training (we stop when the validation loss does not decrease for ten consecutive epochs). The model is trained end-to-end using Adam optimizer with learning rate 0.0005, $\beta_1 = 0.75$, $\beta_2 = 0.999$ and batch size 8. We found that with 16 latent dimensions and a weight of 10 for regression loss (after normalizing the scales of all three losses) we got the best forward prediction accuracy with good reconstructions. More details of architecture, training and hyperparameter tuning appear in the supplementary material (please refer to the arxiv version of the paper).

Forward Inference. Table 1 shows accuracy in the prediction of C_{11} using different methods. MAPE is the mean absolute percentage error ($MAPE = 1/n\Sigma_i|y_i - \hat{y}_i|/\bar{y}$, where $\bar{y}$ is the mean observed value and $\hat{y}$ are the predictions) and R^2 is the coefficient of determination. The first block corresponds to two physics-based methods, the numbers reproduced from [21]. The next block corresponds to VAE and Gaussian process regression (GPR) trained separately. Supplementary material contains similar results obtained using other regression models such as support vector regression. The second last row corresponds to a state-of-the-art 3D CNN [21] for forward prediction only (we implemented this to reproduce the results), while the last row corresponds to our model. These numbers were obtained by averaging over ten random initializations and

Table 1. Forward Prediction Accuracy. Our method performs much better than traditional methods (first block) and separately trained VAE and regression (second block), and is as accurate as the state-of-the-art fwd model (Reg-Only).

Method	MAPE	R^2
V-R-H Avg [21]	46.66	–
2pt Stats+Reg [21]	6.79	–
VAE + GPR	7.23 (±0.15)	0.97 (±1e–3)
Reg-Only [21]	3.69 (±0.17)	0.99 (±4e–4)
VAE-Reg (ours)	**3.50** (±0.24)	0.99 (±5e–4)

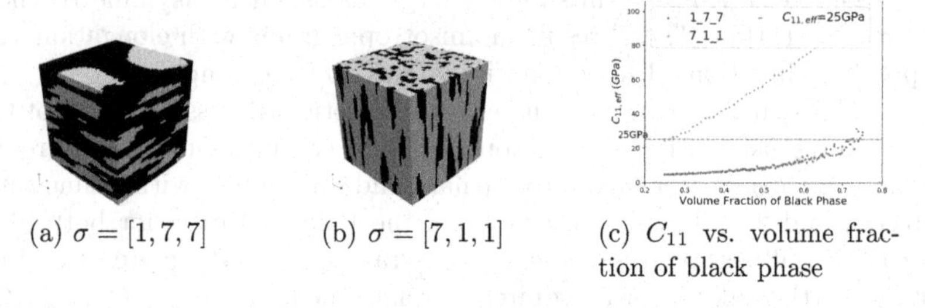

(a) $\sigma = [1, 7, 7]$ (b) $\sigma = [7, 1, 1]$ (c) C_{11} vs. volume fraction of black phase

Fig. 2. Two microstructures with $C_{11} \approxeq 25\,\text{GPa}$. The volume fraction of black phase in (a) is 74.49% and in (b) is 26.02%, as shown by the green line in (c)

train-test splits (standard deviations in brackets). While our model is much more accurate than separately trained VAE and regression, it is comparable to the state-of-the-art regression-only model (considering the standard deviations) and additionally provides direct inverse inference.

Inverse Inference. The effective elastic stiffness property C_{11} is largely dependent on the volume fractions. However, this function changes with the morphology. This is apparent from Fig. 2. Figure 2a and 2b show microstructures with $C_{11} \approxeq 25GPa$, generated from $\sigma = [1, 7, 7]$ and $\sigma = [7, 1, 1]$ having black phase volumes 74.49% and 26.02%, respectively. Figure 2c, shows C_{11} against the volume fraction of the black phase in the microstructures from these two morphologies. Thus, a given value of C_{11} can be achieved by multiple microstructures, possibly each coming from a different morphology with a different volume fraction of the black phase. This motivates the use of a multi-modal conditional prior.

To demonstrate the effectiveness of multi-modal inverse inference, we chose a subset of microstructures from the data, corresponding to $\sigma = [1, 1, 7]$ and $\sigma = [7, 1, 1]$, which have remarkably different morphologies (see Fig. 2a and 2b for an example) and are well-separated in the property space, but with some overlap at the extremes, as shown in Fig. 2c. We perform inverse inference for six target values of C_{11} spread across the complete range observed in the data. For each target C_{11} value, we obtain the mean of the conditional prior (for the multi-modal prior, the means of each component) and decode it. For illustration, we discuss the inverse inference for $C_{11} = 30\,\text{GPa}$. Figure 3a shows the real (top) and inferred microstructures from the uni-modal Gaussian prior (middle) and the two components of the Gaussian mixture prior. Note that the uni-modal prior tends to infer an average of the two possible solutions. Whereas the Gaussian mixture prior learns the solutions separately under different mixture components, with suitable weights.

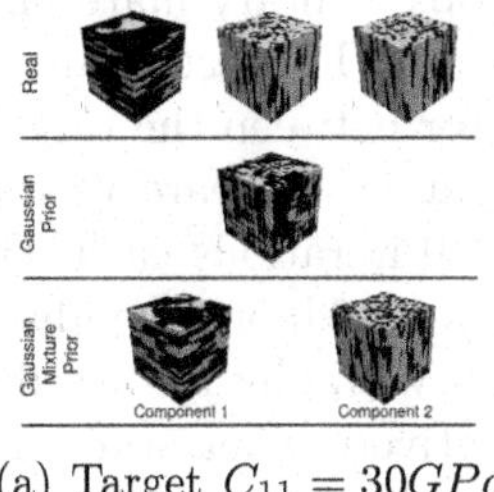

(a) Target $C_{11} = 30GPa$

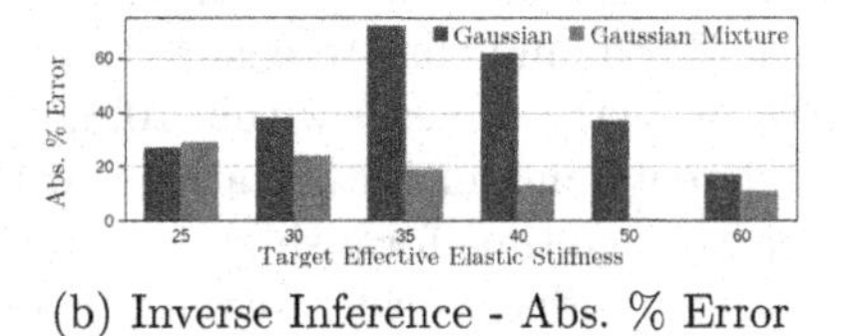

(b) Inverse Inference - Abs. % Error

Fig. 3. (a) Inverse inference for target $C_{11} = 30\,\text{GPa}$. Top row shows real microstructures with the target C_{11}. The microstructures inferred using a uni-modal Gaussian prior (middle row) tend to be like an average, whereas the multi-modal Gaussian mixture prior learns the multiple possible solutions under separate mixture components (last row). (b) Evaluation of inverse inference through FEM simulations to get the achieved C_{11} and compute the absolute % error between target and achieved C_{11}. The multi-modal prior learns multiple solutions under separate components unlike the uni-modal Gaussian prior, leading to more accurate inference (error within 20% in most cases).

Further, we validated the inferred microstructures through FEM simulations to estimate the achieved properties. Figure 3b shows the absolute percentage error between target and achieved properties using the mean solutions under uni-modal and multi-modal priors for a range of C_{11} values. The solutions inferred from the Gaussian mixture prior achieve the target properties better than those inferred from the uni-modal Gaussian prior. The difference is more pronounced for target C_{11} values in the middle (e.g., 35 GPa), where both the morphologies are likely. The average absolute error using Gaussian mixture prior is about 16%, with an R^2 value of 0.97.

Comparison with Optimization

As discussed in Sects. 1 and 3, some recent works have proposed optimization in the latent space of VAE or GAN for inverse inference. However, exploring the latent space efficiently and finding multiple optima can be difficult [7]. Searching from multiple random initial points may be required to find good solutions. In contrast, our inverse inference method directly provides some good initial candidates in the optimal regions. Detailed search can then be efficiently performed in the neighborhood to reach high-quality solutions.

To demonstrate this, we implemented simulated annealing optimization in the VAE latent space. The objective function is implemented using the forward prediction model referred to as Reg-only in Table 1. For each target property value, we performed the optimization starting at multiple random initial points and those inferred by our method. As an example, Fig. 4a shows the results for target $C_{11} = 35\,\text{GPa}$. The top row shows two real microstructures with $C_{11} \approxeq 35\,\text{GPa}$, the middle row shows two distinct solutions found from optimization runs starting with 5 random initial points and the last row shows the result of optimization starting with the points inferred by our method. Similar results for other target property values are shown in the supplementary material. The solutions are validated through FEM simulations as before. The results are shown in Fig. 4c. We found that to ensure that the mean error between the target and achieved property values is within 10%, we had to run at least 5 searches starting from different random initializations, with at least 200 iterations each. Due to this, the average time for a single inverse inference was ∼1.5 h on a Nvidia V100 GPU, using a reduced-order forward model. Doing this with physics-based FEM simulation model would be clearly infeasible. Comparatively, when starting from the points inferred by our method, we could get to similar accuracy (i.e. mean error <10%) within 10 iterations, which took ∼2 min on the same hardware.

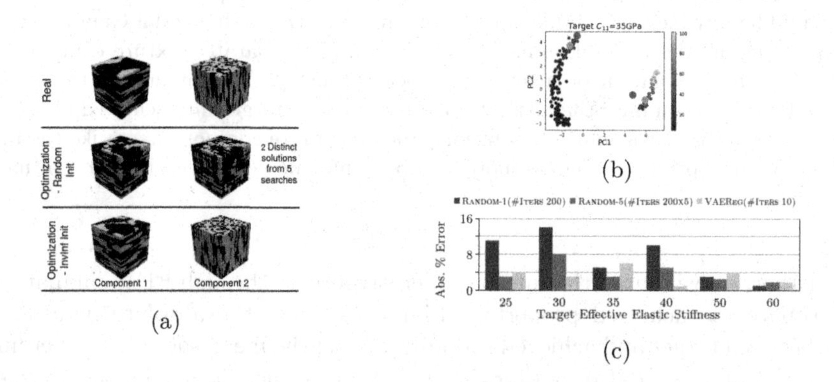

Fig. 4. Optimization based inverse inference (a) For target $C_{11} = 35$ GPa, real microstructures (top row), those obtained through optimization by starting at 5 random initial points of which two were distinct (middle row), and at the points inferred by our method (last row). (b) Visualization in latent space. Optimization starting from the points inferred by our method (red) converges to the orange points in 10 iterations, leading to high-quality solutions (c) The absolute % error for a range of target C_{11} values. When starting at random points, at least 5 searches, 200 iterations each, are needed to achieve mean error <10%. Starting at points inferred by our method, this is achieved in just 10 iterations. (Color figure online)

This can perhaps be explained as follows: The VAE latent space has regions of high and low probabilities, with high probability regions being quite sparse. The optimization algorithm guided only by the forward-prediction model has no knowledge of this distribution and hence spends considerable amount of time exploring low probability regions. In our approach, the conditional prior used for inverse inference is learned jointly with VAE and property prediction which ensures the conditional prior picks up the true distribution of the latent space. Figure 4b shows for target $C_{11} = 35\,\text{GPa}$, how the latent space points inferred by our method are just nudged right to improve the error from $\sim$20% (please see Fig. 3b) to $\sim$6%. The plot is a visualization of the latent space in 2D using Principal Components Analysis (PCA). The points are colored by the C_{11} value. The two large red circles are the points inferred by VAE-Reg multi-modal prior (component 1 and 2), which are fed as initial points for optimization. The large orange circles are the points found by optimization after 10 iterations. It can be seen that the points are pushed towards the right high-probability regions, leading to microstructures with C_{11} very close to 35 GPa.

Extension to Multiple Properties. Our model can be easily extended for the case when c is a vector, by suitably changing the sizes of the regressor's output and the generator's input layers. To show this, we extended the dataset by estimating the entire elastic stiffness tensor C through FEM simulations and applied our method for forward and inverse inference involving the vector C. These experiments are discussed in the supplementary material.

Discussion and Limitations. As discussed already, the number of mixture components K in the prior is treated as a hyperparameter presently. A good value of K could be set using Bayesian global optimization, or can be learned using a suitable hyper-prior for K. In some cases, materials scientists may provide a heuristic about the number of practically feasible solutions for a target property value based on domain knowledge of the specific material system under consideration.

5 Summary and Conclusions

We have developed a model for forward and inverse structure-property linkages in materials science by combining VAE and regression. For forward prediction, the combined model performs better than separately trained VAE and regression and is comparable to the state-of-the-art regression-only model. Whereas for inverse inference, the candidate microstructures inferred using our model achieve the target properties reasonably accurately and a local optimization search around these candidates using a reduced-order model quickly reaches target accuracy. Thus a detailed exploration in the small optimal region using physics-based simulations or experiments becomes feasible.

References

1. Blei, D.M., Kucukelbir, A., McAuliffe, J.D.: Variational inference: a review for statisticians. J. Am. Stat. Assoc. **112**(518), 859–877 (2017)
2. Cang, R., Li, H., Yao, H., Jiao, Y., Ren, Y.: Improving direct physical properties prediction of heterogeneous materials from imaging data via convolutional neural network and a morphology-aware generative model. Comput. Mater. Sci. **150**, 212–221 (2018)
3. Cecen, A., Dai, H., Yabansu, Y.C., Kalidindi, S.R., Song, L.: Material structure-property linkages using three-dimensional convolutional neural networks. Acta Mater. **146**, 76–84 (2018)
4. Dilokthanakul, N., et al.: Deep Unsupervised Clustering with Gaussian Mixture Variational Autoencoders. arXiv e-prints (2016)
5. Fernandez-Zelaia, P., Yabansu, Y.C., Kalidindi, S.R.: A comparative study of the efficacy of local/global and parametric/nonparametric machine learning methods for establishing structure-property linkages in high-contrast 3D elastic composites. Integrat. Mater. Manuf. Innov. **8**(2), 67–81 (2019)
6. Gatys, L.A., Ecker, A.S., Bethge, M.: Image style transfer using convolutional neural networks. In: Proceedings of the IEEE Conference on Computer Vision and Pattern Recognition (2016)
7. Griffiths, R.R., Hernández-Lobato, J.M.: Constrained bayesian optimization for automatic chemical design using variational autoencoders. Chem. Sci. **11**, 577–586 (2020)
8. Gómez-Bombarelli, R., Wei, J.N., Duvenaud, D., Aspuru-Guzik, A.: Automatic chemical design using a data-driven continuous representation of molecules. ACS Cent. Sci. **4**, 268–276 (2018)
9. Hershey, J.R., Olsen, P.A.: Approximating the kullback leibler divergence between gaussian mixture models. In: 2007 IEEE International Conference on Acoustics, Speech and Signal Processing - ICASSP 2007, vol. 4, pp. IV-317–IV-320 (2007)
10. Joy, T., Schmon, S., Torr, P., Siddharth, N., Rainforth, T.: Capturing label characteristics in vaes. In: International Conference on Learning Representations (2021)
11. Joy, T., Shi, Y., Torr, P., Rainforth, T., Schmon, S.M., Siddharth, N.: Learning multimodal VAEs through mutual supervision. In: International Conference on Learning Representations (2022)
12. Kim, Y.: Exploration of optimal microstructure and mechanical properties in continuous microstructure space using a variational autoencoder. Mater. Des. **202**, 109544 (2021)
13. Kingma, D.P., Rezende, D.J., Mohamed, S., Welling, M.: Semi-supervised learning with deep generative models. In: Proceedings of the 27th International Conference on Neural Information Processing Systems, NIPS 2014, vol. 2, pp. 3581-3589. MIT Press (2014)
14. Kingma, D.P., Welling, M.: Auto-encoding variational bayes. In: Bengio, Y., LeCun, Y. (eds.) 2nd International Conference on Learning Representations, ICLR 2014 (2014)
15. Lavda, F., Gregorová, M., Kalousis, A.: Data-dependent conditional priors for unsupervised learning of multimodal data. Entropy **22**(8), 888 (2020)
16. Li, X., Zhang, Y., Zhao, H., Burkhart, C., Brinson, L.C., Chen, W.: A transfer learning approach for microstructure reconstruction and structure-property predictions. Sci. Rep. **8**, 13461 (2018)

17. Mao, Y., et al.: Generative adversarial networks and mixture density networks-based inverse modeling for microstructural materials design. Integrat. Mater. Manuf. Innov. **11**, 637–647 (2022)
18. Sardeshmukh, A., Reddy, S., P., G.B., Bhattacharyya, P.: TextureVAE: learning interpretable representations of material microstructures using variational autoencoders. In: Proceedings of the AAAI 2021 Spring Symposium on Machine Learning for Physical Sciences. CEUR Workshop Proceedings, vol. 2964 (2021)
19. Serway, R., Jewett, J.: Physics for Scientists and Engineers. Cengage Learning (2013)
20. Simonyan, K., Zisserman, A.: Very deep convolutional networks for large-scale image recognition. In: Bengio, Y., LeCun, Y. (eds.) 3rd International Conference on Learning Representations, ICLR 2015 (2015)
21. Yang, Z., et al.: Deep learning approaches for mining structure-property linkages in high contrast composites from simulation datasets. Comput. Mater. Sci. **151**, 278–287 (2018)
22. Zhao, Q., Adeli, E., Honnorat, N., Leng, T., Pohl, K.M.: Variational AutoEncoder for regression: application to brain aging analysis. In: Shen, D., et al. (eds.) MICCAI 2019. LNCS, vol. 11765, pp. 823–831. Springer, Cham (2019). https://doi.org/10.1007/978-3-030-32245-8_91
23. Yang, Z., Li, X., Catherine Brinson, L., Choudhary, A.N., Chen, W., Agrawal, A.: Microstructural materials design via deep adversarial learning methodology. J. Mech. Des. **140**(11), 111416 (2018)

A Weighted Cross-Modal Feature Aggregation Network for Rumor Detection

Jia Li[1], Zihan Hu[1], Zhenguo Yang[1], Lap-Kei Lee[2](✉), and Fu Lee Wang[2]

[1] Guangdong University of Technology, Guangzhou, China
2112205148@mail2.gdut.edu.cn, yzg@gdut.edu.cn
[2] Hong Kong Metropolitan University, Hong Kong, China
{lklee,pwang}@hkmu.edu.hk

Abstract. In this paper, we propose a Weighted Cross-modal Aggregation network (WCAN) for rumor detection in order to combine highly correlated features in different modalities and obtain a unified representation in the same space. WCAN exploits an adversarial training method to add perturbations to text features to enhance model robustness. Specifically, we devise a weighted cross-modal aggregation (WCA) module that measures the distance between text, image and social graph modality distributions using KL divergence, which leverages correlations between modalities. By using MSE loss, the fusion features are progressively closer to the original features of the image and social graph while taking into account all of the information from each modality. In addition, WCAN includes a feature fusion module that uses dual-modal co-attention blocks to dynamically adjust features from three modalities. Experiments are conducted on two datasets, WEIBO and PHEME, and the experimental results demonstrate the superior performance of the proposed method.

Keywords: Rumor detection · Adversarial training · Cross-modal alignment

1 Introduction

Rumors refer to false information that is widely spread on social media platforms, news media and other channels. It disrupts society and causes panic. Rumor detection is a task to stop the spread of rumors quickly and accurately. The purpose of the rumor detection scenario is to protect the public's right to know and the safety of their interests, deeply prevent false information from negatively affecting people's thoughts and behaviors. At the same time, rumor detection also helps maintain the normal order of public opinion, promotes clearer and more truthful information flow, and helps improve the intelligence level and public quality of society.

Traditional rumor detection methods rely on text to classify rumors. With the development of multimedia, the form of rumor dissemination transforms into images coupled with text. As shown in Fig. 1, user comments and images can

D.-N. Yang et al. (Eds.): PAKDD 2024, LNAI 14650, pp. 42–53, 2024.
https://doi.org/10.1007/978-981-97-2266-2_4

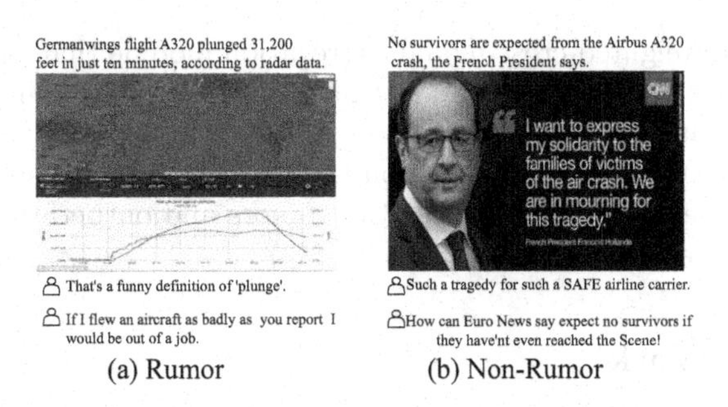

(a) Rumor (b) Non-Rumor

Fig. 1. An illustration of a piece of fake news and related user comments, which can be used for extracting useful information to assist rumor detection.

provide additional information that can help with the rumor classification task. To address this issue, many multimodal rumor detection methods have been proposed. For example, Yan et al. [1] propose a heterogeneous image attention network and use two-way information propagation to achieve cross-modal fusion. However, it cannot effectively utilize inter-modal information. Qian et al. [2] propose a hierarchical multimodal contextual attention network to effectively use multimodal information, which is susceptible to noise and sample imbalance of the dataset. At the same time, some researchers pay attention to adversarial training in the rumor detection domain. Sun et al. [3] first introduce adversarial learning to the rumor detection task and use contrastive learning to capture features that are invariant to rumor events, which can not model inter-modality interactions well. Zheng et al. [4] propose a novel Multi-modal feature-enhanced Attention Networks (MFAN) for rumor detection, which considers both the complement and alignment relationships between different modalities to achieve better fusion. However, the modal information may be imbalanced in the attention mechanism based on similarity calculation.

In this paper, we propose a Weighted Cross-modal Aggregation network (WCAN) for rumor detection in order to aggregate strongly correlated data in different modalities and provide a uniform representation. For the purpose of making the model more resistant to disturbances in the dataset, WCAN uses an adversarial training method to add perturbations on text embeddings. Additionally, we develop a weighted cross-modal aggregation (WCA) module. Taking text features as the main focus at first. Then calculate the correlation weights of the three modal features of text, image, and social graph. After that, combining the correlation weights with the original modality to obtain multimodal features. The fused features and the original features of the image and social graph are drawn closer by MSE loss, and the information of each mode is comprehensively considered. Moreover, WCAN offers a feature fusion module that dynamically modifies features from three modalities using dual-modal co-attention blocks.

The contributions of this paper are summarized below:

- We develop a weighted cross-modal aggregation module that can take advantage of the correlation between different modalities and obtain a comprehensive representation.
- We conduct extensive experiments on the PHEME and Weibo datasets. Our proposed model can effectively identify rumors and outperform the state-of-the-art baselines.

2 Related Work

In this section, we investigate the related works of rumor detection and adversarial training.

2.1 Rumor Detection

Since the rise of the Internet, a large number of rumors spread in social media. Researchers propose a number of rumor detection methods. Among them, many scholars learn the representations of rumor propagation to effectively prevent the spread of rumors. For example, Wiegmann et al. [5] model the social media communication process and use time series analysis to determine the origin and propagation path of rumors. However, the aforementioned method does not take into account the role of image features in rumor classification. At the same time, some scholars use adversarial learning method in many rumor detection scenarios. Sun et al. [3] first use contrastive learning to identify rumors, and employ contrastive learning to capture traits that are invariant to rumor events. This approach ignores the correlation between different modes. Many scholars attempt to use multimodal information in rumor detection. Ye et al. [6] propose a cross-modal self-attentive network that can span multiple visual modalities and capture the correlation between them. Conversely, the above method performs poorly for data with inter-modal imbalance and uncertainty.

2.2 Multimodal Alignment

Multimodal alignment refers to the alignment and fusion of data from modalities, and one of the first scenarios in which multimodal alignment techniques are used in medical imaging. Nowadays, multimodal alignment has been widely used in the field of deep learning. For instance, Wu et al. [7] use a combination of self-attention mechanisms and fully connected layers to learn to align visual and textual information in cross-modal retrieval task. This method is not ideal for small data volumes. Ma et al. [8] propose a cross-modal alignment method to align different modalities to the same feature space using a similarity measure function in reinforcement learning task. This method has limited alignment effect for other tasks. In addition, in terms of visual-language relationship alignment, Pandey et al. [9] propose a cross-modal attention consistency regularization method. But the method requires a large amount of annotated data for model training.

3 Methodology

3.1 Overview of WCAN

The overall WCAN model framework is illustrated in Fig. 2, which consists of three main modules, the multimodal feature extraction module, the multimodal feature alignment module, and the feature fusion module. Specifically, the multimodal feature extraction module contains three feature extractors to extract text features, image features, and social graph features, respectively. The weighed cross-modal aggregation module acquires features with high correlation across modalities and obtains a representation of the same feature space. The feature fusion module fuses the features of the three different modalities to obtain the fused feature representations.

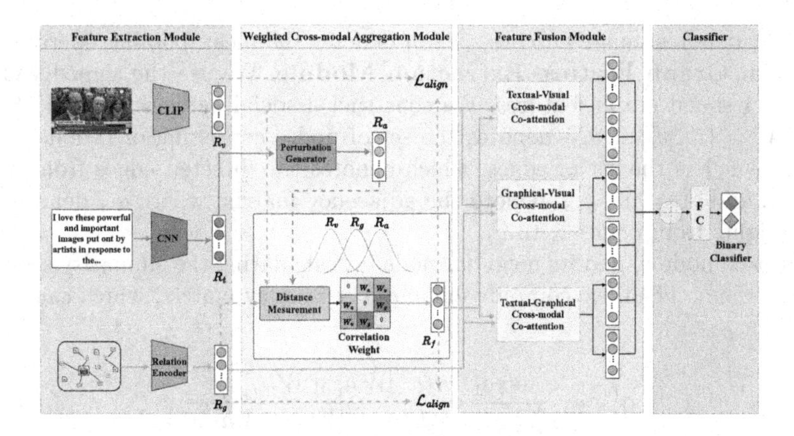

Fig. 2. The architecture of WCAN.

3.2 Feature Extraction

1) Text Feature Extraction Module. We use the TextCNN model as the text feature extractor. Given an input text sequence x, we first convert each word into a corresponding word vector representation. After that, we use multiple convolutional kernels of different sizes to capture local features in the text. Then, we perform the maximum pooling operation for each feature map. Finally, we input the pooled vector into a fully connected layer to get the text features:

$$R_t = concat(P_{k=3},\ P_{k=4},\ P_{k=5}). \tag{1}$$

where k represents the size of the convolution kernel. P denotes the CNN model.

After getting a text feature R_t, the generated perturbation is calculated as follows:

$$\epsilon = \varepsilon \cdot \operatorname{sign}\left(\nabla L\left(R_t\right)\right) \tag{2}$$

where $\nabla L(R_t)$ denotes the gradient of the text feature and ϵ is a hyperparameter.

Then, the adversarial sample is calculated as:

$$R_a = R_t + \epsilon. \tag{3}$$

where ϵ denotes the perturbation, R_t denotes the text feature.

2) Visual Feature Extraction Module. We use CLIP [10] to extract image features due to its ability to extract image features with rich semantic information. For each image region i, the corresponding vector X_i is selected as its visual embedding in the output tensor of CLIP. For the whole image, all visual embeddings are merged by mean pooling to obtain a visual feature vector R_v with 512 dimensions.

$$R_v = Mean(X_1, X_2, ..., X_n). \tag{4}$$

where X_i denotes image patches, $Mean$ denotes the mean pooling operation.

3) Social Graph Feature Extraction Module. We use the signedGAT [4] to extract social graph features. We construct a social network $G = (V, E, A)$, where $V = \{v_1, v_2, ..., v_n\}$ denotes the set of nodes consisting of n nodes, $E = \{e_{ij} | v_i \rightarrow v_j\}$ is the set of edges, which denotes the directed edges from nodes v_i to v_j, and $A = [a_{ij}]_{n \times n}$ denote the adjacency matrix, where a_{ij} denotes the edge weights from nodes v_i to v_j.

For each node v_i and its neighbor node v_j, calculating the similarity score s_{ij} between them. Then, we compute enhanced adjacency matrix, which calculated as follows:

$$\alpha'_{ij} = \frac{\exp(LR(a^T[Wh_i \parallel Wh_j]))}{\sum_{k \in N'_i} \exp(LR(a^T[Wh_i \parallel Wh_k]))} \tag{5}$$

where h_i and h_j denotes the feature vectors of nodes i and j, LR denotes LeakyReLU, and N'_i denotes the set of neighboring nodes.

After obtaining the attention coefficients, we calculate the connected feature vector $\tilde{h}'_i$, which are calculated as follows:

$$\tilde{h}'_i = \sum_j \alpha'_{ij} W h_j \tag{6}$$

where W is the learnable linear transformation matrix. $\tilde{h}'_i$ belongs to $\tilde{h}^+_i$ and $\tilde{h}^-_i$.

Then, we obtain the final representations $\hat{h}_i = [\tilde{h}^+_i \parallel \tilde{h}^-_i]$ of the current node i. Finally, We concatenate the updated node embeddings of each head together as the overall graph feature:

$$R_g = \parallel_{i=1}^{H} \sigma\left(\hat{h}_i\right) \tag{7}$$

where H denotes the number of heads. $\parallel$ means concatenation operation.

3.3 Weighted Cross-Modal Aggregation Module

Inspired by previous work [4], this method did not consider the modal imbalance between text and image features. In addition, the text contains many interfering words, which have a strong ability to interfere with the model. Therefore, we propose a weighted cross-modal aggregation module. In addition, the model will add perturbed text, image, and social graph features to calculate KL divergence pairwise, with text as the dominant factor. Firstly, the association weights between text and image, text and social graph, and image and social graph will be obtained. Then, the weighted sum of the associated weights and the original features is used to obtain the enhanced text features that fuse the three modalities. Then, MSE loss is used to narrow the distance between the enhanced text features, image features, and social graph features. Figure 3 shows the details of the weighted aggregation module.

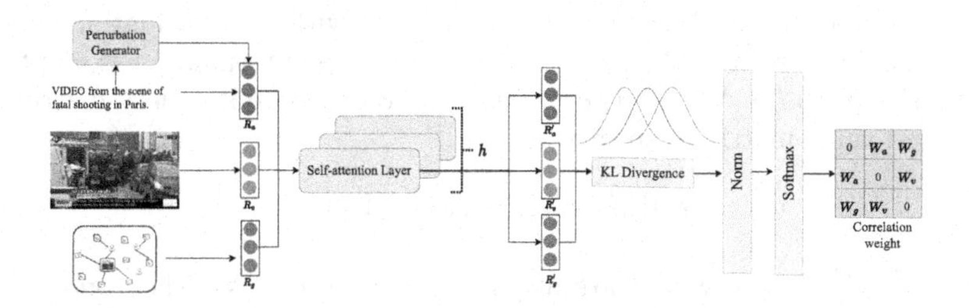

Fig. 3. The details of proposed WCA module.

Specifically, the KL divergence, which is defined as the difference between the entropy of q with respect to q and the cross-entropy of p, is calculated as follows:

$$D_{KL}(p||q) = \sum_i P(i) \log \frac{P(i)}{q(i)} \tag{8}$$

where p and q denote two probability distributions, respectively.

Given three embedding vectors R_a, R_v, R_g, which represent text, image and social graph features, respectively. Then, the three features are normalized, and the similarity between the features is obtained after the KL divergence operation between the two processed features. For example, for the text feature R_t and image feature R_v, we calculate the KL divergence between two distributions as:

$$W_a = KLDiv(logsoftmax(R_a), softmax(R_v)) \tag{9}$$

where $KLDiv$ denotes the KL divergence operation, $logsoftmax$ and $softmax$ denote the log normalization and normalization operation, respectively. And W_a

denotes the KL divergence between text features and image features. Then, the representations are normalized as follows:

$$W'_m = \frac{\exp(-W_m)}{\exp(-W_a) + \exp(-W_g) + \exp(-W_v)} \tag{10}$$

where W'_m denotes the weights after normalization, m belongs to W_a, W_v, W_g.

Finally, the three embedding vectors are weighted and averaged according to the corresponding weights to obtain the fusion vector R_f:

$$R_f = W'_a \cdot R_a + W'_v \cdot R_v + W'_g \cdot R_g \tag{11}$$

where R_a, R_v, R_g represent text, image and social graph features, respectively.

3.4 Multimodal Feature Fusion

We use the co-attention mechanism [11] to interact with the information between different modalities and obtain a richer feature representation. For example, for the fusion feature R_f. We use Q_f, K_f and V_f to calculate its query matrix, key matrix and value matrix, respectively:

$$Q_f = R_f W_f^Q,\ K_f = R_f W_f^K,\ V_f = R_f W_f^V \tag{12}$$

where $W_f^Q, W_f^K, W_f^V \in \mathbb{R}^{d \times \frac{d}{H}}$ are linear transformations. We then produce the multi-head self-attention feature of the text modal as:

$$Z_f = \left(\|_{h=1}^{H} \operatorname{softmax}\left(\frac{Q_f K_f^T}{\sqrt{d}} \right) V_f \right) W_f^O \tag{13}$$

where h denotes the h-th head, W_t^O is the output linear transformation. We perform the same operations on R_v and R_g to obtain the features Z_v and Z_g.

Then we produce the multi-modal features by using co-attention mechanism. Similar to the self-attention mechanism, but we replace R_f with Z_v to get the query matrix Q_v, and replace R_f with Z_f to get the key matrix K_f and value matrix V_f. Then we obtain the cross-modal enhanced feature Z_{vf} as:

$$Z_{vf} = \left(\|_{h=1}^{H} \operatorname{softmax}\left(\frac{Q_v K_f^T}{\sqrt{d}} \right) V_f \right) W_{vf}^O \tag{14}$$

where W_{vf}^O is the output linear transformations. Z_{vf} represents the enhanced text feature with the image feature.

For each pair of the three modal features, we again perform the cross-modal co-attention mechanism mentioned above. Then we get six cross-modal enhancement features: Z_{vf}, Z_{fv}, Z_{vg}, Z_{gv}, Z_{gf}, Z_{fg}. We then concatenate them together as the final multimodal features:

$$\tilde{Z} = concat(Z_{vf}, Z_{fv}, Z_{vg}, Z_{gv}, Z_{gf}, Z_{fg}) \tag{15}$$

Finally, we feed $\tilde{Z}$ into the fully connected layer FC for affect predictions ŷ, which can be formulated as:

$$\hat{y} = FC\left(\tilde{Z}\right) \tag{16}$$

3.5 Objective Function

Our proposed method uses two losses, including MSE loss and cross-entropy loss. The MSE loss is to shorten the distance between two probability distributions.

$$\mathcal{L}_{align} = \frac{1}{N}\sum_{i=1}^{N}(y_i - \hat{y}_i)^2 \tag{17}$$

where y_i and $\hat{y}_i$ denotes two different probability distributions.

We use the cross-entropy loss function as:

$$\mathcal{L}_{classify} = -y\log(\hat{y}_i) - (1-y)\log(1-\hat{y}_i) \tag{18}$$

where $\hat{y}_i$ denotes the predicted probability of p_i being a rumor.

The final loss can be written as follows:

$$\mathcal{L} = \lambda_c\mathcal{L}_{classify} + \lambda_a\mathcal{L}_{align} \tag{19}$$

where λ_c and λ_a are two hyperparameters that are used to balance the two losses.

4 Experiments

In this section, we present the experiments to evaluate the effectiveness of WCAN.

4.1 Datasets

1) PHEME Dataset [12]: The PHEME dataset is a large-scale real-world dataset for studying rumor propagation in social media. It consists of over 5000 tweets collected from Twitter, covering rumors, non-rumors, and related information from 10 different events.
2) Weibo Dataset [13]: The Weibo dataset is a collection of social media content from one of the largest microblogging platforms in China, Sina Weibo. It comprises a vast amount of user-generated data, including text, images, videos, and other multimedia content.

Table 1. Results of different models on PHEME and Weibo datasets.

Dataset	Methods	Accuracy	Fake news			Real news		
			Precision	Recall	F1	Precision	Recall	F1
PHEME	EANN [14]	0.681	0.685	0.664	0.694	0.701	0.750	0.747
	TextGCN [15]	0.828	0.775	0.735	0.737	0.827	0.828	0.828
	MVAE [16]	0.852	0.806	0.719	0.760	0.871	0.917	0.893
	SpotFake [17]	0.823	0.743	0.745	0.744	0.864	0.863	0.863
	MFAN [4]	0.887	**0.856**	**0.841**	**0.849**	0.884	0.872	0.871
	HMCAN [2]	0.881	0.830	0.838	0.834	**0.910**	0.905	0.907
	WCAN	**0.887**	0.855	0.734	0.790	0.898	**0.949**	**0.923**
Weibo	EANN [14]	0.782	0.827	0.697	0.756	0.752	0.863	0.804
	TextGCN [15]	0.787	**0.975**	0.573	0.727	0.712	**0.985**	0.827
	MVAE [16]	0.824	0.854	0.769	0.809	0.802	0.875	0.837
	SpotFake [17]	0.869	0.877	0.859	0.868	0.861	0.879	0.870
	MFAN [4]	0.903	0.873	0.865	0.868	0.905	0.897	0.898
	HMCAN [2]	0.885	0.920	0.845	0.881	0.856	0.926	0.890
	WCAN	**0.918**	0.910	**0.910**	**0.910**	**0.926**	0.926	**0.926**

4.2 Baselines

We compare our model with a series of representative baselines, such as EANN [14], TextGCN [15], MVAE [16], SpotFake [17] and HMCAN [2].

Table 1 summarizes the performance of the approaches on PHEME and Weibo, from which we have three-fold observations: 1) Our proposed model achieves the highest accuracy, demonstrating that the use of multimodal alignment and adversarial training improves the performance of the model. 2) MFAN is more accurate than other baseline models as it uses co-attention mechanisms to focus on important parts of features. Our model surpasses MFAN, showing that the WCA module can obtain features with high correlation between modalities, improving the effectiveness of the model. 3) In the case of fake news, the recall and F1 of MFAN are the highest on the PHEME dataset while the precision of TextGCN is the highest on the Weibo dataset. And in the real news, the precision of HMCAN is higher than our model on PHEME dataset. The results indicate that combining text context information has a positive effect on rumor detection.

4.3 Ablation Experiment

Our method includes an adversarial training (AT) method and a weighted cross-modal aggregation (WCA) module. To verify their validity, we evaluate different variants of the model, including WCAN without AT (WCAN_T) and WCAN_A without WCA (WCAN_A). The performance of the variations is summarized in Table 2, from which we have two observation: 1) WCAN performs better than WCAN_T, indicating that adversarial training enhances the robustness of the

Table 2. Results of the variations of WCAN on PHEME and Weibo datasets.

Dataset	Methods	Accuracy	Precision	Recall	F1
PHEME	WCAN_T	0.873	0.871	0.873	0.872
	WCAN_A	0.855	0.851	0.855	0.850
	WCAN	**0.887**	**0.885**	**0.887**	**0.884**
Weibo	WCAN_T	0.871	0.872	0.871	0.871
	WCAN_A	0.885	0.885	0.885	0.884
	WCAN	**0.918**	**0.910**	**0.910**	**0.910**

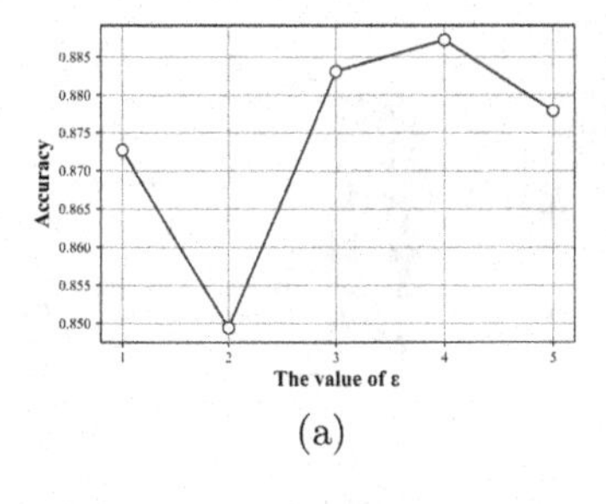

(a)

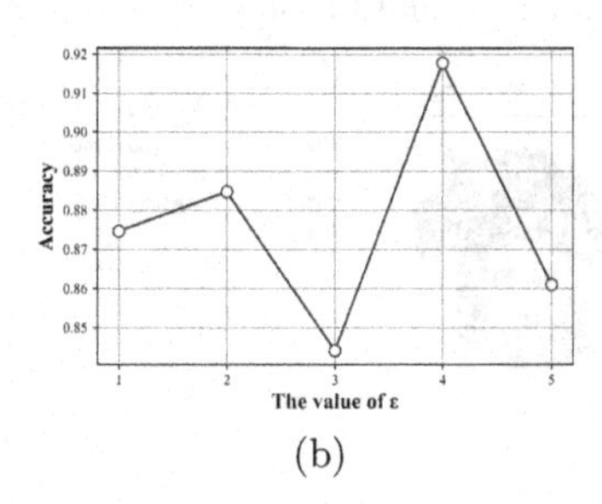

(b)

Fig. 4. Impact of the value of ε for the accuracy of fake news on two datasets. The left figure shows the result on the PHEME dataset, and the right figure shows the effect on the WEIBO dataset.

model and improves the accuracy of the model for datasets with fewer data. 2) WCAN performs better than WCAN_A, it shows that the weighted cross-modal aggregation module proposed by us can combine the features with high correlation between different modalities, and can obtain better feature representations. WCAN solves the problem of multimodal feature heterogeneity and provides more effective information for classification tasks.

4.4 Hyper-parameter Analysis

During the training process, we introduce the FGSM adversarial attack method. In the FGSM algorithm, hyperparameter ε affects the performance of the model. We conduct a series of experiments by setting different hyperparameter values and observing their effect on the accuracy of the model, as shown in Fig. 4. It turns out that when $\varepsilon = 4$, the model achieves the best performance.

4.5 Visualization on the Representations

We visualize multimodal representations on PHEME and Weibo datasets in Fig. 5, from which we can observe that the two-aspect representations have been disentangled clearly, while the single representations is tight to each other in the feature space, indicating the effectiveness of the feature disentanglement from two aspects.

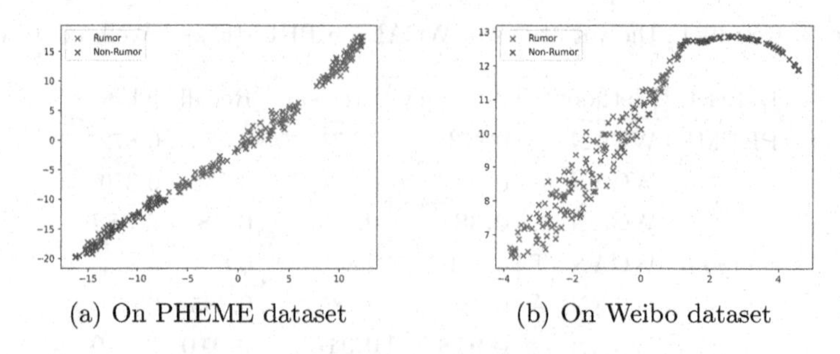

(a) On PHEME dataset (b) On Weibo dataset

Fig. 5. Visualization of representations on PHEME and Weibo dataset.

Fig. 6. Failure samples have a high degree of matching between image and text, but are misclassified. The image or text can provide very little information to prove that it is a rumor.

4.6 Case Study

Figure 5 shows two failed cases. Case (a) is a failure case that the model misclassified it as a rumor. The model is disturbed because the image content is in an exaggerated style and the text and image content have high correlation. Case (b) is a rumor but the WCAN model incorrectly classifies it as a non-rumor. The image content can only provide little evidence to prove that they are rumors, and the correlation between the text and image is high, so the model incorrectly distinguishes them as non-rumors (Fig. 6).

5 Conclusions

In this paper, we propose a Weighted Cross-modal Aggregation network (WCAN) for rumor detection, which uses information from three modalities: text, image, and social graph to classify rumors. In addition, We propose the WCA module, which can measure the correlation weights between different modal distributions and ultimately obtain a multimodal feature representation dominated by text. Simultaneously, the distance between multimodal features, image features, and social graph features is narrowed through MSE loss. The experiments are conducted on two rumor datasets, PHEME and Weibo, demonstrating the effectiveness of our proposed WCAN.

Acknowledgement. This work is supported by the Youth Talent Support Programme of Guangdong Provincial Association for Science and Technology (No. SKXRC202305), and a grant from the RGC of the HKSAR, China (UGC/FDS16/E17/21).

References

1. Yan, M., Yang, W., Sun, B., Zhu, Y.: Heterogeneous graph attention networks with bi-directional information propagation for rumor detection. In: ICBDA, pp. 236–242 (2022)
2. Qian, S., Wang, J., Hu, J., Fang, Q., Xu, C.: Hierarchical multi-modal contextual attention network for fake news detection. In: ACM SIGIR, pp. 153–162 (2021)
3. Sun, T., Qian, Z., Dong, S., Li, P., Zhu, Q.: Rumor detection on social media with graph adversarial contrastive learning (2022)
4. Zheng, J., Zhang, X., Guo, S., Wang, Q., Zang, W., Zhang, Y.: MFAN: multi-modal feature-enhanced attention networks for rumor detection. In: IJCAI, pp. 2413–2419 (2022)
5. Wiegmann, M., Khatib, K.A., Khanna, V., Stein, B.: Analyzing persuasion strategies of debaters on social media. In: 29th International Conference on Computational Linguistics, pp. 6897–6905 (2022)
6. Ye, L., Rochan, M., Liu, Z., Wang, Y.: Cross-modal self-attention network for referring image segmentation. In: CVPR, pp. 10502–10511 (2019)
7. Lu, W., Chenyu, W., Guo, H., Zhao, Z.: A cross-modal alignment for zero-shot image classification. IEEE Access **11**, 9067–9073 (2023)
8. Ma, J., Wu, F., Chen, Y., Ji, X., Ding, Y.: Effective multimodal reinforcement learning with modality alignment and importance enhancement. arXiv preprint arXiv:2302.09318 (2023)
9. Pandey, R., Shao, R., Liang, P.P., Salakhutdinov, R., Morency, L.P.: Cross-modal attention congruence regularization for vision-language relation alignment. arXiv preprint arXiv:2212.10549 (2022)
10. Radford, A., et al.: Learning transferable visual models from natural language supervision. In: International Conference On Machine Learning, pp. 8748–8763. PMLR (2021)
11. Lu, Y.J., Li, C.T.: GCAN: graph-aware co-attention networks for explainable fake news detection on social media. arXiv preprint arXiv:2004.11648 (2020)
12. Zubiaga, A., Liakata, M., Procter, R.: Exploiting context for rumour detection in social media. In: 9th International Conference on Social Informatics, pp. 109–123 (2017)
13. Song, C., Yang, C., Chen, H., Tu, C., Liu, Z., Sun, M.: CED: credible early detection of social media rumors. IEEE Trans. Knowl. Data Eng. **33**, 3035–3047 (2019)
14. Wang, Y., et al.: EANN: event adversarial neural networks for multi-modal fake news detection. In: ACM SIGKDD, pp. 849–857 (2018)
15. Yao, L., Mao, C., Luo, Y.: Graph convolutional networks for text classification. In: AAAI, pp. 7370–7377 (2019)
16. Khattar, D., Goud, J.S., Gupta, M., Varma, V.: MVAE: multimodal variational autoencoder for fake news detection. In: WWW, pp. 2915–2921 (2019)
17. Singhal, S., Shah, R.R., Chakraborty, T., Kumaraguru, P., Satoh, S.: SpotFake: a multi-modal framework for fake news detection. In: BigMM, pp. 39–47 (2019)

Texts, Web, Social Network

Quantifying Opinion Rejection: A Method to Detect Social Media Echo Chambers

Kushani Perera(✉) and Shanika Karunasekera

University of Melbourne, Melbourne, VIC 3010, Australia
{kushani.perera,karus}@unimelb.edu.au

Abstract. Social media echo chambers are known to be common sources of misinformation and harmful ideologies that have detrimental impacts on society. Therefore, techniques to detect echo chambers are of great significance. Reinforcement of supporting opinions and rejection of dissenting opinions are two significant echo chamber properties that help detecting them in social networks. However, existing echo chamber detection methods do not capture the opinion rejection behaviour, which leads to poor echo chamber detection accuracy. Measures used by them do not facilitate quantifying both properties simultaneously while preserving the connectivity between echo chamber members. To address this problem, we propose a new measure, *Signed Echo* (*SEcho*) that quantifies opinion reinforcement and rejection properties of echo chambers and an echo chamber detection algorithm, *Signed Echo Detection Algorithm* (*SEDA*) based on this measure, which preserves the connectivity among echo chamber members. The experimental results for real-world data show that *SEDA* outperforms the state-of-the-art echo chamber detection methods in detecting the communities with echo chamber properties, such as reinforcement of supporting opinions, rejection of dissenting opinions, connectivity between community members, spread of mis/disinformation and emotional contagion.

Keywords: Echo chambers · Community detection algorithms · Social networks · Quantitative metrics

1 Introduction

Echo chambers are identified as a major culprit for promoting mis/disinformation and harmful ideologies in social media, resulting in dangerous repercussions in society [2]. For example, social media echo chambers have immensely contributed to promoting anti-vaccine ideology, resulting in the spread of diseases [7,13] and political polarization [12], spreading hatred in the society. Therefore, finding methods to detect social media echo chambers is crucial.

An *echo chamber* can be roughly defined as *a network of users with the same opinion regarding a given topic whose users frequently reinforce the content that supports their pre-existing opinions while discrediting and excluding dissenting*

D.-N. Yang et al. (Eds.): PAKDD 2024, LNAI 14650, pp. 57–69, 2024.
https://doi.org/10.1007/978-981-97-2266-2_5

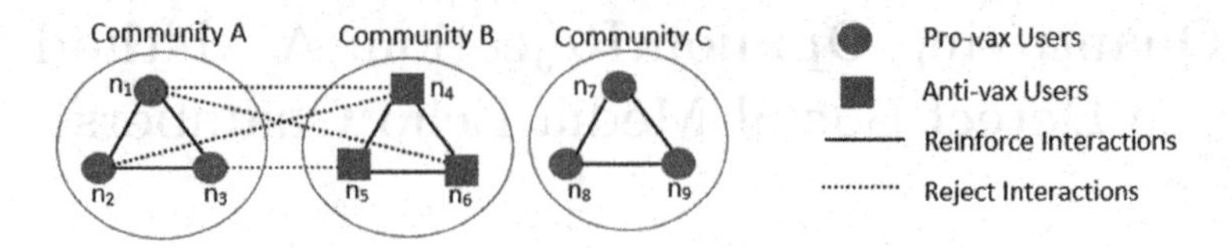

Fig. 1. Echo chambers and filter bubbles in a social network.

opinions [2]. Core properties that define echo chambers are *opinion polarization, homophily of the users, connectivity between members, reinforcement of supporting opinions* and *active rejection of the dissenting opinions.* Properties correlated with echo chambers include *spread of mis/disinformation, amplification of the content,* and *the emotional contagion of the users* [2,5,17]. These properties are helpful in detecting echo chambers in social networks.

Filter bubbles are also user groups with most of the echo chamber properties, except the *active rejection of dissenting opinions* [2,14]. They get formed because of users' ignorance about dissenting opinions, yet echo chambers due to the cognitive bias of the users. Therefore, filter bubbles are isolated communities with no interactions with external users, while echo chambers are well-connected communities with frequent rejection interactions with external users. Both have frequent reinforcing interactions between community members.

Figure 1 shows echo chambers and filter bubbles in a social network with pro-vax (support vaccination) and anti-vax (oppose vaccination) users. In it, all communities have reinforce interactions between their members. Community A and B, echo chambers, have opinion rejection interactions between their members. Community C, a filter bubble, has no interactions with the rest of the network.

Echo chambers are more abundant than filter bubbles in today's highly connected social media platforms, and methods to detect them are required. However, existing echo chamber detection methods, which do not consider the *rejection of dissenting opinions* property, mostly detect filter bubbles instead of echo chambers [14]. Furthermore, we identify that none of the existing echo chamber measures incorporates both opinion polarization and network polarization aspects to enable the detection method to consider both at the community detection phase (please refer to Sect. 2). Therefore, existing echo chamber detection methods, which use these measures, cannot detect communities based on both opinion reinforcement and rejection aspects.

A possible solution for the above problem is to model the echo chamber detection problem as a community detection problem in a signed user network whose positive/negative edges are the reinforce/rejection interactions and use existing community detection measures [19] to detect the echo chambers. These measures facilitate incorporating both opinion and network polarization aspects and quantifying both opinion reinforcement and rejection properties of communities, yet do not guarantee the connectedness of the community members, which is also a significant property of echo chambers (please refer to Sect. 4). To address this problem, first, we propose *SEcho*, a metric that quantifies opinion reinforcement and rejection properties of the communities and the connectivity among the community members. Next, we propose an echo chamber detection method whose objective is to find a set of communities with maximal average *SEcho* value. As this is an NP-hard problem, to obtain a near-optimal solution, we propose a greedy algorithm, *SEDA* whose computational complexity is $O(Nlog(N))$ where N is the user count. The experiments on real-world data show that compared to the baselines, *SEDA* detects communities, which demonstrate strong core and correlated echo chamber properties. Following are our contributions.

- Propose a measure to quantify the echo chamber properties: opinion reinforcement, opinion rejection and connectivity among members.
- Propose an algorithm which utilises the proposed measure to detect echo chambers in social networks.
- Experimentally validate that the proposed method outperforms the state-of-the-art methods in detecting communities with echo chamber properties.

2 Related Work

Echo chamber detection methods detect user networks that possess one or more echo chamber properties. The properties are either core [2,9] or correlated [5,17]. We focus on the methods based on core properties. Both quantitative [7,8,18] and qualitative measures [18] used to measure these properties are of two types: (1) *opinion polarization measures* [7,8] and (2) *network polarization measures* [8,17,18]. *None of the existing echo chamber measures incorporates both opinion polarization and network polarization aspects.* Some echo chamber detection methods use *opinion polarization measures* to measure the opinion polarization and user homophily properties of echo chambers [3,6,10]. Some use *network polarization measures*, which are the same measures used in community detection algorithms [4,15,16], to measure opinion reinforcement property of echo chambers. *None of the methods considers the "opinion rejection" property of the echo chambers.* Therefore, existing echo chamber measures, which do not incorporate both opinion and network polarization aspects cannot be used to quantify both opinion reinforcement and rejection behaviour simultaneously. Sequential application of network and opinion polarization measures to detect and validate echo chambers does not solve this problem [7,18]. Therefore, to improve the echo chamber detection accuracy, new measures to quantify both opinion reinforcement and rejection properties of echo chambers are required.

3 Preliminaries

This section and Table 1 introduce the frequently used terms in the paper. Given a matrix $X \in \mathbb{R}^{n \times m}$ with n rows and m columns, X_i is the ith row of X and $X_{i,j}$ is the element in i^{th} row and j^{th} column. Given a set X, $|X|$ is the cardinality of X. Given a scalar value, x, $|x|$ is the absolute value of x. G is the social network graph. The adjacency matrix of G is $W \in \mathbb{R}^{|N| \times |N|}$ and $W_{i,j}$ is the edge weight between i and j. Given that G_A/G_D is the resulting graph after removing the edges with negative/positive weights in G, A/D is the adjacency matrix of G_A/G_D such that $A_{i,j} = max\ (0, W_{i,j})$ and $D_{i,j} = min\ (0, W_{i,j})$.

Table 1. Frequently used terms.

N	Set of nodes in G	i, j	Nodes in G, $i, j \in N$
E	Set of edges in G	$e_{i,j}$	Edge between node i and j
C	Set of communities in G	$W/A/D$	Adjacency matrix of $G/G_A/G_D$
c	A community, $c \in C$, $c \subseteq N$	$X_{i,j}$	Edge weight between node i and j

Definition 1. *Given that* i, j, c, W, A, D *and* $X_{i,j}$ *are as defined in Table 1, the total weight of the edges of graph* $G/G_A/G_D$ *that lie within community* c, $In(X)_c = \frac{1}{2} \sum_{i,j \in c} X_{i,j}$ *where* $X = W, A, D$.

Definition 2. *Given that* i, j, c, W, A, D *and* $X_{i,j}$ *are as defined in Table 1, the total weight of the edges of graph* $G/G_A/G_D$ *that lie between the nodes in* c *and the nodes in other communities,* $Out(X)_c = \sum_{i \in c} \sum_{j \notin c} X_{i,j}$ *where* $X = W, A, D$.

Definition 3. *Given that* i, c, W, A, D *and* $X_{i,j}$ *are as defined in Table 1, the total weight of the edges of graph* $G/G_A/G_D$ *between node* i *and nodes in* c, $k(X)_{i,c} = \sum_{j \in c} X_{i,j}$ *where* $X = W, A, D$.

Definition 4. *Given that* i, c, W, A, D *and* $X_{i,j}$ *are as defined in Table 1, the total weight of the edges of graph* $G/G_A/G_D$ *between node* i *and nodes that are not in* c, $k'(X)_{i,c} = \sum_{j \notin c} X_{i,j}$ *where* $X = W, A, D$.

Definition 5. *Given that* i, j *and* $W_{i,j}$ *are as defined in Table 1, the total weight of the edges of graph* G, $w = \frac{1}{2} \sum_{\forall i,j \in N} |W_{i,j}|$.

4 Echo Chamber Detection in Signed Networks

The types of reinforce/reject interactions we identify in social networks include a user likes/dislikes another's content, one user replies agreeing/disagreeing with another, two users in a discussion chain with the same/opposite stance. We model these interactions between user pairs as the positive/negative signed edges of an undirected user graph. The sign of the edge represents the type of the interaction (reinforce/agreement or reject/disagreement). The edge weight represents the strength of the relationship (number of interactions). Given two user nodes in G, we count the total number of agreements (*acount*) and disagreements (*dcount*) between the users and compute the edge weight between the two nodes using the equation, $edgeweight = acount - dcount$. We define an edge whose

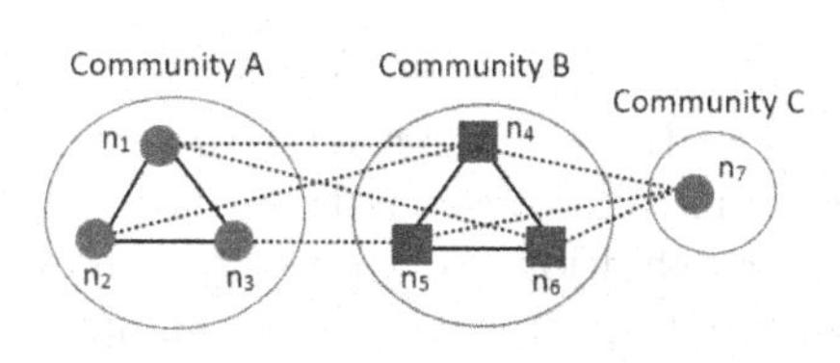

(a) Scenario 1: n_7 in its own community

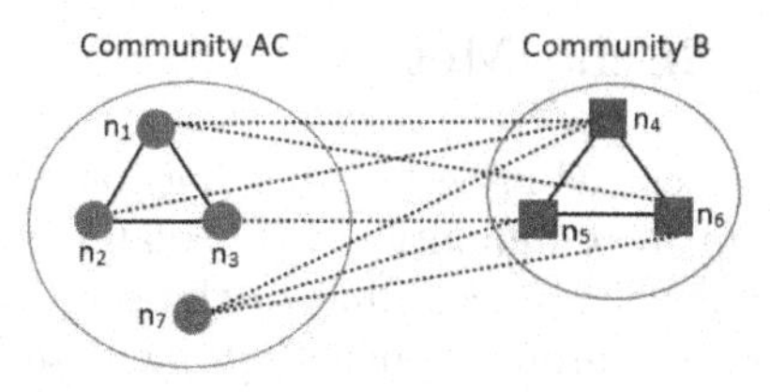

(b) Scenario 2: n_7 added to Community A

Fig. 2. Assigning a community to a node with no agree and many disagree edges.

edgeweight > 0 as an *agreement edge* and an edge whose *edgeweight* < 0 as an *disagreement edge.* The echo chamber detection problem is equivalent to finding a set of sub-networks in this graph with maximal intra/inter-community agree/disagree edges and minimal intra/inter-community disagree/agree edges. We discuss the ability of using existing community detection measures for signed graphs to solve the problem, using *Signed Modularity* (*SMod*) [19] as an example.

Given that all terms are as defined in Sect. 3 and $Tot(X)_c = \sum_{i \in c} \sum_{j \in N} X_{i,j}$ $(X = A, D)$, *SMod* of c, $Q_c = \left[\left(\frac{In(A)_c}{w}\right) - \left(\frac{Tot(A)_c}{2w}\right)^2\right] - \left[\left(\frac{In(D)_c}{w}\right) - \left(\frac{Tot(D)_c}{2w}\right)^2\right]$. *SMod* objective is to find the graph partitions, C, that maximise Q where $Q = \sum_{c \in C} Q_c$. According to the equations, the nodes that have high inter-community disagreements yet no/few intra-community agreements can increase Q_c by increasing $(\frac{Tot(D)_c}{2w})^2$. As shown in Definition 6, we define the nodes with no/few intra-community agreements as isolates. As an echo chamber is defined as a network of users, isolates cannot be considered as echo chamber members and the communities with isolates are not strong echo chambers. Therefore, *SMod detects communities that do not comply with the echo chamber definition and result in false positives.* We show this by applying *SMod* to Example 1.

Definition 6. *Given that i, c and A are as defined in Table 1,* $i \in c$, $k(A)_{i,c}$ *is the total weight of the agree edges between i and other nodes in c and* $\alpha \in \mathbb{R}_0^+$ *is a user defined parameter, i is an* isolate *if and only if* $|c| > 1$ *and* $k(A)_{i,c} \leq \alpha$.

Example 1: Figure 2 shows a network of pro-vax and anti-vax users whose edges are reinforcement and rejection interactions. We assume the weight of each edge is 1. Scenario 1 (S1) and Scenario 2 (S2) are two graph partitions where *S1:* A (n_1, n_2, n_3), B (n_4, n_5, n_6), C (n_7). *S2:* AC (n_1, n_2, n_3, n_7), B (n_4, n_5, n_6).

SMod of S1, $Q_{s1} = Q_A + Q_B + Q_C$, where $Q_A = 1/13\ (3 - 6^2/52 - 0 + 4^2/52) = 0.2$, $Q_B = 1/13\ (3 - 6^2/52 - 0 + 7^2/52) = 0.25$ and $Q_C = 1/13\ (0 - 0 - 0 + 3^2/52) = 0.01$. Therefore, $Q_{s1} = 0.46$. *SMod* of S2, $Q_{s2} = Q_{AC} + Q_B$, where $Q_{AC} = 1/13\ (3 - 6^2/52 - 0 + 7^2/52) = 0.25$ and $Q_B = 1/13\ (3 - 6^2/52 - 0 + 7^2/52) = 0.25$. Therefore, $Q_{s2} = 0.5$. As $Q_{s1} < Q_{s2}$, S2 is selected over S1, resulting in Community AC, with an isolate, n_7.

5 SEcho Method

As discussed in Sect. 4, existing community detection measures for signed graphs result in communities with isolates, which cannot be considered as echo chamber members according to the echo chamber definition. To address this problem, we propose metric, *SEcho*, to measure the strength of a community based on the aspects: opinion reinforcement, opinion rejection and connectivity between community members. We also propose an echo chamber detection algorithm, *SEDA*, that uses *SEcho* to detect communities with echo chamber properties: high intra/inter-community agreements/disagreements, low intra/inter-community disagreements/agreements and minimal isolates.

5.1 SEcho Metric

Definition 7 defines *SEcho* of a community c, SE_c. *SEcho* increases with $In(A)_c$ and $Out(D)_c$ and decreases with $In(D)_c$ and $Out(A)_c$. $I(c)$, indicates whether c includes an isolate. Minimum SE_c is 0, which is achieved when c includes an isolate. Theorem 1 discusses the minimum and maximum values of SE_c.

Definition 7. *Given that i, c, A, $In(A)_c$, $Out(A)_c$, $In(D)_c$, $Out(D)_c$, are as defined in Sect. 3, $\alpha \in \mathbb{R}_0^+$ is a user defined parameter and $I(c)$ is an indicator function such that $I(c) = 1$ if $\forall i \in c \; \exists \; j \neq i \in c \; A_{i,j} > \alpha$, $I(c) = 0$ otherwise, the SEcho of community c, SE_c, is computed using Eq. (1).*

$$SE_c = \left(1 + \frac{In(A)_c - Out(A)_c - In(D)_c + Out(D)_c}{w}\right) I(c) \tag{1}$$

Theorem 1. *Given that c is a community, SE_c and an isolate are as defined in Definition 7 and 6 respectively, the minimum and maximum values of SE_c are 0 and 2, respectively. Minimum is achieved when c contains an isolate in it.*

Proof. Please refer to this link[1] for the proof.

As shown in Eq. (2), the objective of *SEcho* based community detection is to find the set of non-overlapping communities, C, which achieves the maximum average *SEcho* value of the graph, SE.

$$\underset{C}{\operatorname{argmax}} \, SE \text{ where } SE = \frac{\sum_{c \in C} SE_c}{|C|} \tag{2}$$

5.2 Greedy Optimisation

As the optimisation problem of Eq. (2) is NP-hard, we propose a greedy method, *SEDA*, to reach a local maximisation to achieve it. *SEDA* uses the same iterative approach as Louvain algorithm [4], which consists of two phases (1) community

[1] https://sites.google.com/view/kushani/publications.

Algorithm 1: Community Detection Phase of SEDA

input : Adjacency matrix G (W), Maximum iterations ($maxIter$)
output: Dictionary of neighbour communities ($neighcomms$)

1 $allNodes \leftarrow$ All nodes in W ; $incr \leftarrow -1$; $iter \leftarrow 0$; $commID \leftarrow 1$;
2 **foreach** $i \in allNodes$ **do**
3     $neighcomms[i] \leftarrow commID$; $commID{+}{+}$;
4 **end**
5 **while** $(incr! = 0)$ & $(iter \leq maxIter)$ **do**
6     $incr \leftarrow 0$;
7   **foreach** $i \in allNodes$ **do**
8         $iNN \leftarrow$ Neighbours of i in W ;
9         $commNN \leftarrow [neighcomms[iNN[0]], neighcomms[iNN[1]], \cdots$;
10         $ucommNN \leftarrow \{ commNN \}$; $mods = []$;
11   **foreach** $c \in ucommNN$ **do**
12             $mods[c] \leftarrow \Delta SE_{(i,c)}$; (Theorem 2, Eq. (3))
13   **end**
14         $\Delta X_{max} \leftarrow \max(mods)$;
15   **if** $\Delta X_{max} > 0$ **then**
16             $maxComm \leftarrow$ Community with ΔX_{max} ;
17             $neighcomms[i] \leftarrow maxComm$; $incr{+}{+}$;
18   **end**
19   **end**
20     $iter{+}{+}$;
21 **end**

detection and (2) graph construction. The main difference from Louvain is the quantity maximised at a given iteration (Line 12 of Algorithm 1) at the community detection phase, $\Delta SE_{i,c}$. Given that i is any node in the graph, c' is the current community of i, c is the community to which i is to be added, $SE_{i,x}$ is the *SEcho* of the graph when i is in community x, we define $\Delta SE_{i,c} = SE_{i,c} - SE_{i,c'}$. As described in Algorithm 1, at each iteration of the community detection phase, *SEDA* assigns a node to a community to achieve the objective in Eq. (3). As shown in Theorem 2, $\Delta SE_{i,c}$ can be efficiently computed by considering only the edges connected to i. From the final set of detected communities, we select the ones with no isolates (*SEcho*> 0) as echo chambers. If none such communities are found, we conclude that the dataset does not have echo chambers.

$$\underset{c \in C}{\operatorname{argmax}}\, \Delta SE_{i,c} \tag{3}$$

Theorem 2. *Given that $k(A)_{i,c}$, $k(D)_{i,c}$, $k'(A)_{i,c'}$, $k'(D)_{i,c'}$ and w are as defined in Sect. 3 and SE_c and $SE_{c'}$ are as defined in Eq. (1),* SEcho *gain achieved by removing node i from community c' and adding to community c, $\Delta SE_{i,c} = \frac{2}{w}\left(k(A)_{i,c} - k(A)_{i,c'} - k(D)_{i,c} + k(D)_{i,c'}\right)$.*

Proof. To prove this, we use the fact that all edges except the edges of i remain constant for $SE_{i,c}$ and $SE_{i,c'}$. Please refer to this link[2] for the detailed proof.

Example 1 Revisited: We solve the community detection problem in Example 1 (in Sect. 4) again, using *SEcho* measure. According to Eq. (1) and (2), *SEcho* of the $s1$, $SE_{s1} = (\sum SE_A + SE_B + SE_C)/3$ where $SE_A = [1+(3-0-0+4)/13)].1 = 1.54$, $SE_B = [1+(3-0-0+7)/13)].1 = 1.77$ and $SE_C = [1+(0-0-0+3)/13)].1 = 1.23$. Therefore, $SE_{s1} = 1.51$. *SEcho* of the $s2$, $SE_{s2} = (\sum SE_{AC} + SE_B)/2$ where $SE_{AC} = [1+(3-0-0+7)/13)].0 = 0$ and $SE_B = [1+(3-0-0+7)/13)].1 = 1.77$. Therefore, $SE_{s2} = 0.88$. As $SE_{s1} > SE_{s2}$, in the *SEcho* based detection, $S1$ is selected over $S2$. This shows that *SEcho* selects communities with minimal isolates, which comply with the echo chamber definitions.

Computational Complexity: Given that N is the number of graph nodes, the computational complexity of *SEDA* is $O(Nlog(N))$, which is same as Louvain [4].

6 Experiments

Datasets: We evaluate the performance of *SEcho* method using real-world data. D_1, D_2 and D_3 datasets include ~4, ~9, ~36 million English language tweets related to vaccination topic, published during the periods, 1st of June–31st of August 2020, 1st of September–30th of November 2020, 1st of December–1st of February 2021, respectively. They were created from a Twitter dataset [11] whose tweets were collected by a vaccination related keyword search. The stance of each tweet is labelled as pro-vax, anti-vax or neutral. The labelling is performed using OpenAPI's GPT transformer-based stance detection tool [1], trained on 42,000+ manually labelled tweets. D_4 dataset includes ~1 million English language tweets related to Russo-Ukrainian war, collected using a keyword search for the period 1st–31st of August, 2022. The stance of each tweet is manually labelled as pro-Russian, pro-Ukrainian or neutral. We construct the reply network among the users for each dataset. The edge weights are computed as discussed in Sect. 4. A reply with the same/different stance as the original tweet is considered an agreement/disagreement. Neutral replies are assigned a zero weight. D_4's reply network includes only the users who have tweeted more than 50 times and the users mentioned in their tweets.

Baselines: As we address a community detection problem, we select existing community detection algorithms as baselines. We select an SMod based algorithm [19], which is designed for signed networks. We also select Louvain [4] and Infomap [16], which are popular community detection algorithms that use two different metrics: Modularity and Walk Metric. As Louvain is not applicable for signed networks, when using Louvain, reply count is used as edge weight.

[2] https://sites.google.com/view/kushani/publications.

Table 2. Core echo chamber properties of detected communities.

		IEF_A	OEF_A	IEF_D	OEF_D	ISF
D_1	SEDA	**0.82**	**0.18**	**0.006**	**0.994**	**0**
	SMod	0.801	0.2	0.008	0.992	0.04
	Louvain	**0.885**	**0.115**	0.871	0.129	0.64
	Infomap	0.71	0.29	0.667	0.333	0.65
D_2	SEDA	**0.797**	**0.203**	**0.016**	**0.984**	**0**
	SMod	0.777	0.223	0.018	0.982	0.04
	Louvain	**0.88**	**0.12**	0.866	0.133	0.6
	Infomap	0.714	0.286	0.674	0.326	0.61
D_3	SEDA	**0.942**	**0.058**	**0**	**1**	**0**
	SMod	0.932	0.068	0.002	0.998	0.02
	Louvain	**0.986**	**0.014**	0.99	0.009	0.55
	Infomap	0.928	0.072	0.95	0.052	0.55
D_4	SEDA	**0.734**	**0.266**	**0**	**1**	**0**
	SMod	0.657	0.343	0.007	0.992	0.03
	Louvain	**0.736**	**0.266**	0.875	0.125	0.25
	Infomap	0.412	0.588	0.765	0.235	0.25

Evaluation Metrics: In this section, we define the metrics used to evaluate the detected communities. $InEdgeFrac_A$ (IEF_A), $InEdgeFrac_D$ (IEF_D), $OutEdgeFrac_A$ (OEF_A), $OutEdgeFrac_D$ (OEF_D) and $IsolateFraction$ (ISF) measure the core echo chamber properties. Suspend Neighbour Fraction (SNF), Suspend Reply Fraction (SRF), Suspend Agree Fraction (SAF) and Suspend Disagree Fraction (SDF), Positive Sentiment (S_p) and Negative Sentiment (S_n) measure the correlated echo chamber properties.

IEF_A/IEF_D is the fraction of intra-community agree/disagree edges. OEF_A/OEF_D is the fraction of inter-community agree/disagree edges. Given that $e_{i,j}$ is as defined in Table 1, C is the set of echo chambers, $c \in C$, E is the set of edges connected to the nodes in echo chambers, $w_{i,j}$ is the weight of $e_{i,j}$, $X_c = \{e_{i,j} | \, i, j \in c\}$ and $X'_c = \{e_{i,j} | \, i \in c, j \notin c\}$, $w^p_{i,j} = |\sum \max(0, w_{i,j})|$ and $w^n_{i,j} = |\sum \min(0, w_{i,j})|$, $IEF_A = \frac{\sum_{c \in C} \sum_{e_{i,j} \in X_c} w^p_{i,j}}{|\sum_{e_{i,j} \in E} w^p_{i,j}|}$ and $OEF_A = \frac{\sum_{c \in C} \sum_{e_{i,j} \in X'_c} w^p_{i,j}}{2|\sum_{e_{i,j} \in E} w^p_{i,j}|}$. $IEF_D = \frac{\sum_{c \in C} \sum_{e_{i,j} \in X_c} w^n_{i,j}}{|\sum_{e_{i,j} \in E} w^n_{i,j}|}$ and $OEF_D = \frac{\sum_{c \in C} \sum_{e_{i,j} \in X'_c} w^n_{i,j}}{2|\sum_{e_{i,j} \in E} w^n_{i,j}|}$. ISF is the fraction of isolate nodes in detected communities. Given that i_c is the number of isolates in c, $ISF = \frac{1}{|C|} \sum_{c \in C} \frac{i_c}{|c|}$. IEF_A, OEF_A quantify the opinion reinforcement IEF_D, OEF_D quantify the opinion rejection property. ISF quantifies the connectedness of the community nodes in terms of the community's isolate count. The higher the IEF_A, OEF_D and the lower the OEF_A, IEF_D and ISF, the stronger the echo chamber effect.

Table 3. Correlated echo chamber properties of detected communities.

	D_1		D_2		D_3		D_4			
	S_p	S_n	S_p	S_n	S_p	S_n	SRF	SNF	SAF	SDF
SEDA	**0.2**	**−0.09**	**0.19**	**−0.09**	**0.15**	**−0.07**	**0.25**	**0.25**	**0.25**	**0**
SMod	0.15	−0.01	0.11	−0.05	0.11	−0.05	0.16	0.18	0.16	0.08
Louvain	0.13	−0.06	0.13	−0.06	0.12	−0.06	0.24	0.24	0.22	0.07
Infomap	0.15	−0.01	0.15	−0.07	0.11	−0.06	0.08	0.1	0.06	0.1

SNF, SRF, SAF and SDF measure the echo chamber members' interactions with suspended users who are assumed to be mis/disinformation spreaders. Given that S is the set of suspended users, and S_c is the set of suspend users in c, $SNF = \frac{1}{|C|}\sum_{c\in C}\frac{|S_c|}{|c|}$, $SRF = \frac{1}{|C|}\sum_{c\in C}\frac{\sum_{i\in c,j\in S}|w_{i,j}|}{\sum_{i\in c,j\in N}|w_{i,j}|}$, $SAF = \frac{1}{|C|}\sum_{c\in C}\frac{\sum_{i\in c,j\in S}w^p_{i,j}}{\sum_{i\in c,j\in N}w^p_{i,j}}$, $SDF = \frac{1}{|C|}\sum_{c\in C}\frac{\sum_{i\in c,j\in S}w^n_{i,j}}{\sum_{i\in c,j\in N}w^n_{i,j}}$. The higher the SRF, SNF and SAF and the lower the SDF, the higher the vulnerability of the community members to mis/disinformation. S_p and S_n are the average positive and negative sentiment of the tweets shared by the community members. Given that $t_{i,j}$ is a tweet shared between user i and j and $s_{i,j}$ sentiment of $t_{i,j}$, $s^p_{i,j} = |\sum \max(0, s_{i,j})|$ and $s^n_{i,j} = |\sum \min(0, s_{i,j})|$ and c and C are as defined in Table 1, T_c is the total number of tweets sent or received by the members of c, $S_x = \frac{1}{|C|}\sum_{c\in C}\frac{\sum_{i\in c}s^x_{i,j}}{T_c}$ where $x = p, n$. The higher the S_p and S_n the higher intensity of the emotions of the tweets.

Experimental Setup: We apply *SEDA* and baselines to the the reply network of the datasets and analyse the properties of the detected communities using the evaluation metrics. We set $\alpha = 0$. The objective of Experiment 1 is to validate to which extent the detected communities possess core echo chamber properties. The objective of Experiment 2 and 3 is to validate to which extent the detected communities possess correlated echo chamber properties.

Experiment 1: We analyse the communities using the evaluation metrics: IEF_A, OEF_A, IEF_D, OEF_D and ISF. **Experiment 2:** As promotion of emotional content is an echo chamber property [18], we analyse the communities using the evaluation metrics: S_p and S_n. The sentiment score of the tweets are available in D_1, D_2 and D_3. A qualitative analysis on D_4 showed clear evidence of aggressive content circulating within *SEDA* echo chamber members, originated by suspended users. However, we do not report them due to Twitter privacy policies. **Experiment 3:** As spread of mis/disinformation is an echo chamber property, we analyse the communities using the evaluation metrics: SRF, SNF, SAF and SDF. We report the results for the top 5 echo chambers (ranked according to the metric value used to detect them) of D_4. As Walk Metric does not measure echo chamber strength, *SMod* is used to rank Infomap communities.

Experimental Results: As shown in Table 2, *SEDA* has achieved zero ISF for all datasets. None of the baselines has achieved this in any dataset. That is, unlike *SEDA*, baselines detect communities with isolates. For all datasets, *SEDA* shows the highest OEF_D and the lowest IEF_D. In D_3 and D_4, *SEDA* shows a zero IEF_D. That is, *SEDA* communities have the highest inter-community disagreements and lowest intra-community disagreements. *SEDA* also has the second highest IEF_A and second lowest OEF_A. That is, *SEDA* communities have high intra-community agreements and low inter-community agreements. Louvain shows the highest IEF_A and the lowest OEF_A. However, it shows poor performance in terms of IEF_D, OEF_D and ISF. In contrast, *SEDA shows high performance in terms of all metrics* in Table 2. According to Table 3, *SEDA* communities show the highest S_p and S_n. This shows that *SEDA* community members share emotional content compared to other community members. Compared to the baselines, *SEDA* communities contain the highest SRF, SNF and SAF values and the lowest SDF value, which is zero. This shows that *SEDA* communities have frequent interactions with suspended users and agree with the mis/disinformation shared by them. Louvain communities' SRF, SNF values are only slightly lower than *SEDA*. However, low SAF and high SDF, show that Louvain community members do not agree with the suspended users.

Evaluation Insights: The results for inter/intra community agreement/disagreement edge fractions show that, unlike the baselines, *SEcho is suitable to measure both opinion rejection and reinforcement properties of echo chambers. SEDA* communities with no isolates *are more compatible with the user network concept of the echo chamber definition.* High performance in terms of all the metrics in Table 2 shows that *SEDA communities demonstrate core echo chamber properties and are more compatible with echo chamber definitions.* Frequent interactions and agreements with suspended users, shows that *SEDA* communities promote mis/disinformation. The high sentiment of the content shared in *SEDA* communities is an evidence of the emotional contagion of their users. These results show that *in addition to core echo chamber properties, SEDA communities also demonstrate correlated echo chamber properties such as spread of mis/disinformation and emotional contagion of users.*

7 Conclusion

In this paper, we proposed a measure, *SEcho*, to quantify the social media echo chamber properties and an echo chamber detection algorithm, *SEDA*, based on *SEcho*. Unlike existing echo chamber measures, *SEcho* incorporates both opinion polarization and network polarization aspects and facilitates quantifying opinion rejection and opinion reinforcement behaviours of community members. It also ensures that the members of the detected communities are well connected, as required by the echo chamber definition. The experiments on real-world data show that compared to the communities detected by baselines, *SEDA* communities are well connected and have high intra/inter-community opinion reinforces/rejections and low intra/inter-community opinion rejections/reinforces.

Comparatively, they also show higher vulnerability to mis/disinformation and emotional content. These results demonstrate that compared to baselines, *SEDA* detects communities with strong core and correlated echo chamber properties. In future, we plan to improve *SEcho* and *SEDA* to incorporate neutral interactions between users in addition to reinforcement and rejection interactions.

Acknowledgments. This study was funded by Defence Science and Technology Group (DSTG), Australia, under the grant number, MyIP 11113.

Disclosure of Interests. Author, Kushani Perera is funded by DSTG, Australia.

References

1. Transfer learning in NLP for tweet stance classification. https://towardsdatascience.com/transfer-learning-in-NLP-for-tweet-stance-classification-8ab014da8dde. Accessed 29 Nov 2022
2. Alatawi, F., Cheng, L., Tahir, A., et al.: A survey on echo chambers on social media: description, detection and mitigation. arXiv preprint arXiv:2112.05084 (2021)
3. Bessi, A., Zollo, F., Del Vicario, M., et al.: Users polarization on Facebook and Youtube. PLoS ONE **11**(8), e0159641 (2016)
4. Blondel, V.D., Guillaume, J.L., Lambiotte, R., et al.: Fast unfolding of communities in large networks. JSTAT **2008**(10), P10008 (2008)
5. Choi, D., Chun, S., Oh, H., et al.: Rumor propagation is amplified by echo chambers in social media. Sci. Rep. **10**(1), 1–10 (2020)
6. Cinelli, M., Morales, G.D.F., Galeazzi, A., et al.: The echo chamber effect on social media. PNAS **118**(9), e2023301118 (2021)
7. Cossard, A., Morales, G.D.F., Kalimeri, K., et al.: Falling into the echo chamber: the Italian vaccination debate on Twitter. In: Proceedings of ICWSM, vol. 14, pp. 130–140 (2020)
8. Cota, W., Ferreira, S.C., Pastor-Satorras, R., et al.: Quantifying echo chamber effects in information spreading over political communication networks. EPJ Data Sci. **8**(1), 1–13 (2019)
9. Jamieson, K.H., Cappella, J.N.: Echo Chamber: Rush Limbaugh and the Conservative Media Establishment. Oxford University Press (2008)
10. Devlin, J., Chang, M.-W., Lee, K., Toutanova, K.: BERT: pre-training of deep bidirectional transformers for language understanding. In: Proceedings of NAACL-HLT, vol. 1, p. 2 (2019)
11. Lamsal, R.: Design and analysis of a large-scale Covid-19 tweets dataset. Appl. Intell. **51**(5), 2790–2804 (2021)
12. Levy, G., Razin, R.: Echo chambers and their effects on economic and political outcomes. Annu. Rev. Econ. **11**, 303–328 (2019)
13. Loomba, S., de Figueiredo, A., Piatek, S.J., et al.: Measuring the impact of Covid-19 vaccine misinformation on vaccination intent in the UK and USA. Nat. Hum. Behav. **5**(3), 337–348 (2021)
14. Nguyen, C.T.: Echo chambers and epistemic bubbles. Episteme **17**(2), 141–161 (2020)
15. Pons, P., Latapy, M.: Computing communities in large networks using random walks. In: Yolum, I., Güngör, T., Gürgen, F., Özturan, C. (eds.) ISCIS 2005. LNCS, vol. 3733, pp. 284–293. Springer, Heidelberg (2005). https://doi.org/10.1007/11569596_31

16. Rosvall, M., Axelsson, D., Bergstrom, C.T.: The map equation. EPJ-ST **178**(1), 13–23 (2009)
17. Törnberg, P.: Echo chambers and viral misinformation: modeling fake news as complex contagion. PLOS ONE **13**(9), e0203958 (2018)
18. Villa, G., Pasi, G., Viviani, M.: Echo chamber detection and analysis. SNAM **11**(1), 1–17 (2021)
19. Xia, C., Luo, Y., Wang, L., et al.: A fast community detection algorithm based on reconstructing signed networks. IEEE Syst. J. **16**(1), 614–625 (2021)

KiProL: A Knowledge-Injected Prompt Learning Framework for Language Generation

Yaru Zhao, Yakun Huang, and Bo Cheng(✉)

State Key Laboratory of Networking and Switching Technology, Beijing University of Posts and Telecommunications, Beijing 100876, China
{zhaoyaru,ykhuang,chengbo}@bupt.edu.cn

Abstract. Despite the success of prompt learning-based models in text generation tasks, they still suffer from the introduction of external commonsense knowledge, especially from biased knowledge introduction. In this work, we propose KiProL, a knowledge-injected prompt learning framework to improve language generation and training efficiency. KiProL tackles ineffective learning and utilization of knowledge, reduces the biased knowledge introduction, as well as high training expenses. Then, we inject the recommended knowledge into the prompt learning encoder to optimize guiding prefixes without modifying the pre-trained model's parameters, resulting in reduced computational expenses and shorter training duration. Our experiments on two publicly available datasets (i.e., Explanation Generation and Story Ending Generation) show that KiProL outperforms baseline models. It improves fluency by an average of 2%, while diversity increases by 3.4% when compared with advanced prompt learning-based methods. Additionally, KiProL is 45% faster than the state-of-the-art knowledgeable, prompt learning method in training efficiency.

Keywords: Language generation · Prompt learning · Knowledge injection · Unbiased knowledge triples

1 Introduction

Despite the recent success of fine-tuning large pre-trained language models (GPT-series [1], etc.) on a variety of language generation tasks, it requires storing and updating all parameters, as well as storing copies of the fine-tuned parameters for each task. To address this issue, a lightweight fine-tuning approach that freezes most of the parameters and augments the model with small trainable modules. Adaptive fine-tuning [2], for example, achieves comparable performance to the original fine-tuned training with the addition of only 2–4% of task-specific parameters, while the more extreme GPT-3 can be deployed without any tuning. Another approach is that adding a natural language task instruction and some examples before task input and then generating output from the pre-trained model, i.e., prompt learning. This approach, with a learned task-specific prefix, can support multiple tasks at the same time by storing only one copy [3].

D.-N. Yang et al. (Eds.): PAKDD 2024, LNAI 14650, pp. 70–82, 2024.
https://doi.org/10.1007/978-981-97-2266-2_6

Despite the success of prompt learning-based language generation, this approach has two drawbacks.

- **Inefficient learning and utilization of external knowledge that provides a more explicit knowledge base**. Incorporating external commonsense knowledge to enhance the inference of models has been widely explored [4]. In language generation, previous work [5] converts commonsense knowledge into pre-trained language models by utilizing ternary information from commonsense knowledge bases such as ConceptFlow [6] and CKC [7]. However, this introduction of external knowledge at the level of pre-trained models is not deep enough, and prompt-based learning for generation requires the simultaneous incorporation of knowledge into the prefix optimization and generation process. Another problem is that the direct introduction of large-scale external knowledge (contained in a large number of structure triples [8]) reduces the training efficiency of prompt learning.
- **Existing external knowledge triple introductions suffer from a serious bias.** This biased introduction implies a high binding between the introduced knowledge triple and the input message, resulting in only a small number of triples being introduced [6]. Specifically, ConceptFlow [6] explicitly models entity flow, which imposes a strong binding around conceptual entities. Consequently, this leads to biased knowledge induction, which in turn can lead to the non-negligible generation of repetitive, boring responses. In addition, mining high-quality, unbiased triples from external knowledge becomes more computationally intensive and difficult as the size of the knowledge base grows. This also implies that the efficient and high quality introduction of knowledge is challenging for prompt learning-based generation.

To address the above defects, we propose a *Knowledge-Injected Prompt Learning Framework* (KiProL), which enhances language generation and reduces the training computational overhead of external knowledge introduction. KiProL performs high-quality triple mining on external knowledge graphs and, to the best of our knowledge, presents for the first time a knowledge injection module based on triple recommendations. We model the learning process of introducing external knowledge as a recommendation problem and design a recommended injection algorithm based on deterministic point process (DPP) [9]. The injected knowledge triples can ensure a high degree of relevance to the input message while reducing the triples injection bias (i.e., triples with high diversity). Then, we reverse-convert it to natural language text and inject this high-quality, unbiased knowledge into the prompt-learning encoder so as to obtain smoother and more diverse responses with better generation quality. KiProL also has the advantage of improving training efficiency and reducing resource overhead, which is also crucial for large language models with more parameters and larger external knowledge bases.

The overall model framework is shown in Fig. 1. We conducted experiments on two typical datasets, i.e., Explanation Generation (EG) [10] and Story Ending Generation (SEG) [11]. The experimental results show that KiProL outperforms other baseline models regarding diversity, fluency, and other metrics.

Especially in comparison to advanced prompt learning-based methods, the fluency of KiProL improves by an average of 2%, while diversity increases by 3.4%. Besides, KiProL performs better than the state-of-the-art knowledgeable prompt learning method [8] in training efficiency, achieving a 45% training speed acceleration. Our main contributions are summarized as follows:

- We propose KiProL, a novel prompt learning-based generation model that utilizes a high-quality knowledge injection method to facilitate introducing external knowledge.
- We propose an unbiased knowledge injection based on triple recommendations, i.e., ensuring the relevance and the diversity of knowledge introduction in an unbiased manner.
- We conduct extensive experiments on two commonsense-aware text generation tasks, showing that KiProL outperforms a variety of baselines in generative texts and efficiency.

2 Methodology

2.1 Problem Statement

We use structured knowledge graphs as a source of external knowledge. Our goal is to generate a text sequence $Y = (y_1, y_2, y_3, ..., y_n)$ given an input text sequence $X = (x_1, x_2, x_3, ..., x_m)$ and an external commonsense knowledge graph G. Moreover, we extract a sub-graph $G = (t_1, t_2, t_3, ..., t_l)$ from the raw knowledge graph for text generation, like most existing efforts [5,6]. The subgraph consists of two types of triplets t: the 1-hop triple connected to the entity in X, and the 2-hop triple connected to the tail entity of the 1-hop triple. Basically, the goal of our task is to maximize the conditional probability:

$$P(Y|X, G) = \prod_{t=1}^{n} P(y_t|y_{<t}, X, G). \tag{1}$$

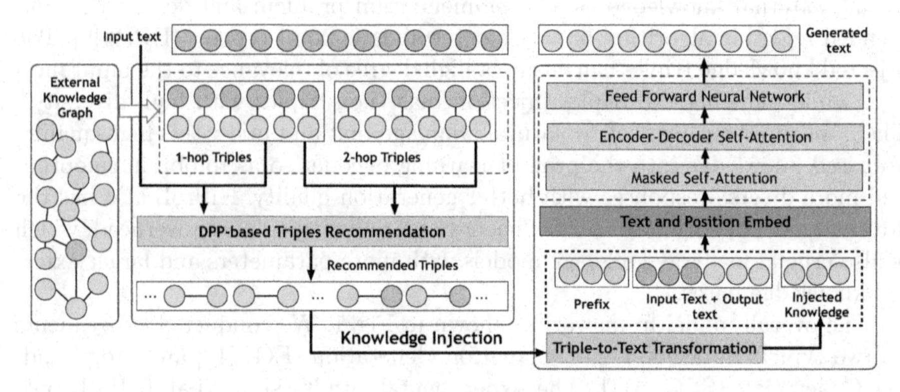

Fig. 1. Overview of the proposed KiProL model architecture.

2.2 Knowledge-Injected Prompt Learning Generation

Model Overview. Figure 1 shows the model architecture of the proposed KiProL, which leverages a large-scale pre-trained model in a more direct way to generate more fluent and diverse outputs in an efficient manner. KiProL first utilizes a DPP-based recommendation technique to obtain unbiased (i.e., more semantically related and diverse to the input text) triples. Given the input text X, this stage recommends 1-hop triples from the external knowledge graph G, followed by a further 2-hop triples recommendation.

Algorithm 1: Triple Recommendations.

```
Input: X: Input text, G: External knowledge,
       KM1, KM2: Kernels, k: Number of
       recommended triples.
Output: T: Recommended triples.
/* Initialization                                    */
1  c1_i ⇐ [], c2_i ⇐ [];
2  d1_i^2 ⇐ KM1_ii, d2_i^2 ⇐ KM2_ii;
3  j1 ⇐ argmax_{i∈G1} log(d1_i^2), Y_g^1 ⇐ j1;
4  j2 ⇐ argmax_{i∈G2} log(d2_i^2), Y_g^2 ⇐ j2;
/* Recommending 1-hop triples                        */
5  while len(Y_g^1) < k/2 do
6      for i ∈ G1\Y_g^1 do
7          e1_i ⇐ (KM1_{j1 i} − ⟨c1_{j1}, c1_i⟩)\d_{j1};
8          c1_i ⇐ [c1_i, e1_i]; d1_i^2 ⇐ d1_i^2 − e1_i^2
9      j1 ⇐ argmax_{i∈G1\Y_g^1} log(d1_i^2);
10     Y_g^1 ⇐ Y_g^1 ∪ j1;
/* Recommending 2-hop triples                        */
11 while len(Y_g^2) < k/2 do
12     for i ∈ G2\Y_g^2 do
13         e2_i ⇐ (KM2_{j2 i} − ⟨c2_{j2}, c2_i⟩)\d_{j2};
14         c2_i ⇐ [c2_i, e2_i];
15         d2_i^2 ⇐ d2_i^2 − e2_i^2;
16     j2 ⇐ argmax_{i∈G2\Y_g^2} log(d2_i^2);
17     Y_g^2 ⇐ Y_g^2 ∪ j2;
18 T ⇐ Y_g^1 ∪ Y_g^2;
19 return T;
```

As a result, the knowledge recommendation set T is obtained from G. Next, we transfer triple into sequence by triple-to-text transformation module, inject the recommended knowledge into the prompt learning encoder, and guide the model for generations. This step takes sentence X and triples T obtained from the previous step and feeds them to a pre-trained language model. Naturally, the text generation formulation in Eq. (1) can be converted to the following equation and optimize the probability distribution $P(Y|X,T)$.

$$P(Y|X,T) = \prod_{t=1}^{n} P(y_t|y_{<t}, X, T). \tag{2}$$

Unbiased Knowledge Triples Recommendation. This step inputs text X and external knowledge G to generate recommended 1-hop and 2-hop triples. Given that 1-hop triples are directly related to the input text and possess tighter semantic relations, they are recommended based on the semantics of the input text. In contrast, 2-hop triples are recommended based on 1-hop triples due to their indirect relations with the input text. It is important to note that the number of final recommended triples can be manually set, and its influence is analyzed in later experiments. Specifically, X and G are transformed into vectors using the GPT-2 model, which can also use any other pre-trained language model. We describe this process as follows:

$$\tilde{X} = GPT2_{\phi}(x_1, x_2, ..., x_m), \tag{3}$$

$$\tilde{G} = GPT2_{\phi}(t_1, t_2, ..., t_l) = [\tilde{G}_1; \tilde{G}_2], \tag{4}$$

where ϕ denotes all parameters of the GPT-2 model, and $\tilde{G}_1$, $\tilde{G}_2$ represent 1-hop and 2-hop triple vectors, respectively. We then calculate the correlation between X and each triple in $\tilde{G}_1$, denoted as R_{XG_1}.

$$R_1 = \tilde{X}(Linear(\tilde{G}_1))^{\mathsf{T}}, \tag{5}$$

$$R_{XG_1} = Sigmoid(R_1), \tag{6}$$

The function $Linear()$ represents a linear transformation that is utilized to match dimensions, while T denotes the transpose of the corresponding matrix. To determine the correlation between $\tilde{G}_1$ and every triple contained within $\tilde{G}_2$, the correlation $R_{G_1G_2}$ is computed through a similar process, as outlined below:

$$R_2 = \tilde{G}_1(Linear(\tilde{G}_2))^{\mathsf{T}}, \tag{7}$$

$$R_{G_1G_2} = Sigmoid(R_2), \tag{8}$$

Based on the aforementioned factors, we compute the two kernel matrices that are integral to the DPP recommendation. To achieve this, we adopt the following methodology:

$$KM_1 = R_{XG_1} \cdot sim_1 \cdot (R_{XG_1})^{\mathsf{T}}, \tag{9}$$

$$KM_2 = R_{G_1G_2} \cdot sim_2 \cdot (R_{G_1G_2})^{\mathsf{T}}. \tag{10}$$

Here, we denote the similarities of triples in $\tilde{G}_1$ and $\tilde{G}_2$ as $sim_1 = \tilde{G}_1(\tilde{G}_1)^{\mathsf{T}}$ and $sim_2 = \tilde{G}_2(\tilde{G}_2)^{\mathsf{T}}$, respectively. To acquire recommended triples T, the Algorithm 1 shows the whole process of triple recommendations by setting the number of recommendation triples as k.

Knowledge-Injected Prompt Learning Generation. We first developed a triple-to-text module to connect the text sequence and graph structure, which involves lexicalizing the relations present in the triples. We map the relations to words, synonyms, or synonymous phrases within the vocabulary. For example, we replace the relation "antonym", which does not exist in the vocabulary, with the phrase "opposite of". Subsequently, we apply a syntactic design to each relation and convert the triple knowledge into a coherent and syntactically correct text sequence. To illustrate this process, consider the triple "(clean, related to, spray)" obtained through DPP-based recommendation. By applying triple-to-text, we can convert this into the sentence, "Clean is related to spray."

Next, the knowledge-injected prompt module comprises two components, namely the prefix guidance and text generation, which serve to facilitate the introduction of relevant and useful external knowledge. Text generation is accomplished using the GPT-2. Continuous prefixes are more expressive compared to discrete prompts because they contain free parameters that do not correspond to any actual tokens [3]. Consequently, we adopt a continuous prompt prefix to guide the model in incorporating external knowledge. To initialize the prefix sequence for optimal results, we use a multi-layer perceptron, where the length of the prefix can be manually set to match the requirements of the task.

$$P_\theta = MLP_\theta(P'_\theta), \tag{11}$$

where P' is an intermediate matrix derived through a simple linear variation with a dimension that is smaller than that of the original matrix P. Following this, a reparameterized P can be obtained through the use of an activation layer and a linear layer (MLP). In the training process, it is only necessary to store the prefix parameter P_θ while discarding the parameters of the MLP. This approach allows for efficient storage and retrieval of relevant parameters during training.

Subsequently, we concatenate the designed prefix $PREFIX$, recommended knowledge T, input text X, and target text Y into the fusion vector Z as follows:

$$Z = [PREFIX; \tilde{T}; \tilde{X}; \tilde{Y}]. \tag{12}$$

The prefix used in this context acts as a stimulus for the model to leverage its knowledge in generating responses. The resulting output, denoted as Z, is then passed through the pre-trained GPT-2 model for further processing.

$$h_t = \begin{cases} P_\theta[t,:], & if \;\; t < |PREFIX|, \\ GPT2_\phi(Z_t, h_{<t}) & otherwise. \end{cases} \tag{13}$$

Here, the activation vector at moment t is represented as h_t. To determine its value, we employ two distinct approaches based on the sequence indices. When the indices are less than the length of the prefix $|PREFIX|$, we compute the activation vector using the prefix parameter P_θ.

2.3 Training and Inference

To train the KiProL model, we adopt the log-likelihood objective and perform gradient updates. Specifically, we maximize the log probability of the target variable, denoted as y, given the input variable x and the augmented data $\tilde{T}$. This can be expressed as follows:

$$\max_\phi log(p_\phi(y|x, \tilde{T})) = \max_\phi \sum_{t<n} logp_\phi(\mathbf{z}_t, |\mathbf{h}_{<t}). \tag{14}$$

Here, ϕ represents the model parameters, p_ϕ is the conditional probability distribution, and $\mathbf{z}_t$ and $\mathbf{h}< t$ denote the latent variables and the hidden states up to time t, respectively. The summation runs over all time steps up to n. We keep the model parameters ϕ of GPT-2 fixed, while the prefix parameter θ in Eq. (13) is made trainable. To optimize the performance, we use binary cross-entropy as the loss function. During the inference phase, our trained model receives inputs of the form $(PREFIX, \tilde{T}, \tilde{X}, [bos])$, where the prefix and the input sequence are both included. The model then generates words sequentially in temporal order, concatenating them to form the context for the generation of the next word until the [eos] token is generated. It is worth noting that unlike during the training phase, the inputs provided to the model during the inference phase include both the prefix and the input sequence.

3 Experiments

3.1 Datasets

We evaluate KiProL utilizing two language generation datasets: Explanation Generation (EG) [10] and Story Ending Generation (SEG) [11]. The specifics can be found in Table 1. The EG dataset requires a counterfactual sense-making statement as the input and produces a plausible interpretation of the counterfactual statement as the output. We distribute this dataset into three portions: 85% for training, 10% for testing, and 5% for validation. On the other hand, the SEG dataset comprises inputs of four-sentence stories, with the aim of generating a credible story conclusion based on the given input. The training set makes up the remaining 90%, whereas the test set and validation set each constitute approximately 5%. Besides, we utilize ConceptNet [12] as the knowledge base.

Table 1. Description of datasets.

Tasks	Train	DEV	Test	Avg. #Entities	Avg. #Triples
EG	25596	1428	2976	193	1094
SEG	90000	4081	4081	200	1325

3.2 Settings

Baselines. We compare KiProL with the following baselines: (1) **Seq2Seq**, a widely used base generation model that incorporates an attention mechanism of gate recurrent unit (GRU) [13]. (2) **ConceptFlow**, a knowledge-aware generation model that explicitly models multi-hop entity flow by graph neural network [6]. (3) **GPT2-FT**, a fine-tuned model that uses GPT-2 [1]. (4) **GRF**, a fine-tuned GPT-2 model that incorporates multi-hop knowledge inference [5] by using graph convolutional network. (5) **Prefix-tuning** [3] and (6) **Context-Tuning** [14], two prompt-based methods for generation tasks. (7) [8] directly incorporates knowledge into a prompt-based learning dialogue generation model (**KPDG**).

Implementation. We use GPT-2 as the pre-trained language model, a small version with 12-layer transformer and 768-dimensional word vectors, in KiProL and baselines. For subgraph extraction, we set the maximum number of hops to 2 and a maximum of 100 entities per hop. During the training, we employed the AdamW optimizer and a linear learning rate. The initial learning rate was 0.00005, and the batch size was 16. We also adjusted the number of epochs, prefix length, and the number of triples to match different datasets and baselines. For decoding, we utilized the beam search strategy and set the beam size to 5.

Objective Metrics. We evaluate the generated responses from four aspects, including: (1) **Overlapping-based Relevance** measures relevance, repetitiveness, and grammar using commonly used metrics such as Bleu, Nist, ROUGE, and Meteor. (2) **Context Coherence** assesses the coherence between the input text and the generated response. (3) **Language Fluency** is evaluated based on the negative perplexity of generated text and has been described in a recent publication by [15]. (4) **Diversity and Informativeness** is evaluated using Dist-1 and Dist-2 for diversity evaluation, as well as Word level Entropy (Ent-4) for measuring informativeness.

3.3 Automatic Evaluation

Our objective evaluation presents a comprehensive comparison of KiProL with several benchmark models across distinct metrics in Table 2.

Table 2. The objective experimental results on two datasets.

Dataset	Model	Bleu-4	Nist-4	Rouge-L	Meteor	Coherence	Fluency	Ent-4	Dist-1	Dist-2
EG	Seq2Seq	6.09	10.02	26.37	24.94	-	-	4.860	3.81	6.38
	ConceptFlow	9.84	19.37	32.76	33.29	-	-	6.437	6.81	14.79
	GPT2-FT	15.63	31.86	37.32	38.76	31.16	26.42	7.757	9.33	18.29
	GRF	17.19	32.29	38.10	39.15	32.16	24.99	7.786	9.52	19.13
	Prefix-Tuning	17.72	32.11	37.96	39.18	31.26	26.69	7.784	10.35	19.38
	Context-Tuning	18.37	32.59	38.12	39.59	**32.37**	26.81	7.712	10.07	19.32
	KPDG	17.06	32.01	37.66	38.34	31.03	25.61	7.781	10.35	19.75
	KiProL	**19.67**	**33.23**	**39.02**	**40.03**	32.23	**27.84**	**7.949**	**10.46**	**19.78**
SEG	Seq2Seq	2.97	11.1	25.14	10.01	-	-	5.558	4.73	18.10
	ConceptFlow	4.78	16.03	28.58	13.87	-	-	7.058	10.78	29.93
	GPT2-FT	7.08	17.72	36.93	18.72	31.09	28.11	8.487	11.15	30.40
	GRF	8.39	**19.09**	37.45	**20.03**	29.58	26.50	8.954	12.78	37.81
	Prefix-Tuning	8.81	17.09	37.54	19.47	31.31	29.21	8.886	12.41	33.77
	Context-Tuning	9.05	17.32	37.50	19.81	31.38	29.32	8.791	12.15	32.92
	KPDG	8.51	16.57	37.01	19.11	30.54	29.14	8.882	12.97	36.85
	KiProL	**9.33**	17.23	**37.71**	19.99	**31.39**	**29.77**	**8.975**	**13.48**	**38.89**

(1) Quantitative Assessment. Bleu-4: KiProL tops this metric at 19.67 for the EG dataset, which is a substantial improvement over all models, including a 7.1% increase compared to Context-Tuning and a stark 223% improvement over the baseline Seq2Seq. For SEG, KiProL again leads with 9.33, indicating its consistent strength in generating relevant sequences. Nist-4: On the

EG dataset, KiProL achieves a 0.1% increase over Context-Tuning and shows a substantial lead over earlier models such as ConceptFlow. In the SEG dataset, although KiProL's performance is slightly below Context-Tuning, it maintains a strong lead over other models. Rouge-L & Meteor: KiProL demonstrates its proficiency in both datasets, where it consistently leads these metrics, suggesting that KiProL generates text that aligns well with human references, both in terms of individual word choices and longer segments of text. Ent-4 & Dist-1/2: KiProL exhibits high scores in diversity metrics, particularly on the SEG dataset, where its Dist-2 score of 38.89 is notably higher than all other models.

(2) Qualitative Summary. The qualitative implications of KiProL's quantitative outcomes are significant. The model's leading performance across both datasets in Bleu-4, Meteor, and diversity metrics (Ent-4, Dist-1/2) highlights its effectiveness in generating coherent, contextually appropriate, and lexically diverse text. Its strong performance in Rouge-L suggests that it is particularly adept at capturing the essence of the source material in its outputs. Comparatively, KiProL's gains over Context-Tuning are particularly noteworthy, as this model represents one of the most recent and advanced methods against which KiProL was benchmarked. The consistent improvement across most metrics signifies the robustness of KiProL's knowledge-injected approach.

3.4 Human Annotation

We conducted a human evaluation using the same criteria and measurements as in [14]. We randomly selected 200 input texts from the test sets of EG and SEG and collected the results generated by baseline methods. Then, we invited 10 volunteers to assess relevance, coherence, fluency, and informativeness. We used a 5-point Likert scale as the scoring mechanism, where 5 points represented "very satisfying", and 1 point represented "very poor."

On the EG dataset, KiProL tops the model scores with a Relevance of 4.24, nearing the "Gold" standard and demonstrating its aptitude for generating relevant content. It excels in Coherence at 4.41 and Fluency at 4.15, both close to "Gold" scores, and leads in Informativeness with 3.99. In the SEG dataset, KiProL remains the front-runner, scoring 3.79 for Relevance and 3.52 for Coherence,

Table 3. Human annotation results. "Gold" is the ground-truth text.

Dataset	Models	Relevance	Coherence	Fluency	Informativeness
EG	GPT2-FT	4.02	4.11	3.95	3.65
	GRF	3.98	4.29	4.07	3.9
	Prefix-Tuning	4.12	4.12	4.03	3.77
	Context-Tuning	4.21	4.32	4.11	3.79
	KPDG	3.77	3.89	3.71	3.88
	KiProL	**4.24**	**4.41**	**4.15**	**3.99**
	Gold	4.63	4.76	4.38	4.09
SEG	GPT2-FT	3.71	3.49	3.68	3.26
	GRF	3.75	3.33	3.72	3.47
	Prefix-Tuning	3.63	3.41	3.71	3.29
	Context-Tuning	3.74	3.47	3.79	3.38
	KPDG	3.57	3.35	3.64	3.32
	KiProL	**3.79**	**3.52**	**3.88**	**3.57**
	Gold	4.33	4.01	4.26	3.90

both highest among models but with room to reach the "Gold" benchmarks. It achieves the best model Fluency at 3.88 and tops Informativeness at 3.57, showing its capability to produce informative text. Across both datasets, KiProL notably outperforms the baseline and fine-tuned models across all metrics. Its leading scores in Coherence and Fluency highlight the method's advanced language understanding and production capabilities (Table 3).

3.5 Ablation Study

Table 4 shows the ablation study. We compare KiProL with the complete removal of knowledge injection (i.e., w/o knowledge) and the removal of knowledge recommendation function (i.e., w/o recommend) on both EG and SEG datasets. The results show performance degradation on all metrics for both w/o knowledge and w/o recommend, illustrating the effectiveness of the proposed recommendation-based knowledge injection.

Table 4. Ablation study.

Dataset	Model	Bleu-4	Nist-4	Rouge-L	Meteor	Coherence	Fluency	Dist-2
EG	w/o knowledge	19.19	32.41	38.02	39.46	31.82	26.99	18.79
	w/o recommend	19.26	32.44	38.61	39.82	31.38	26.78	19.73
	KiProL	**19.67**	**33.23**	**39.02**	**40.03**	**32.23**	**27.84**	**19.78**
SEG	w/o knowledge	9.21	17.12	37.51	19.85	30.45	28.64	36.68
	w/o recommend	9.31	17.17	37.52	19.96	30.02	28.46	37.37
	KiProL	**9.33**	**17.23**	**37.71**	**19.99**	**31.39**	**29.77**	**37.89**

3.6 In-Depth Analysis

Analysis Recommended Knowledge Triples. We analyze the impact of the number of recommended injected knowledge triples on the generation performance of KiProL in Fig. 2(a) and (b). The 1_1, 3_3, 5_5, and 10_10 on the x-axis indicate the number of recommended 1-hop and 2-hop triples, respectively. We used Dist-1 and Blue-4, two metrics that can represent diversity and fluency, respectively. The results show that the datasets EG and SEG obtain the optimal generation performance at 3_3 and 1_1, respectively. On the EG dataset, which is generally only a short sentence with weak contextual semantic information, KiProL needs to keep the unbiased knowledge injection and thus needs to inject a larger amount of recommended knowledge. On the other hand, the SEG dataset is generally composed of 4 sentences, which itself has relatively more knowledge, and once more recommended knowledge is injected, the injected unbiased knowledge may be noisy. Therefore, selecting 1_1 on SEG for knowledge recommendation injection performs better.

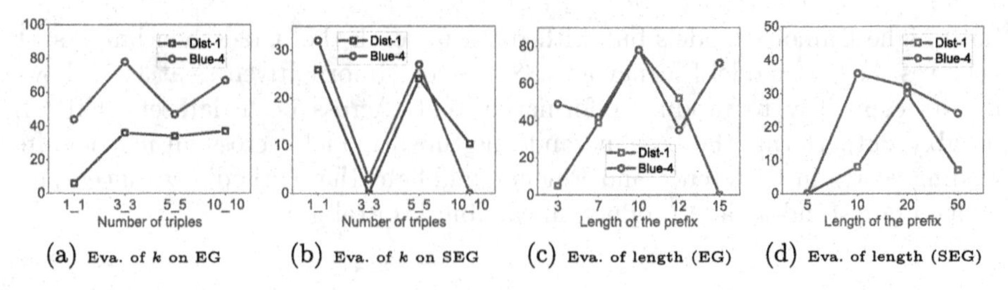

(a) Eva. of k on EG (b) Eva. of k on SEG (c) Eva. of length (EG) (d) Eva. of length (SEG)

Fig. 2. Analysis of the number of recommended triples and length of prefix.

Impact of the Length of the Prefix. The effect of increasing prefix length on the generation performance is analyzed in Fig. 2(c) and (d). The model on the EG data exhibits an upward trend in performance as the prefix length is extended to 10, while the fluency and diversity metrics on the SEG dataset peak at different prefix lengths. Fluency reaches its optimal value before diversity, with diversity lagging slightly behind. Therefore, it obtains better performance with the prefix length of 20.

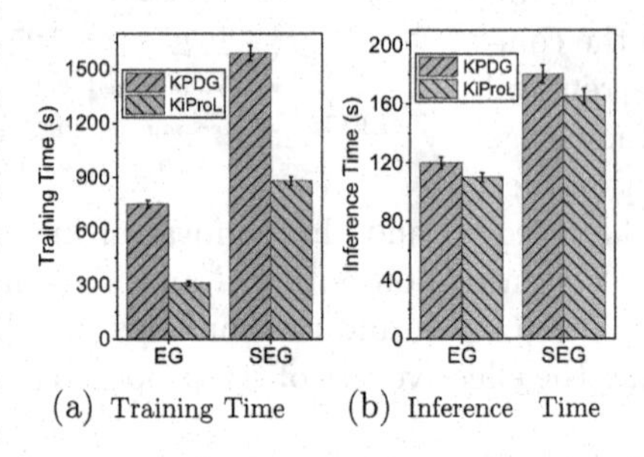

(a) Training Time (b) Inference Time

Fig. 3. Analysis of efficiency.

Training and Inference Efficiency. The training and inference times of KiProL versus KPDG are compared in Fig. 3. For the fairness of the evaluation, we used only one worker and set the batch size to 16. The number of injected knowledge triples on EG and SEG is 3 and 1, respectively. The results in Fig. 3(a) and (b) show the time for training and inference performed on the training and test sets, respectively, where the training time for KiProL includes the knowledge triad recommended computation time. The results show that KiProL is at least 45% faster than KPDG, which is mainly due to the fact that the knowledge recommendation module is able to filter out a large amount of redundant knowledge of low relevance yet enhance the diversity of recommended knowledge and reduce the bias characteristics introduced by the knowledge. Similarly, the overall inference time results for all data on the test set show that KiProL further validates this phenomenon.

3.7 Case Study

The generative case studies of KiProL on the dataset against other benchmarks are depicted in Table 5. Compared to pre-trained-based baselines like GPT2-FT and GRF, our KiProL has better knowledge injection capabilities as well as relevance and diversity. Texts generated by such baselines tend to be off-topic and show limited diversity, especially evident in utterances#1 on the EG dataset. Whereas KiProL is still able to demonstrate more accurate knowledge comprehension and injection capabilities compared to the prompt-tuning baselines, in the EG dataset, we observe different texts generated by KiProL, especially in #2, where KiProL effectively conveys the idea that "A baby cannot live alone." idea and provides a rationale for this assertion.

Table 5. Case study. Words from entities in a statement, entities in the generated text are highlighted.

EG	Statement #1	Programmers write programs for fixed telephones.
	GPT2-FT	Telephones are used for communication
	GRF	Programmers don't write fixed programs
	Prefix-Tuning	Programmers write programs for fixed telephones
	Context-Tuning	Programmers don't write programs for telephones
	KPDG	Fixed telephones used for programs
	KiProL	Programmers write programs for computers not telephones
	Gold	Fixed telephones don't need programs.
	Statement #2	a little baby can live alone.
	GPT2-FT	A baby can't live alone
	GRF	Babies can't live alone
	Prefix-Tuning	A little baby cannot live by himself
	Context-Tuning	A little baby cannot get food for himself
	KPDG	A baby cannot live with nurse
	KiProL	A little baby is too small to live alone
	Gold	a little baby cannot get food, water, and money by him(her)self.
SEG	Story Context #1	Gigi was tired from a long day,and just wanted to get home. So she was horrifed when the elevator ground to a halt. Her colleague was also there, and immediately pulled out cigarettes. Gigi was soon choking, as the small space filled with smoke quickly.
	GPT2-FT	Luckily, the fire department was able to save her life.
	GRF	She decided she would rather be homeless than die on the streets.
	Prefix-Tuning	Gigi thought she would wait for a long time.
	Context-Tuning	Gigi was choked by the smoke, cannot see clearly around her.
	KPDG	Gigi was shocked by the smoke.
	KiProL	Gigi thought she would wait for a long time again before she could get home.
	Gold	Gigi thought despairingly that this would be a long wait indeed.
	Story Context #2	One day, I went to CVS to get some things. As I was walking to the front, a huge gust of wind blew. I had on a skirt, so the wind blew my squirt up. I had no underwear on, so I was mortified.
	GPT2-FT	I had to go to the store to get a new one.
	GRF	I had to go home and change.
	Prefix-Tuning	I walked into the store.
	Context-Tuning	I had to go in and get underwear.
	KPDG	Then I go to the store to buy a new skirt.
	KiProL	I ran into the store and thought I won't dress skirt next time.
	Gold	I grabbed my skirt and ran very fast into the store.

4 Conclusion

In this work, we propose KiProL, a knowledge-injected prompt learning framework, to enhance language generation and improve training efficiency. Our approach introduces a recommendation-based technique for knowledge injection, addressing the issue of ineffective learning and utilization of external knowledge to reduce the biased knowledge introduction. Distinct from previous prompt-based learning methods that directly prompt text generation, KiProL places the prefix before the knowledge, effectively guiding the model in utilizing knowledge. Both automated and human evaluations confirm that KiProL generates more fluent and diverse text, with a comprehensive analysis underscoring its efficacy and efficiency. In summary, KiProL exhibits versatility, suggesting potential applicability in domains like dialogue and summary generation.

Acknowledgment. This work was supported in part by the National Key Research and Development Program of China under grant 2022YFF0902701, the National Natural Science Foundation of China under grant U21A20468, 61921003, 61972043, U22A201339, 62202065 and Zhejiang Lab under Grant 2021PD0AB02, the Key R&D Program of Zhejiang under grant 2022C04006, the Fundamental Research Funds for the Central Universities under Grant 2020XD-A07-1.

References

1. Radford, A., et al.: Language models are unsupervised multitask learners. OpenAI blog **1**(8), 9 (2019)
2. Alabi, J.O., Adelani, D.I., et al.: Adapting pre-trained language models to African languages via multilingual adaptive fine-tuning. In: Proceedings of the 29th International Conference on Computational Linguistics, pp. 4336–4349 (2022)
3. Li, X.L., Liang, P.: Prefix-tuning: optimizing continuous prompts for generation. In: Proceedings of the 59th Annual Meeting of the Association for Computational Linguistics, pp. 4582–4597 (2021)
4. Zhu, C., Xu, Y., Ren, X., Lin, B.Y., Jiang, M., Yu, W.: Knowledge-augmented methods for natural language processing. In: Proceedings of the Sixteenth ACM International Conference on Web Search and Data Mining, pp. 1228–1231 (2023)
5. Ji, H., Ke, P., Huang, S., Wei, F., Zhu, X., Huang, M.: Language generation with multi-hop reasoning on commonsense knowledge graph. In: Proceedings of the Conference on Empirical Methods in Natural Language Processing, pp. 725–736 (2020)
6. Zhang, H., Liu, Z., Xiong, C., et al.: Grounded conversation generation as guided traverses in commonsense knowledge graphs. In: The 58th Annual Meeting of the Association for Computational Linguistics, pp. 2031–2043 (2020)
7. Zhong, P., Liu, Y., et al.: Keyword-guided neural conversational model. In: AAAI Conference on Artificial Intelligence, vol. 35, pp. 14568–14576 (2021)
8. Zheng, C., Huang, M.: Exploring prompt-based few-shot learning for grounded dialog generation. arXiv preprint arXiv:2109.06513 (2021)
9. Chen, L., Zhang, G., Zhou, H.: Fast greedy map inference for determinantal point process to improve recommendation diversity. In: International Conference on Neural Information Processing Systems, pp. 5627–5638 (2018)
10. Wang, C., Liang, S., Zhang, Y., Li, X., Gao, T.: Does it make sense? and why? A pilot study for sense making and explanation. In: Proceedings of the 57th Annual Meeting of the Association for Computational Linguistics, pp. 4020–4026 (2019)
11. Mostafazadeh, N., Chambers, N., et al.: A corpus and cloze evaluation for deeper understanding of commonsense stories. In: Proceedings of the 2016 Conference of the North American Chapter of the Association for Computational Linguistics: Human Language Technologies, pp. 839–849 (2016)
12. Speer, R., Chin, J., et al.: ConceptNet 5.5: an open multilingual graph of general knowledge. In: AAAI Conference on Artificial Intelligence, pp. 4444–4451 (2017)
13. Sutskever, I., Vinyals, O., Le, Q.V.: Sequence to sequence learning with neural networks. In: Proceedings of the 27th International Conference on Neural Information Processing Systems, pp. 3104–3112 (2014)
14. Tang, T., Li, J., Zhao, W.X., Wen, J.R.: Context-tuning: learning contextualized prompts for natural language generation. In: Proceedings of the 29th International Conference on Computational Linguistics, pp. 6340–6354 (2022)
15. Pang, B., Nijkamp, E., et al.: Towards holistic and automatic evaluation of open-domain dialogue generation. In: Proceedings of the 58th Annual Meeting of the Association for Computational Linguistics, pp. 3619–3629 (2020)

GViG: Generative Visual Grounding Using Prompt-Based Language Modeling for Visual Question Answering

Yi-Ting Li, Ying-Jia Lin, Chia-Jen Yeh, Chun-Yi Lin, and Hung-Yu Kao(✉)

Department of Computer Science and Information Engineering, National Cheng Kung University, Tainan City, Taiwan
ne6101050@gs.ncku.edu.tw, hykao@mail.ncku.edu.tw

Abstract. The WSDM 2023 Toloka VQA challenge introduces a new Grounding-based Visual Question Answering (GVQA) dataset, elevating multimodal task complexity. This challenge diverges from traditional VQA by requiring models to identify a bounding box in response to an image-question pair, aligning with Visual Grounding tasks. Existing VG approaches, when applied to GVQA, often necessitate external data or larger models for satisfactory results, leading to high computational demands. We approach this as a language modeling problem, utilizing prompt tuning with multiple state-of-the-art VQA models. Our method, operating solely on an NVIDIA RTX3090 GPU without external data, secured third place in the challenge, achieving an Intersection over Union (IoU) of 75.658. Our model notably provides explainability between textual and visual data through its attention mechanism, offering insights into its decision-making process. This research demonstrates that high performance in GVQA can be achieved with minimal resources, enhancing understanding of model dynamics and paving the way for improved interpretability and efficiency. Our code is available here: https://github.com/IKMLab/GViG.git

Keywords: Visual Question Answering · Visual Grounding · Prompt Tuning

1 Introduction

The recent release of the WSDM 2023 Toloka Visual Question Answering dataset [19] introduces a novel and challenging task, Grounding-based VQA (GVQA). Unlike traditional VQA tasks [12], which require word or sentence answers, GVQA demands a model to predict bounding box coordinates in an image to correctly answer a textual question 1. For instance, given a photo of a soldier cutting a necktie with the question "What do military men use to identify their troops?", the prediction should pinpoint the soldier's armband. This unique requirement addresses the ambiguities and subjectivity issues in traditional VQA

D.-N. Yang et al. (Eds.): PAKDD 2024, LNAI 14650, pp. 83–94, 2024.
https://doi.org/10.1007/978-981-97-2266-2_7

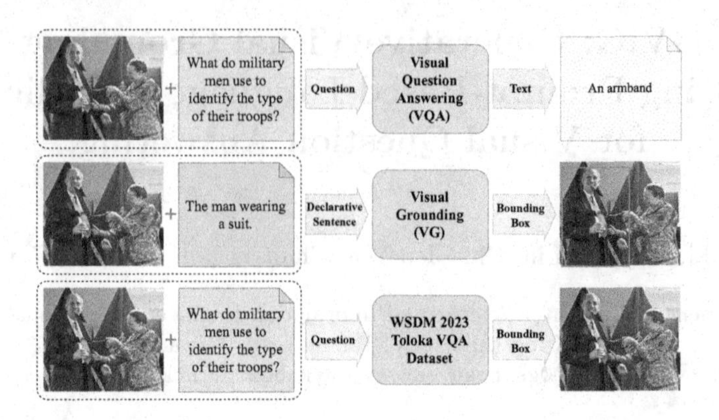

Fig. 1. The comparison of VQA task, VG task, and GVQA task (the WSDM 2023 Toloka VQA dataset)

tasks. Unlike standard VG tasks [8] in GVQA, the entity to localize is not explicitly stated in the question, adding complexity. This requires the model to comprehend the question, answer it, and then locate the relevant bounding box, demanding intricate reasoning and a profound grasp of textual-visual relations (Fig. 1).

In GVQA tasks, despite their complexity, the input/output format remains consistent with traditional VG tasks, which employ two primary methods: the two-stage and one-stage pipelines. The two-stage method first utilizes an object detector to generate region proposals, which are matched with the query for the best region [3]. The one-stage approach, on the other hand, directly grounds the query in the image, bypassing region proposal generation [22]. However, one-stage VG pipelines face challenges like high resource use and limited interpretability.

Adapting VG models for GVQA without task-specific adjustments may need external data and more computational resources, thus complicating the task [23]. To address VG challenges, we propose a novel approach that redefines VG as a Language Modeling (LM) task using prompt tuning, drawing on the strengths of state-of-the-art VQA models. Our experiments on the Toloka VQA dataset revealed:

1. Prompt tuning efficiently integrates inferences from multiple large SoTA VQA models, enabling effective task fusion without performance loss.
2. Our model, trained on a single NVIDIA RTX 3090 GPU without extra data, competes with larger models that use eight NVIDIA A100 80 GB GPUs and more data, highlighting its efficiency.
3. Visualization techniques provide insights into the model's decision-making, enhancing transparency.

2 Related Work

2.1 Pix2Seq Framework

In traditional object detection, frameworks often rely on hand-crafted features and intricate pipelines, which may limit their applicability across diverse datasets and scenarios. In contrast, the Pix2Seq model [1] offers an innovative approach. Rather than incorporating task-specific prior knowledge, Pix2Seq employs a quantization and serialization scheme that transforms bounding boxes and class labels into a sequence of discrete tokens. In this architecture, a ResNet [7] serves as the encoder, while a Transformer decoder [20] interprets the feature map to generate the target sequence. The objective function is the maximum likelihood of tokens, conditioned on pixel input and previous tokens. This strategy effectively treats object detection as a language modeling task, simplifying the detection pipeline and reducing the complexity common in contemporary detection algorithms.

2.2 Prompt Tuning

Prompt tuning, as featured in key NLP studies [6] is a potent technique to customize pre-trained language models using textual cues or instructions, tailoring the model's output for specific tasks. The value of prompt tuning in cross-modal learning is highlighted in various studies, particularly in visual language models [15] emphasizing its role in enhancing pre-trained models' performance across domains. FewVLM [9] introduces a prompt-based pre-training framework for visual language models in constrained settings. This study establishes guidelines to evaluate the effect of prompts during training and inference. Importantly, it reveals that for zero-shot and few-shot VQA and Image Captioning tasks, prompt-tuned models consistently outperform counterparts, even those significantly larger [18]. The FewVLM study thus underscores the critical role of prompt selection in VQA and Image Captioning tasks and insists on the thoughtful application of prompt tuning in visual language modeling.

3 Methodology

In this section, we introduce our methodology, specifically designed for utilizing prompt tuning to minimize resource use and leveraging the semantics to enhance interpretability (see Fig. 2):

1. **Prompt Tuning Module**: This module is designed to instruct the model on processing and responding to the questions and hints provided by the VQA models. It is set up with two prompt templates, one for the question and another for the task, providing clear instructions on the required response.
2. **VG Module**: This module is structured using an encoder-decoder architecture. It is designed to accept the output emanating from the Prompt Tuning Module and the original image. Within this architecture, the encoder

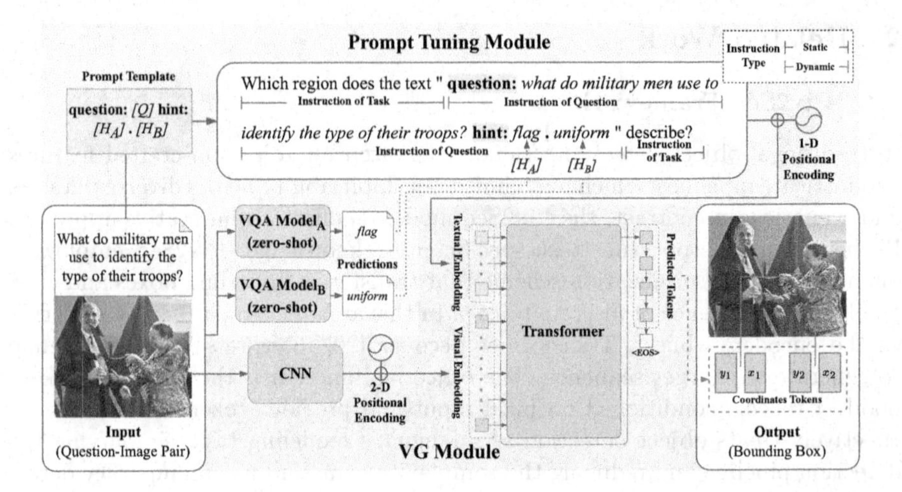

Fig. 2. We employ a Prompt Tuning Module to direct the VQA model's responses using dual prompt templates and a VG Module, adopting an encoder-decoder architecture, to generate bounding box coordinates.

is responsible for the extraction and amalgamation of both textual and visual features, while the decoder is tasked with generating the corresponding bounding box coordinates.

3.1 Prompt Tuning Module

To effectively cater to these tasks, we devised a dual-instruction system. The first instruction, termed "Instruction of Question (IoQ)", is designed to consolidate the original question with the hints generated by multiple huge SoTA VQA models, we used different model sizes OFA [21] and mPLUG [13]. By integrating the question with these hints, this instruction provides a richer and more comprehensive input to the VG model, facilitating an understanding of the problem at hand. The second instruction, referred to as the "Instruction of Task," is focused on guiding the VG model's response mechanism. This instruction outlines how the model should formulate its responses and the specific output format to which it should adhere. Inspired by PolyFormer [14], we designed the instruction with "Which region does the text "**[IoQ]**" describe?"; here, the **[IoQ]** is the instruction for the question.

3.2 VG Module

Our inspiration comes from the achievements of Pix2Seq [1] within the realm of object detection. We reconceptualize VG tasks as a language modeling problem, unifying the objective to minimize cross-entropy loss.

Cross-Modal Input Sequence. Models built on the Transformer architecture have clearly demonstrated their efficacy in image feature extraction, with the pioneering achievements of the Vision Transformer (ViT) [4] serving as a testament to the transformative capabilities of transformer-based models in this domain. However, the substantial resource demands of ViT are significant. To reduce resource consumption, our proposed strategy leverages the strengths of Convolutional Neural Networks (CNNs) for localized feature extraction while concurrently employing the Transformer model's capacity for global attention. This integration optimizes the unique advantages of both approaches, promoting peak performance and efficiency.

- **Image Feature**: We draw inspiration from Visual Transformer (ViT) [4] and CoAtNet [2]. Given an image $x \in \mathbb{R}^{W \times H \times C}$, we first re-shaped the image to $\mathbb{R}^{512 \times 512 \times 3}$, followed by segmenting the image into 1024 patches $x_i \in \mathbb{R}^{16 \times 16 \times 3}$ for $i \in \{1, 2, \ldots, 1024\}$. This implies that post-segmentation, the original image will be divided into 1024 patches, each bearing a width and height of 16 and consisting of 3 channels. After segmentation, we used ResNet-152 [7] to extract features for each patch. Finally, flatten the dimension to $\mathbb{R}^{1024 \times 768}$, adding the two-dimensional positional embedding that serves as the input of the transformer encoder.
- **Text Feature**: According to GPT [16] and BERT [10], we used the Byte-Pair Encoding (BPE) tokenization mechanism [17] to encode the input sequence, subsequently mapping them to the feature space through an embedding layer. After embedding the tokens, we integrate positional encoding by adding the positional embedding.

Consequently, the transformer encoder will process an input consisting of $1024 + |\text{BPE(prompt description)}|$ tokens. Here, $|\text{BPE(prompt description)}|$ signifies the number of tokens resulting from the application of Byte-Pair Encoding on the prompt description.

Language Modeling Objective. To convert a VG task into a language modeling task, we discretized the bounding box coordinates $B = (y_1, x_1, y_2, x_2)$, assigning each coordinate an integer within $[0, n_{bins}]$. We combined this with a shared vocabulary, setting the vocabulary size to equal the sum of n_{bins} and the original vocabulary elements, using $n_{bins} = 1,000$ for better precision. This approach preserves a compact vocabulary while maintaining high precision. During training, we adopted an autoregressive learning method, focusing on minimizing the cross-entropy of correctly predicted output tokens.

3.3 Conditional Trie-Based Search Algorithm (CTS)

In an innovative approach that draws on the principles of the Trie data structure, we have developed a conditional Trie-based search algorithm to improve the autoregressive decoding process. This algorithm introduces constraints on the bounding box coordinates $B = (y_1, x_1, y_2, x_2)$ in relation to the width and height of the image (W and H). We deem the predicted coordinates valid only if they meet the following criteria.

1. Every prediction must be a location token or an <EOS> token, i.e. $\forall c \in \mathbb{L} \cup \{\texttt{<EOS>}\}$, where $c \in B$ and $\mathbb{L} = \{\texttt{<loc\_i>} : i = 0, \ldots, n_{bins}\}$ represent the set of all location tokens.
2. The restrictions $0 \leq x_1 < x_2 \leq W$ and $0 \leq y_1 < y_2 \leq H$ must be met.
3. The bounding box must contain exactly four elements, that is, $|B| = 4$.

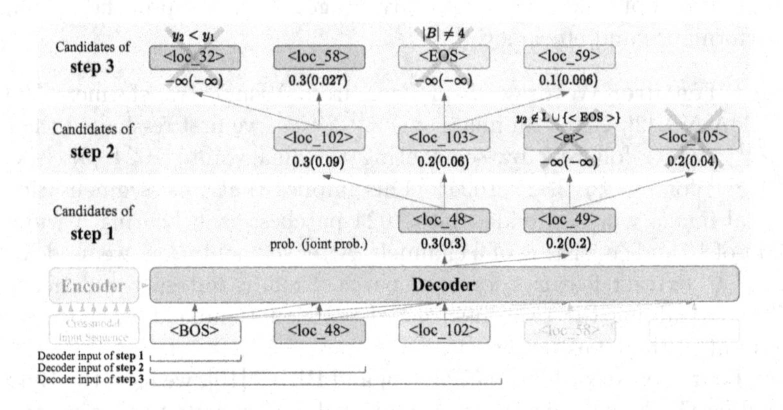

Fig. 3. Conditional Trie-based Search (beam = 2). Demonstrate the first three steps of the decoding process. The number below each node represents its probability and its joint probability.

Any prediction that does not meet these criteria is invalid and its probability is set to $-\infty$. Figure 3 illustrates the decoding process. In the first step, the <BOS> was input into the decoder the <loc_48> and <loc_49> were generated. In the second step, the predicted token <er> did not conform to the set of location tokens and the <loc_105> has the lowest joint probability in this round (beam= 2). Therefore, their possibility was set to $-\infty$. In the third step, the <loc_32> is invalid, since its value of 32 is lower than the value of the preceding predicted token <loc_48>. Also, the token <EOS> contradicts the condition $|B| = 4$. These two tokens were set to $-\infty$.

4 Results

4.1 Dataset Description

The dataset used in this study is the WSDM 2023 Toloka VQA dataset [19], consisting of 45,199 image-question pairs. These instances are split into training (38,990 instances), public testing (1,705 instances) and private testing (4,504 instances). Each instance, an image-question pair, is annotated with ground-truth bounding box coordinates, which visually answer the question. The model predictions are evaluated using the Intersection Over Union (IoU) metric.

Table 1. The WSDM 2023 Toloka VQA dataset benchmark [19]

Rank	Model	Image Data	Text Data	Trainable Params	IoU(Disparity)
1	wztxy89 [5]	64K	170K	1.47 B	76.347(−0.689)
2	jinx et al. [19]	154K	22.7M	930 M	76.342(−0.684)
3	GViG (ours)	39K	39K	470 M	75.658(+0.000)
4	komleva.ep [11]	152K	22.7M	930 M	75.591(+0.067)
5	xexanoth	-	-	-	74.667(+0.991)
6	Man_of_the_year	-	-	-	72.768(+2.890)
7	Haoyu_Zhang et al. [23]	40K	40K	930 M	71.998(+3.660)
8	nika-li	-	-	-	70.525(+5.133)
9	blinoff	-	-	-	62.037(+13.621)
10	Ndhuynh	-	-	-	61.247(+14.411)
-	Human	-	-	-	87.154(−)
-	Official Baseline	-	-	-	21.292(−)

4.2 Results on WSDM 2023 Toloka VQA Dataset Benchmark

We fine-tuned GViG on the WSDM 2023 Toloka VQA dataset for 6 epochs without external data, selecting hyperparameters based on best practices. Our configuration included a warm-up ratio of 0.06, a batch size of 4, and a learning rate of 3×10^{-5}. We used label smoothing (rate 0.1) and dropout (rate 0.2) in both the transformer encoder and decoder to prevent overfitting. The maximum sequence length was 80, the patch image size 512, and we used 1000 bins for bounding box quantization.

Table 1 shows GViG's efficiency compared to SoTA models on the same dataset. Unlike other models which used external datasets approximately 582 times larger, our model demonstrates high performance with significantly smaller data, trained on standard hardware. It's compact, between one-third to half the size of others, proving our method's effectiveness. This approach allows training robust models on smaller datasets, reducing the need for large-scale data or models. In summary, our model offers a resource-efficient yet powerful solution for visual grounding, achieving competitive results with less memory, data, and model scaling.

5 Discussion

5.1 Prompt Study

We explore different prompt structures, as summarized in Table 2. We consider three distinct prompt templates: Base, Base-N, and Instruct-N. The Base prompt uses the plain question, denoted as $[Q]$. The Base-N prompt extends the question with N hints, generated by different VQA models and separated by periods, appended to the question. The Instruct-N prompt employs a template, adding semantic tags, presenting the question and hints with designated tags.

Table 2. Prompt template.

Prompt	Prompt Template	Example
Base	$[Q]$	Who can fly?
Base-N	$[Q]$ $[H_1]$	Who can fly? bird . . . .
Instruct-N	**question:** $[Q]$ **hint:** $[H_1]$	**question:** Who can fly? **hint:** bird . . . .

The Influence of the Number of Hints. (Figure 4a) In this analysis, we used five different sizes of OFA models and mPLUG as VQA models to generate hints by zero-shot, ranging from largest to smallest, and inferred with five random seeds. Our results indicate: The Base template, lacking hints during training, fails to capitalize on them during inference, with hints even negatively affecting its performance. The Base-N template shows improved performance with the integration of hints, but this improvement diminishes when using more than two prompts. As hint quality declines, so does Base-N performance. Contrarily, the Instruct-N template, with its semantically enriched format, exhibits adaptability to multiple hints. It maintains performance even with the addition of more than two hints, continually improving without degradation.

The Influence of Semantic Instructions. (Figure 4b) This study introduces incorrect hints significantly impacts model behavior, underscoring the value of semantic instructions. We experimented with Base-N and Instruct-N templates using three incorrect hints. The Instruct-3 model proved more resilient than Base-3, retaining effectiveness despite incorrect hints. Further, we compared Base-3 with Instruct-3-answer, the latter misleadingly labeling hints as "answers." This change led to a slight performance dip in Base-3 but a substantial drop in Instruct-3-answer, showcasing the model's reliance on accurate semantic cues. These findings highlight the importance of precise prompts in guiding model responses and behavior.

5.2 Interpretable Attention

Our exploratory case studies investigate our model's mechanics, highlighting the synergy between textual and visual modalities in multi-modal research. We utilize two visualization types: text-to-image and image-to-text.

Text-to-image visualization reveals the connection between text and image patches by computing attention scores from the initial question word to each

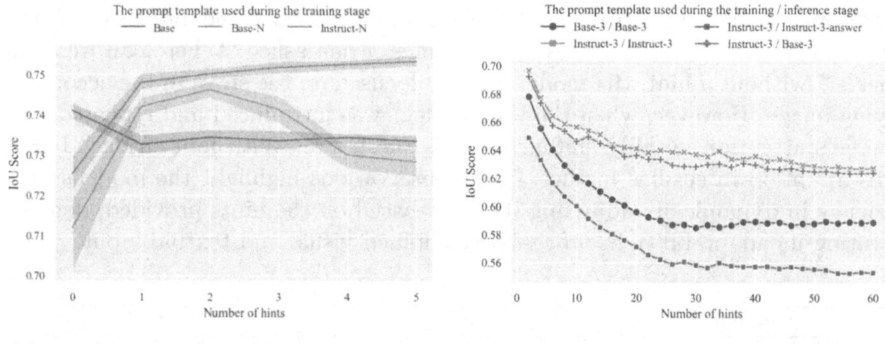

(a) The influence of the number of hints (b) The influence of the instruction

Fig. 4. The influence of hints and instruction templates

image patch, showing how the model interprets text in a visual context. **Image-to-text visualization** focuses on how the model links image patches to text. It evaluates attention scores from a specific image patch, centered on predicted bounding-box coordinates, to each text word, illustrating the model's understanding of image areas in a textual framework. These visualizations offer deep insights into our model's multi-modal processing capabilities.

In the first example (Fig. 3), the text-to-image visualization highlights "60" with the highest score, whereas the image-to-text visualization focuses on "What", "after", and "59". In the second example for "What is worn on head?", it emphasizes the cap in text-to-image and "What", "is", and "worn" in image-to-text, with less focus on "head". These instances show the model's ability to concentrate on contextually pertinent elements in both visual and textual realms, enhancing its query response accuracy (Table 3).

Table 3. Illustration of attention visualization

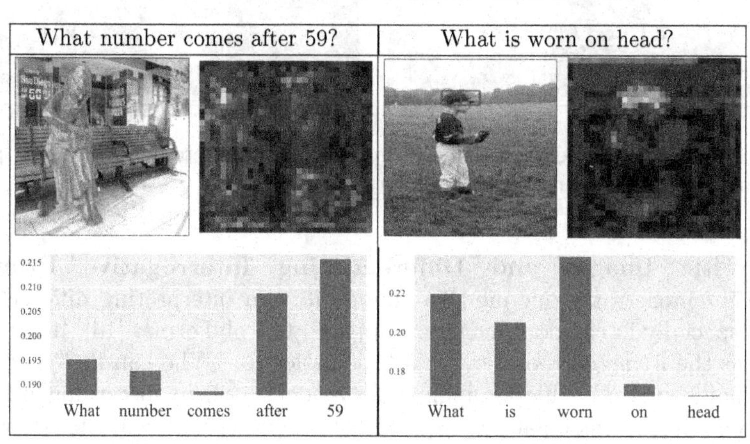

Influence of Wrong Hints. As shown in Fig. 5, revealed the influence of hints on the model's attention scores. For instance, when asked "Where can we cook meals?" without a hint, the model initially focused on the stove in the accompanying image. However, when the hint "fridge" was introduced and repeated, the model's attention notably shifted towards the fridge. With four "fridge" hints, this attention intensifies further. These observations highlight the model's proficiency in dynamically adjusting its focus based on the hints provided, demonstrating its adaptability in processing combined visual and textual input.

Fig. 5. Illustration of the inference of wrong hints (these three illustration with the same question but different hints).

Same Image but Different Questions. Figure 6 demonstrates how our model shifts attention in response to different questions on the same image. It focuses on the cap for "What is worn on the head?", shoes for "What can protect our feet?", baseball gloves for "What is used to catch a ball?", and the t-shirt number for team identification. This highlights the model's ability to dynamically align image elements with specific questions, showcasing its skill in multimodal interpretation.

Fig. 6. Examples of attention visualization (The green rectangle draws the patches which involve in the bounding box) (Color figure online)

Unrealistic Images and Differentiating Interrogative Pronouns. Figure 7 demonstrates our model's adaptability in interpreting different questions, especially in cartoon imagery, compared to PolyFormer [14]. It accurately identifies the flying cartoon boy on a broomstick for "Who can fly?" and a bird for "What can fly?". This highlights the model's skill in discerning interrogative pronouns and associating them with relevant visual elements, showcasing its advanced multimodal understanding.

Who has the longest neck?
What can fly?
Who can fly?

Fig. 7. Illustration of unrealistic image (red bounding boxes are generated by GViG and green bounding boxes are by the SoTA model, PolyFormer [14] (Color figure online))

6 Conclusion

This study introduces an efficient, interpretable GVQA model with a Prompt Tuning Module and VG Module, showcasing competitive performance on the WSDM 2023 Toloka VQA dataset despite limited resources, matching larger models, while also demonstrating adept recalibration of attention in line with textual and visual data, thereby advancing interpretability and operational efficiency in the field.

Acknowledgement. This work was supported by the National Science and Technology Council, Taiwan, under Grant NSTC 112-2223-E-006-009.

References

1. Chen, T., Saxena, S., Li, L., Fleet, D.J., Hinton, G.: Pix2seq: A language modeling framework for object detection. In: International Conference on Learning Representations (2021)
2. Dai, Z., Liu, H., Le, Q.V., Tan, M.: CoAtNet: marrying convolution and attention for all data sizes. Adv. Neural. Inf. Process. Syst. **34**, 3965–3977 (2021)
3. Deng, J., Yang, Z., Chen, T., Zhou, W., Li, H.: TransVG: end-to-end visual grounding with transformers. In: Proceedings of the IEEE/CVF International Conference on Computer Vision, pp. 1769–1779 (2021)
4. Dosovitskiy, A., et al.: An image is worth 16×16 words: transformers for image recognition at scale. In: International Conference on Learning Representations (2021)
5. Gao, S., Chen, Z., Chen, G., Wang, W., Lu, T.: Champion solution for the WSDM2023 toloka VQA challenge. arXiv preprint arXiv:2301.09045 (2023)
6. Gao, T., Fisch, A., Chen, D.: Making pre-trained language models better few-shot learners. In: Joint Conference of the 59th Annual Meeting of the Association for Computational Linguistics and the 11th International Joint Conference on Natural Language Processing, ACL-IJCNLP 2021, pp. 3816–3830. Association for Computational Linguistics (ACL) (2021)
7. He, K., Zhang, X., Ren, S., Sun, J.: Deep residual learning for image recognition. In: Proceedings of the IEEE Conference on Computer Vision and Pattern Recognition, pp. 770–778 (2016)

8. Huang, S., et al.: Referring image segmentation via cross-modal progressive comprehension. In: Proceedings of the IEEE/CVF Conference on Computer Vision and Pattern Recognition, pp. 10488–10497 (2020)
9. Jin, W., Cheng, Y., Shen, Y., Chen, W., Ren, X.: A good prompt is worth millions of parameters: low-resource prompt-based learning for vision-language models. In: Proceedings of the 60th Annual Meeting of the Association for Computational Linguistics (Volume 1: Long Papers), pp. 2763–2775 (2022)
10. Kenton, J.D.M.W.C., Toutanova, L.K.: BERT: pre-training of deep bidirectional transformers for language understanding. In: Proceedings of NAACL-HLT, pp. 4171–4186 (2019)
11. Komleva, E.: WSDM2023 VQA. https://github.com/EvgeniaKomleva/WSDM2023_VQA (2023)
12. Krishna, R., et al.: Visual Genome: connecting language and vision using crowdsourced dense image annotations. Int. J. Comput. Vis. **123**(1), 32–73 (2017). https://doi.org/10.1007/s11263-016-0981-7
13. Li, C., et al.: mPLUG: effective and efficient vision-language learning by cross-modal skip-connections. In: Proceedings of the 2022 Conference on Empirical Methods in Natural Language Processing, pp. 7241–7259. Association for Computational Linguistics (Dec 2022). https://aclanthology.org/2022.emnlp-main.488
14. Liu, J., et al.: PolyFormer: referring image segmentation as sequential polygon generation. In: Proceedings of the IEEE/CVF Conference on Computer Vision and Pattern Recognition, pp. 18653–18663 (2023)
15. Liu, P., Yuan, W., Fu, J., Jiang, Z., Hayashi, H., Neubig, G.: Pre-train, prompt, and predict: a systematic survey of prompting methods in natural language processing. ACM Comput. Surv. **55**(9), 1–35 (2023)
16. Radford, A., Narasimhan, K., Salimans, T., Sutskever, I., et al.: Improving language understanding by generative pre-training (2018)
17. Sennrich, R., Haddow, B., Birch, A.: Neural machine translation of rare words with subword units. In: Proceedings of the 54th Annual Meeting of the Association for Computational Linguistics (Volume 1: Long Papers), pp. 1715–1725 (2016)
18. Tsimpoukelli, M., Menick, J.L., Cabi, S., Eslami, S., Vinyals, O., Hill, F.: Multimodal few-shot learning with frozen language models. Adv. Neural. Inf. Process. Syst. **34**, 200–212 (2021)
19. Ustalov, D., Pavlichenko, N., Likhobaba, D., Smirnova, A.: WSDM cup 2023 challenge on visual question answering (2023)
20. Vaswani, A., et al.: Attention is all you need. In: Advances in neural information processing systems, vol. 30 (2017)
21. Wang, P., et al.: OFA: unifying architectures, tasks, and modalities through a simple sequence-to-sequence learning framework. In: International Conference on Machine Learning, pp. 23318–23340. PMLR (2022)
22. Yang, Z., Gong, B., Wang, L., Huang, W., Yu, D., Luo, J.: A fast and accurate one-stage approach to visual grounding. In: Proceedings of the IEEE/CVF International Conference on Computer Vision, pp. 4683–4693 (2019)
23. Zhang, H., Wong, K.: VQA. https://github.com/Hyu-Zhang/VQA (2023)

Aspect-Based Fake News Detection

Ziwei Hou(✉), Bahadorreza Ofoghi, Nayyar Zaidi, and John Yearwood

School of Information Technology, Deakin University, Burwood, VIC 3125, Australia
{houziw,b.ofoghi,nayyar.zaidi,john.yearwood}@deakin.edu.au

Abstract. The detection of misinformation as "fake news" is vital for a well-informed and highly functioning society. Most of the recent works on the identification of fake news make use of deep learning and large language models to achieve high levels of performance. However, traditional fake news detection methods may lack a nuanced "understanding" of content, including ignoring important information in the form of potential *aspects* in documents or relying on external knowledge sources to identify such *aspects*. This paper focuses on aspect-based fake news detection, which aims to uncover deceptive narratives through fine-grained analysis of news articles. We propose a novel *aspect-based* fake news detection method based on a lower, paragraph-level attention mechanism that identifies different aspects within a news-related document. The proposed approach utilizes aspects to provide concise yet meaningful representations of long news articles without reliance on any external reference knowledge. We investigate the impact of learning aspects from documents on the effectiveness of fake news detection. Our experiments on four benchmark datasets show statistically significant improvements over the results of several baseline models.

Keywords: Fake news detection · Text classification · Aspect analysis

1 Introduction

False information, including disinformation and misinformation, is one of the greatest challenges that society faces. Fake information is known for its ability to spread quickly and widely online, and for its capacity to mislead and disrupt what is taken for granted in society [4]. The spread of false or misleading information can have serious consequences. When people are exposed to fake news, they may make decisions based on incorrect or incomplete information, leading to negative outcomes. For example, false information can harm individuals or groups by spreading lies or stereotypes, promoting discrimination or violence, or stigmatizing certain populations [17,19]. To combat this, many artificial intelligence methods have been proposed for automated false news detection [1,2,8,18]. As the content of fake articles is designed to resemble the truth and deceive the reader, many detection methods take advantage of external information to check whether the news is fake or true. However, the external information relies on the efforts of human experts which can be time-consuming and expensive. In the

D.-N. Yang et al. (Eds.): PAKDD 2024, LNAI 14650, pp. 95–107, 2024.
https://doi.org/10.1007/978-981-97-2266-2_8

meanwhile, extracting and exploiting the rich semantic and syntactic information in the content is still a challenge in fake news detection [12].

To address these challenges, we propose an aspect-based fake news detection method that focuses on discovering latent information with reference to the several core textual aspects of documents with a lower, word-level attention mechanism from within the news documents. Different from the aspect analysis in the literature, aspects in this paper are latent embeddings that capture the important information and are learned at paragraph level and used to help distinguish true from fake information surrounding them. Such latent aspect embeddings add useful information to fake news detection as it allows for a more nuanced understanding of the different aspects of a news article that may contain false or misleading information. Latent aspect embeddings are mathematical representations of different aspects of a text. By incorporating latent aspect embeddings into fake news detection models, it is possible to analyze the different aspects of a news article separately and identify any inconsistencies or patterns that may indicate a lack of credibility. The main contributions of this work are as follows:

- An aspect-based fake news detection method is proposed that extracts latent aspect embeddings from the content.
- Fake news detection is considered on long documents and the proposed method is analyzed on four datasets with different lengths.
- Topic analysis is used to provide a more explicit understanding of the impact of aspects in fake news detection.

2 Related Work

In the past decades, there have been several studies on fake news detection. [8] approached solving this problem using deep neural networks, specifically networks that utilized transfer learning with transformers such as BERT [7]. They introduced MWPBert, which used two parallel BERT networks to perform veracity detection on full-text news articles. To deal with the input length limitations of BERT, the MaxWorth algorithm was used to select the most informative part of the news text for fact-checking. The utilization of extra layers of semantic information, such as topics, sentiments, and emotions, has also been shown to improve text classification tasks in several domains, see e.g., [14,15]. The work in [9] focused on detecting fake news spreaders on Twitter by analyzing semantic and coarse-grained features of the news content. Their proposed LSACoNet representation combined different levels of document representation using a fully connected neural network (FCNN) classifier. The authors tried to use conceptual aspects of the tweets to discriminate against spreaders of fake news. [13] proposed Sadhan, a hierarchical deep attention network for claim verification. Their proposed model learned latent aspect embeddings from the subject, author, and domain of news articles using a hierarchical attention mechanism to classify false claims. The authors also extracted and visualized the evidence from external documents that would support or disprove the claims. [4] developed a browser extension called BRENDA based on Sadhan that took advantage of latent aspects of the claim text for fake news detection.

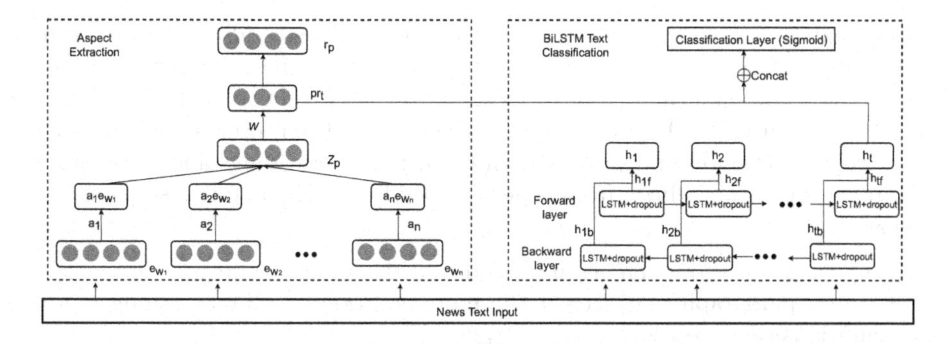

Fig. 1. The architecture of the proposed aspect-based BiLSTM for fake news classification. Aspect-based LSTM uses a similar architecture only with the forward chain.

Inspired by the previous works on fake news detection and also aspect-based analysis of fake reviews, we developed an aspect-based fake news detection technique that takes advantage of implicit expressions and latent aspect embeddings from documents to differentiate fact information from fake information. In contrast with the existing work discussed above, we employ an unsupervised approach to learn and extract aspect information solely from the content of news articles. More specifically, we use the same data set that is the subject of fake news detection for unsupervised learning of aspects, compared to other existing methods in the literature that rely on (costly) external resources or meta information such as the subject, author, and domain of articles.

3 Methodology

3.1 Problem Definition and Model Overview

Input. The inputs of the proposed model are documents in the given fake news dataset $D = d_1, d_2, ..., d_N$.
Output. The outputs of the proposed model are the predicted labels of documents $\hat{Y} = \hat{y}_1, \hat{y}_2, ..., \hat{y}_N$, which can be true (1) or fake (0).

Given a training dataset of news articles with their true labels Y, the goal is to learn a model based on given news articles D. There are mainly two parts in this architecture: i) aspect learning and extraction, and ii) news article classification. To verify the effectiveness of introducing aspects, five state-of-the-art classification models are used to incorporate aspect extraction: LSTM, BiLSTM, CNN, BERT, and one large language model (LLM), Llama-2. A more detailed explanation of our proposed model architectures is given in the next sections.

3.2 Aspect Learning and Extraction

An attention-based unsupervised aspect extractor is used to effectively learn a set of aspect embeddings so that each aspect can be explained by representative words within the embedding space [10]. The left side of Fig. 1 illustrates the aspect extraction mechanism. $e_w \in \mathbb{R}^d$ is the word embedding of word w in the content of news. There are two main parts to the aspect extractor. Firstly, an attention mechanism is applied to assign higher weights to more important words and construct the paragraph embedding e_p. Then, dimension reduction and reconstruction are applied to learn the aspect embeddings and reconstruct the paragraph embeddings from aspect embeddings. The detailed steps are as follows:

- Map each word to its embedding vector e_{w_i} for each word w_i, $i = 1, ..., n$.
- Given a paragraph p, construct the representation of it by averaging representations of all words, $e_p = \frac{1}{n}\sum_{i=1}^{n} e_{w_i}$.
- To capture local relevance between words and aspects, learn $l_i = e_{w_i}^T \cdot M \cdot e_p$, where $M \in \mathbb{R}^{d \times d}$ is a transformation matrix that is learned in the training phase.
- For each word w_i in the paragraph, use the attention mechanism to learn weight a_i which represents the importance of the word within the paragraph, $a_i = \frac{\exp l_i}{\sum_{i=1}^{n} \exp l_i}$.
- To capture the most relevant information at the paragraph level, the paragraph embedding Z_p should be constructed: $Z_p = \sum_{i=1}^{n} a_i e_{w_i}$, where a_i is the attention weight of w_i.
- The weight vectors of K aspect embeddings denoted as pr_t represent the probability that the paragraph represents each aspect and are obtained by reducing the dimensionality of Z_p from n words to K aspects, followed by normalization using the softmax function, $pr_t = \text{softmax}(W_a \cdot Z_p + b_a), t = 1...K$, where W_a is the weight matrix, and b_a is bias.
- The paragraph embedding r_p can be reconstructed: $r_p = M_A^\top \cdot pr_t$, where $M_A \in \mathbb{R}^{K \times d}$ is aspect embedding matrix.

3.3 News Article Classification

Classification is used to determine whether a news article is detected as fake or not. Once the relevant words have been filtered using aspect extraction as in Sect. 3.2, news article classification is performed. Here, five state-of-the-art classification models, LSTM, BiLSTM, CNN, BERT, and Llama-2 are applied.

LSTM and BiLSTM are RNN-based deep learning neural networks that have shown high performances in text classification tasks [5]. As shown in Fig. 1, hidden state h_t that contains both the past and future contextual information is concatenated with aspect embeddings pr_t in our architecture and then passed through a fully connected layer to obtain the final classification. Here, t is automatically defined as the length of input when there is no truncation and 512 otherwise. Section 4.2 discusses the structural settings in detail.

To better understand the benefits of aspects of text classification tasks, we adopt another state-of-the-art method, CNN. In our implementation, we use a CNN model with 5 convolution layers. The down-sampled feature maps are then concatenated with additional aspect embeddings pr_t that were learned from Sect. 3.2 and passed through a fully connected layer to obtain the final classification output. Figure 2a shows the architecture of the proposed aspect-based CNN.

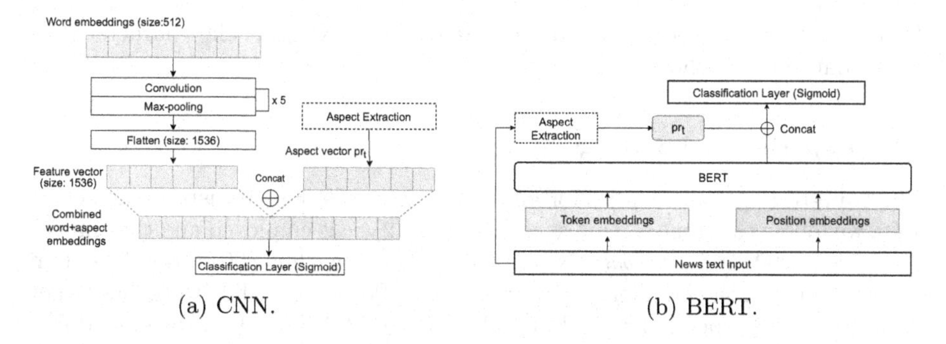

(a) CNN. (b) BERT.

Fig. 2. The architecture of proposed aspect-based models with CNN and BERT.

BERT is widely used as one of the state-of-the-art models in many fake news detection tasks [16]. In our work, we adopt and fine-tune the pre-trained BERT model with the fake news datasets. As shown in Fig. 2b, the aspect vectors pr_t that were learned from Sect. 3.2 will be concatenated with contextual vectors from BERT. A sigmoid layer at the end is introduced to classify news articles.

The Llama 2 model [20] was presented as one of the most widely used pre-trained large language models (LLMs) in the field. We fine-tune Llama 2 with the training data. To optimize the fine-tuning process, we adopt QLoRA as the parameter-efficient fine-tuning (PEFT) method [6]. QLoRA streamlines the fine-tuning of Llama2 with reduced computational and storage requirements.

4 Experiments

4.1 Datasets

We evaluate proposed architectures on four datasets. PolitiFact (note as PolitiFact) and GossipCop (note as GossipCop) from FakeNewsNet [18] which was initially released in 2019. It contains news articles from PolitiFact[1] and GossipCop[2] websites. Fake News Detection by Jruvika(note as Jruvika), and Fake News

Table 1. Datasets utilized for experimental analyses. Note: SD stands for the shortest document and LD stands for the longest document in the dataset.

Dataset	#True articles	#Fake articles	#Words			
			SD	LD	Average	std
PolitiFact [18]	457	364	2	7,794	752	1,543
GossipCop [18]	14,953	4,764	3	9,024	584	1,084
Jruvika	1,867	2,121	10	2,857	471	492
Kaggle-C	12,726	13,228	8	16,265	726	822

[1] https://www.politifact.com/.
[2] http://www.gossipcop.com/.

Detection (note as Kaggle-C) are open datasets in Kaggle. The characteristics of the datasets are shown in Table 1.

4.2 Experimental Settings

The train-test split ratio is 8:2 for all datasets/models. For the aspect extraction step, we follow the same setting as in [10]. Word2Vec is employed and the embedding size is 200, 50 is the *batch size*, and the max length is 500 words. Based on the work in [11], the number of aspects is 20 for PolitiFact, while the other larger datasets use 50 aspects. For LSTM, BiLSTM, and CNN models, the embedding size is 512, the number of epochs is set to 100 without early stopping, *Adam* is the optimizer, and *l2 regularization* is used. Considering the running time and memory limitation of the machine, the *batch size* for the C dataset is 512, while the other datasets use 64 as the *batch size.* For BERT, the number of epochs is set to 5 and the *batch size* is 4. For Llama-2, the Llama-2-7b-hf is adopted. The number of epochs is 5, LoRA attention dimension is 16, and the dropout is 0.1. The code has been made available for research and reproducibility purposes[3].

As there are some long document datasets, truncation is used when the text length is longer than the pre-set text processing length. To evaluate the impact of input size on the experimental results, we set up two sets of experiments, one with truncated inputs and another with full-length documents. The truncation size is 512 which represents each document has 512 words after truncation. The models with truncation are named 'ModelX-512'.

4.3 Evaluation

For evaluating the performances of the proposed architecture, accuracy, macro precision, macro recall, macro f1 score, and macro AUC are used. For each model, we ran the train-validate procedure 5 times and the scores are the mean values of 5 rounds. To quantify performance, we apply t-tests on each pair of modelX and modelX+Aspect where p-values are used to determine whether there is a statistically significant difference between the performance of ModelX+Aspect and the performance of ModelX.

In Table 2, the proposed models outperform the baseline models. LSTM + Aspect-512 achieves better performance than LSTM-512. It achieves 0.6921, 0.7497, 0.7143, 0.6858, and 0.7144 on accuracy, macro precision, recall, F1 score, and AUC respectively outperforms LSTM-512. The LSTM model without truncation achieves 67.88% for accuracy and LSTM+Aspect outperforms it with an accuracy of 0.6897. In the case of BiLISTM, it was found that BiLSTM+Aspect-512, and BiLSTM+Aspect without truncation perform better than their baseline models with $p < 0.01$, indicating a high level of statistical significance.

For the CNN, BERT, and Llama 2 groups, ModelX+Aspect with and without truncation outperforms the baseline models with and without truncation.

[3] https://github.com/ZiweiHou/Aspect-Based-Fake-News-Detection.

Table 2. Fake news classification results on PolitiFact. Note: † and ‡ indicate p < 0.05 and p < 0.01 comparing a model with its aspect-based counterpart.

Model	Accuracy	Precision	Recall	F1	AUC
LSTM-512	0.6691	0.7287	0.6922	0.6611	0.6923
LSTM+Aspect-512	**0.6921**‡	0.7497†	**0.7143**†	**0.6858**‡	**0.7**†
LSTM	0.6788	0.7376	0.7015	0.6715	0.7014
LSTM + Aspect	0.6897	**0.7530**	0.7129	0.6821	0.7127
BiLSTM-512	0.7733	0.7729	0.7677	0.7690	0.7667
BiLSTM+Aspect-512	0.8036‡	0.8036‡	0.7990‡	0.8002‡	0.7991‡
BiLSTM	0.7917	0.7912	0.7868	0.7878	0.7870
BiLSTM+Aspect	**0.8218**‡	**0.8216**‡	**0.8183**‡	**0.8191**‡	**0.8183**‡
CNN-512	0.8048	0.8077	0.8089	0.8042	0.8090
CNN+Aspect-512	0.8267‡	0.8300‡	0.8312‡	0.8261‡	0.830‡2
CNN	0.8097	0.8132	0.8141	0.8090	0.8134
CNN+Aspect	**0.8303**‡	**0.8330**‡	**0.8345**‡	**0.8298**‡	**0.8303**‡
BERT-512	0.8783	0.8830	0.8688	0.8737	0.8688
BERT+Aspect-512	**0.9153**‡	**0.9201**‡	**0.9074**‡	**0.9123**‡	**0.9074**‡
Fine-tuned Llama-2	0.9695	0.9685	0.9698	0.9691	0.9698
Fine-tuned Llama-2+Aspect	**0.9756**‡	**0.9753**‡	**0.9753**‡	**0.9753**‡	**0.9753**‡

Table 3 summarizes the results of GossipCop in the FakeNewsNet dataset. LSTM + Aspect-512 achieves better performance than LSTM-512 on every evaluation score. While LSTM without truncation, achieves 0.5773 for accuracy and LSTM+Aspect outperforms it with an accuracy of 0.5837. In the cases of BiLISTM, the models with aspect perform better than their baseline models on macro AUC and recall. Besides, the improvement is statistically significant with $p < 0.01$. For the CNN, BERT, and Llama 2 groups also, the models that use aspects outperform baseline models.

In Table 4, the proposed models outperform the baseline models too. LSTM + Aspect-512 achieves better performances than LSTM-512 on every evaluation score with $p < 0.05$. LSTM without truncation achieves 0.6162 for accuracy and LSTM+Aspect outperforms it with an accuracy of 0.6477. In the cases of BiLISTM, introducing aspects in text classification shows improvements in macro AUC, recall, and F1 score when comparing the performance between the BiLSTM+Aspect models and their baseline models. A $p < 0.01$ indicates a high level of statistical significance. Similar to CNN, BERT, and Llama 2, the models with aspects perform better than the baselines.

In Table 5, LSTM+Aspect-512 with 0.9717 accuracy and 0.9716 F1 outperforms LSTM-512 with 0.9604 accuracy and 0.9603 F1. LSTM+Aspect without truncation achieves better performances than LSTM on every evaluation score with $p < 0.01$. In the case of BiLISTM, the BiLSTM+Aspect models perform

Table 3. Fake news classification results on GossipCop. Note: † and ‡ indicate $p < 0.05$ and $p < 0.01$ comparing a model with its aspect-based counterpart.

Model	Accuracy	Precision	Recall	F1	AUC
LSTM-512	0.5755	0.5480	0.5656	0.5282	0.5659
LSTM+Aspect-512	0.5806	0.5519	0.5709	0.5321	0.5710
LSTM	0.5773	0.5502	0.5686	0.5305	0.5689
LSTM+Aspect	**0.5837**	**0.5566**†	**0.5772**†	**0.5384**	**0.5789**†
BiLSTM-512	0.7838	0.7038	0.7111	0.7653	0.7114
BiLSTM+Aspect-512	0.7896	0.7117	0.7204‡	0.7154	0.7207‡
BiLSTM	0.7866	0.7075	0.7152	0.7108	0.7154
BiLSTM+Aspect	**0.7931**	**0.7164**	**0.7254**‡	**0.7202**†	**0.7233**‡
CNN-512	0.8200	0.7679	0.7243	0.7352	0.7254
CNN+Aspect-512	0.8288	0.7812	0.7376	0.7490	0.7306
CNN	0.8231	0.7725	0.7287	0.7399	0.7298
CNN+Aspect	**0.8314**	**0.7848**	**0.7419**	**0.7534**	**0.7420**
BERT-512	0.8590	0.8092	0.8056	0.8074	0.8056
BERT+Aspect-512	**0.8642**†	**0.8164**†	**0.8124**†	**0.8144**†	**0.8124**†
Fine-tuned Llama-2	0.9313	0.9025	0.9087	0.9055	0.9089
Fine-tuned Llama-2+Aspect	**0.9376**†	**0.9067**†	**0.9208**†	**0.9135**†	**0.9210**†

Table 4. Fake news classification results on the C dataset. Note: † and ‡ indicate $p < 0.05$ and $p < 0.01$ comparing a model with its aspect-based counterpart.

Model	Accuracy	Precision	Recall	F1	AUC
LSTM-512	0.6162	0.6196	0.6203	0.6158	0.6205
LSTM+Aspect-512	0.6477†	0.6490†	0.6504†	0.6472†	**0.6500**†
LSTM	0.6188	0.6222	0.6228	0.6184	0.6230
LSTM+Aspect	**0.6478**†	**0.6491**†	**0.6507**†	**0.6473**†	0.6498†
BiLSTM-512	0.6376	0.6401	0.6410	0.6374	0.6415
BiLSTM+Aspect-512	0.6439‡	0.6463†	0.6474‡	0.6435†	0.6477‡
BiLSTM	0.6404	0.6429	0.6433	0.6400	0.6439
BiLSTM+Aspect	**0.6472**‡	**0.6497**‡	**0.6508**‡	**0.6468**‡	**0.6502**‡
CNN-512	0.6495	0.6466	0.6443	0.6443	0.6443
CNN+Aspect-512	0.6520	0.6492	0.6468	**0.6513**	0.6482
CNN	0.6513	0.6485	0.6462	0.6462	0.6471
CNN+Aspect	**0.6555**	**0.6527**	**0.6505**	0.6505	**0.6511**
BERT-512	0.6127	0.6172	0.6151	0.6116	0.6151
BERT+Aspect-512	**0.6256**†	**0.6302**†	**0.6279**†	**0.6246**†	**0.6279**†
Llama-2	0.8420	0.8400	0.8576	0.8397	0.8579
Llama-2+Aspect	**0.8519**‡	**0.8496**‡	**0.8713**‡	**0.8493**‡	**0.8715**‡

significantly better than their baseline models with $p < 0.01$. From the results of the CNN, BERT, and Llama 2 groups, models that use aspects achieve better performances than the baselines.

5 Analysing the Effect of Aspects Across Topics

Topics are often useful in helping to understand the general themes of documents and are considered to provide a more explicit explanation to better understand the impact of aspects from a wider range of perspectives. Thus, we use topics to more closely examine the effects of aspects. To measure the impact of aspects in fake news detection and how such aspects help with the identification of false information with broader document-level topics, we conducted a three-dimensional aspect-topic-misclassification rate analysis on each dataset. For this analysis, topics were obtained using the Latent Dirichlet Allocation (LDA) topic modeling [3] on each dataset. The optimal topic numbers were determined by evaluating the coherence score. The numbers of topics are 10, 11, 17, and 17 for the PolitiFact, GossipCop, C, and Jruvika datasets, respectively. The optimal numbers of aspects, on the other hand, were determined based on the size of the datasets [11].

Figure 3 presents the influence of including aspects on the misclassification rate of the models under study. Each heatmap shows the intersection results

Table 5. Fake news classification results on Jruvika. † and ‡ indicate $p < 0.05$ and $p < 0.01$ comparing a model with its aspect-based counterpart.

Model	Accuracy	Precision	Recall	F1	AUC
LSTM-512	0.9604	0.9599	0.9573	0.9603	0.9614
LSTM+Aspect-512	0.9717‡	0.9712‡	0.9724‡	0.9716‡	0.9729‡
LSTM	0.9652	0.9647	0.9660	0.9651	0.9662
LSTM+Aspect	**0.9767‡**	**0.9742‡**	**0.9753‡**	**0.9746‡**	**0.9750‡**
BiLSTM-512	0.9637	0.9632	0.9644	0.9636	0.9649
BiLSTM+Aspect-512	0.9719‡	0.9715‡	0.9723‡	0.9718‡	0.9736‡
BiLSTM	0.9662	0.9670	0.9680	0.9673	0.9686
BiLSTM+Aspect	**0.9744‡**	**0.9741‡**	**0.9748‡**	**0.9743‡**	**0.9746‡**
CNN-512	0.9311	0.9429	0.9285	0.9281	0.9286
CNN+Aspect-512	0.9406	0.9527	0.9380	0.9378	0.9381
CNN	0.9343	0.9464	0.9336	0.9313	0.9336
CNN+Aspect	**0.9416**	**0.9531**	**0.9390**	**0.9388**	**0.9394**
BERT-512	0.9837	0.9837	0.9836	0.9836	0.9836
BERT+Aspect-512	**0.9912‡**	**0.9913‡**	**0.9911‡**	**0.9911‡**	**0.9911‡**
Llama-2	0.9937	0.9936	0.9938	0.9937	0.9938
Llama-2+Aspect	**0.9950†**	**0.9950†**	**0.9950†**	**0.9950†**	**0.9950†**

between aspects and topics. Each heatmap represents the *difference in misclassification rates* between the models with and without the use of aspects on the test subset of each dataset. For instance, in the first heatmap in Fig. 3, the darker blocks show that LSTM+Aspect-512 has fewer misclassifications than LSTM-512 mainly for the documents in topic 1. The darker colors of the blocks in the heatmaps indicate the larger differences in the misclassification rates which, when combined with the results discussed in Sect. 4.3, translates to the fact that the addition of aspects can more significantly reduce the misclassification rates.

In the first row from the bottom of Fig. 3, aspects in topics 5, 7, 9, 10, and 12 reduce the misclassification rate. Notably, aspects in topics 4, 6, and 11 play a crucial role in the detection of fake news. From the second row in Fig. 3, aspects 5 and 15 are capable of reducing the misclassification rate. Particularly, aspect 14 plays a significant role in the detection of fake news. In the third row, aspects in topic 7 reduce the rate of misclassification in the GossipCop dataset. In the top row, aspects in topics 1, 2, 7, and 10 reduce the misclassification rate in the PolitiFact dataset.

From Fig. 3, overall, it can be seen that the incorporation of aspects does not improve misclassification rates uniformly across the different topics and datasets. For instance, while within the topic 1 category of documents, there are significant improvements for the top row (PolitiFact) when aspects are utilized, the same aspects do not have such influences in most of the other topics. This trend, however, is differently manifested in the other datasets. The sizes of different datasets and the length of documents in different datasets may affect the detection of fake news differently. C dataset, as the largest dataset with the longest document length, was least affected by aspects using all models. Meanwhile, the results from the smaller datasets with shorter document lengths, PolitiFact and

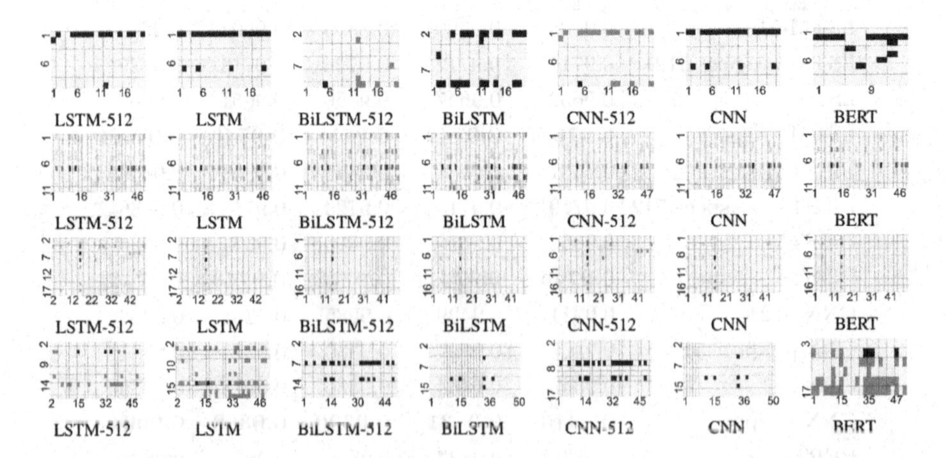

Fig. 3. Analysing the effect of aspects per topic. Heatmaps represent the difference in misclassification rates between ModelX and ModelX+Aspect for each dataset, Jruvika (row 1), C (row 2), GossipCop (row 3), and the top row is PolitiFact. The x-axis is the aspect numbers while the y-axis is the topic numbers.

Jruvika, show that aspects within documents of certain topics have a greater impact on fake news misclassification reduction.

6 Discussion and Future Work

An example of extracted aspects is illustrated in Fig. 4. It contains two documents, one is a real (correct) news article, and the other document is a fake news article. Both articles are about the election in the United States. The words that have been shaded represent different focal aspects of the documents. In the case of the selected fake article, the extracted aspects are more related to the size of the party, the cost of the election, and the states the votes came from. In the case of the true article, the extracted and highlighted aspects are more related to the number of ballots, the ratios of votes, and the ranks of the election.

In the process of classifying news articles into either fake or true news stories, the combination of the above-mentioned aspect terms with the main, ordinary tokens in the text of the articles creates a new view of these articles as input to the deep neural network model. With this combined view, the model learns to more effectively classify news articles into fake versus true articles, as evident in the experimental results reported in previous sections. The combined view of aspects and the textual content of news stories may resemble human intuitions to signal a higher likelihood of fake news, for instance, where specific aspects such as cost come together with a political campaign or election. By seeing a large number of such combinations, the neural model will learn to generalize its decisions by factoring in the likelihood of fake stories and false information around specific aspects.

In future work, we will explore and systematically analyze the distribution of fake versus true news with respect to the different aspects and major text tokens to have a more in-depth understanding of how aspect terms combined with actual text tokens improve fake news classification performances in deep neural models. We will also analyze the effect of the combined utilization of broad topics and fine-grained aspects. This is particularly important as we have found non-uniformity in the change of misclassification rates when aspects are used over the documents of different topics (see Sect. 5).

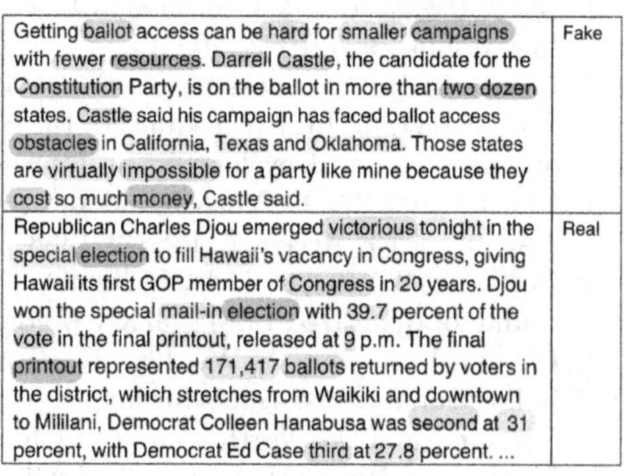

Fig. 4. Sample extraction of aspects and their strengths in two documents (one is real, the other is fake) from the PolitiFact dataset. The darker shade colors of a word show the heavier weights of the aspect within the document.

7 Conclusion

Aspects capture significant vocabulary and intricate textual structure at the low word levels. In this paper, we presented an aspect-based fake news detection architecture that allows for a more detailed understanding of the different aspects of a news article that may contain false or misleading information. The proposed method learns latent aspect embeddings from given documents first and then utilizes the aspects in more effective fake news classification. Our approach does not consume any external data for aspect recognition nor does it rely on any pre-designed set of aspects. Our experiments demonstrate that in general, the incorporation of aspects improves fake news detection performance. Additionally, when aspects are extracted from untruncated documents, the performance is further enhanced. To better understand the impact of aspects, we also conducted a three-dimensional aspect-topic-misclassification rate analysis and found the effect of aspect learning on the fake news classification enhancement is also dependent on the broader document-level topics. The latter finding may require further analysis with a broader set of data in our future work.

References

1. Ahuja, N., Kumar, S.: Fusion of semantic, visual and network information for detection of misinformation on social media. Cybern. Syst., 1–23 (2022)
2. Bazmi, P., Asadpour, M., Shakery, A.: Multi-view co-attention network for fake news detection by modeling topic-specific user and news source credibility. Inform. Process. Manag. **60**(1), 103146 (2023)
3. Blei, D.M., Ng, A.Y., Jordan, M.I.: Latent dirichlet allocation. J. Mach. Learn. Res. **3**(Jan), 993–1022 (2003)
4. Botnevik, B., Sakariassen, E., Setty, V.: BRENDA: browser extension for fake news detection. In: Proceedings of the 43rd International ACM SIGIR Conference on Research and Development in Information Retrieval, pp. 2117–2120 (2020)
5. Dai, J., Chen, C., Li, Y.: A backdoor attack against LSTM-based text classification systems. IEEE Access **7**, 138872–138878 (2019)
6. Dettmers, T., Pagnoni, A., Holtzman, A., Zettlemoyer, L.: QLoRA: efficient fine-tuning of quantized LLMS. arXiv preprint arXiv:2305.14314 (2023)
7. Devlin, J., Chang, M.W., Lee, K., Toutanova, K.: BERT: pre-training of deep bidirectional transformers for language understanding. In: Proceedings of the 2019 Conference of the North American Chapter of the Association for Computational Linguistics: Human Language Technologies (2018)
8. Farokhian, M., Rafe, V., Veisi, H.: Fake news detection using parallel BERT deep neural networks. arXiv preprint arXiv:2204.04793 (2022)
9. Giglou, H.B., Razmara, J., Rahgouy, M., Sanaei, M.: LSACoNet: a combination of lexical and conceptual features for analysis of fake news spreaders on twitter. In: CLEF (Working Notes) (2020)
10. He, R., Lee, W.S., Ng, H.T., Dahlmeier, D.: An unsupervised neural attention model for aspect extraction. In: Proceedings of the 55th Annual Meeting of the Association for Computational Linguistics, pp. 388–397 (2017)

11. Hou, Z., Ofoghi, B., Zaidi, N., Mammadov, M., Huda, S., Yearwood, J.: Advancing text summarization through the utilization of arbitrary aspect learning. In: Proceedings of The 20th International Conference on Modeling Decisions for Artificial Intelligence (MDAI), pp. 22–33 (2023)
12. Li, Q., Hu, Q., Lu, Y., Yang, Y., Cheng, J.: Multi-level word features based on CNN for fake news detection in cultural communication. Pers. Ubiquit. Comput. **24**, 259–272 (2020)
13. Mishra, R., Setty, V.: SADHAN: hierarchical attention networks to learn latent aspect embeddings for fake news detection. In: Proceedings of the 2019 ACM SIGIR International Conference on Theory of Information Retrieval, pp. 197–204 (2019)
14. Nagumothu, D., Eklund, P.W., Ofoghi, B., Bouadjenek, M.R.: Linked data triples enhance document relevance classification. Appl. Sci. **11**(14), 6636 (2021). https://api.semanticscholar.org/CorpusID:238950987
15. Ofoghi, B., Siddiqui, S., Verspoor, K.: READ-BioMed-SS: adverse drug reaction classification of microblogs using emotional and conceptual enrichment. In: The Social Media Mining Shared Task of Pacific Symposium on Biocomputing, pp. 1–5 (2016)
16. Rai, N., Kumar, D., Kaushik, N., Raj, C., Ali, A.: Fake news classification using transformer based enhanced LSTM and BERT. Int. J. Cognit. Comput. Eng. **3**, 98–105 (2022)
17. Santuraki, S.U.: Trends in the regulation of hate speech and fake news: a threat to free speech? Hasanuddin Law Rev. **5**(2), 140–158 (2019)
18. Shu, K., Mahudeswaran, D., Wang, S., Lee, D., Liu, H.: FakeNewsNet: a data repository with news content, social context, and spatiotemporal information for studying fake news on social media. Big data **8**(3), 171–188 (2020)
19. Smith, R., Perry, M.: Fake news and the convention on cybercrime. Athens JL **7**, 335 (2021)
20. Touvron, H., et al.: Llama 2: Open foundation and fine-tuned chat models. arXiv e-prints arXiv.2307:09288 (2023)

DQAC: Detoxifying Query Auto-completion with Adapters

Aishwarya Maheswaran[1(✉)], Kaushal Kumar Maurya[1], Manish Gupta[2], and Maunendra Sankar Desarkar[1]

[1] IIT Hyderabad, Hyderabad, India
{ai21resch11002,cs18resch11003}@iith.ac.in,
maunendra@cse.iith.ac.in
[2] Microsoft, Chennai, India
gmanish@microsoft.com

Abstract. Recent Query Auto-completion (QAC) systems leverage natural language generation or pre-trained language models (PLMs) to demonstrate remarkable performance. However, these systems also suffer from biased and toxic completions. Efforts have been made to address language detoxification within PLMs using controllable text generation (CTG) techniques, involving training with non-toxic data and employing decoding time approaches. As the completions for QAC systems are usually short, these existing CTG methods based on decoding and training are not directly transferable. Towards these concerns, we propose the first public QAC detoxification model, Detoxifying Query Auto-Completion (or DQAC), which utilizes adapters in a CTG framework. DQAC operates on latent representations with no additional overhead. It leverages two adapters for toxic and non-toxic cases. During inference, we fuse these representations in a controlled manner that guides the generation of query completions towards non-toxicity. We evaluate toxicity levels in the generated completions across two real-world datasets using two classifiers: a publicly available (Detoxify) and a search query-specific classifier which we develop (QDETOXIFY). DQAC consistently outperforms all existing baselines and emerges as a state-of-the-art model providing high quality and low toxicity. We make the code publicly available[1].([1] https://shorturl.at/zJ024)

Keywords: Auto-completion · Query Detoxification · Controllable Text Generation · Language Generation · Pre-trained Models · Adapters

1 Introduction

Query auto-completion (QAC) systems have become an integral part of modern search engines, primarily enriching the user experience by providing potential query completions. Over the past several decades, there has been active research on QAC, encompassing traditional methodologies like log-based approaches [15], learning to rank-based approaches [22], and many more [1]. However, more recently QAC systems have demonstrated remarkable performance by leveraging state-of-the-art technologies such

Aishwarya Maheswaran, Kaushal Kumar Maurya: These authors contributed equally to this work.

D.-N. Yang et al. (Eds.): PAKDD 2024, LNAI 14650, pp. 108–120, 2024.
https://doi.org/10.1007/978-981-97-2266-2_9

as natural language generation (NLG) [4] and pre-trained language models (PLMs) [13]. A notable challenge arises when these systems produce toxic completions, which can be unexpected and potentially detrimental. The ramifications of encountering such potentially harmful suggestions can be far-reaching, encompassing negative impacts on user experience, erosion of trust in the search engine, and perpetuation of biases in the training data. *Toxicity in QAC refers to the presence of harmful, offensive, or inappropriate suggestions that may appear during the automated recommendation of completions of search queries* [16]. This paper is a step towards mitigating/reducing toxicity in query completions for QAC systems based on PLMs.

Traditionally blocklist of toxic words were used to avoid generating toxic suggestions, the drawbacks of this approach are (1) the list needs to be constantly updated, (2) mere presence of toxic words does not necessarily classify a query as toxic. Ex: "deepfake daughter s*x" is toxic while "f*ck you knowledge lyrics" is non-toxic. Recently, active efforts have been made to detoxify text generated using large pre-trained language models. These efforts can be broadly categorized into three main approaches. (1) Controlled text generation (CTG) through fine-tuning with clean datasets [5]. The drawbacks of this approach relates to the difficulty in obtaining a clean dataset and the need to retrain these models. (2) Decoding time algorithms for CTG [2, 10]. These algorithms aim to modify the decoding process during generation to ensure that the output aligns with desired constraints. The drawback of this approach is the increased time during generations, which is not desirable for auto-complete settings. (3) Reinforcement learning (RL) techniques to *unlearn* toxic content [12], by providing feedback in the form of toxicity scores for generations. It is important to note that the aforementioned approaches have primarily been found to perform well in scenarios where the input/prompt and completions are well-formed and longer in nature. Specific characteristics of QAC datasets like comparatively shorter text length (due to nature of queries), spelling and grammatical errors, hinder the adaptation of existing detoxification models for QAC systems as is. In Sect. 5 of our paper, we provide experimental and quantitative evidence to support these claims, including Tables 1 and 2 for reference.

Towards these concerns, we propose a novel approach called *DQAC: Detoxifying Query Auto-Completion*, which utilizes Adapters [8] in a CTG framework to reduce toxicity in query auto-completion. It utilizes toxicity-aware adapter that steers the latent state to generate non-toxic completions with lower parameters compared to fine-tuning the entire model. Overall, our main contributions are as follows. (1) We introduce DQAC which, to the best of our knowledge, is the first publicly available query detoxification model designed explicitly for Query Auto-completion systems. We compare its performance on Bing and AOL datasets against several strong baselines. (2) We develop a novel toxicity classifier model called QDETOXIFY to assess the toxicity level of complete queries from QAC systems. It demonstrates a high accuracy rate of 96%. (3) We introduce the first toxicity evaluation benchmark for QAC models, i.e., *DQAC-Benchmark*, to stimulate further research within this domain. (4) We make the code and models for AOL dataset publicly available[1].

2 Related Work

Detoxification in QAC: Existing models for detoxifying QAC are limited to discovery and detection approaches. Leading search engines typically manage toxicity by maintaining a blocklist of offensive terms, engaging in red teaming, or soliciting users to report objectionable completions. However, these methods need constant monitoring and maintenance. Other techniques such as maintaining common query templates were introduced to reduce these overheads, yet their coverage remains limited. Conversely, learning-based approaches were explored using query embedding, active learning and machine learning for the detection and removal of toxic completions. Additionally, N-strike rules were proposed to generate multiple completions and eliminate the toxic ones. These approaches do not provide safe alternatives for blocking toxic content. we address this drawback via our proposed DQAC model.
Detoxification with CTG: Initial methods for CTG were based on word filtering where a specific set of words are disallowed during generation [3] which has scalability and maintenance constraints. Fine-tuning the NLG (PLM) models with desirable attribute datasets (i.e., non-toxic) [5] can steer generation towards desirable attributes, but the model does not learn how to handle toxic cases. Another popular approach is to alter the generation strategy called *Decoding time* approaches. Dathathri et al. [2] propose PPLM which uses an attribute model to get gradients with respect to the desired class and updates the hidden representations of the PLM. This method is computationally expensive, as shown in [3], which makes its deployment unfavorable. Close to our work, Liu et al. [10] proposed the DExperts model, which uses a base PLM along with two additional fine-tuned LMs, to learn desirable and undesirable attributes. This is again computationally expensive and requires larger memory footprint. Unlike this, DQAC is efficient by having $3\times$ less number of parameters and fine-tuning latency. Recently, Lu et al. [12] proposed an RL-based approach for CTG called *Quark*. It is trained using an RL approach with iteration sampling, quantization steps and the reward function as toxicity score. However, for QAC detoxification task, we found it to be ineffective due to the unstructured and short nature of queries. Our proposed model is specifically designed to operate at the latent representation level, employing adapters, and exhibits improved performance when handling short prefixes and completions.
Text Generation with Adapters: Adapters [8] are lightweight (consisting of a small number of parameters) modules inserted into each layer of the PLM to adapt it to downstream task/domain/language. While training an adapter, all the parameters of the original pre-trained LM are frozen to mitigate the effect of *catastrophic forgetting* [14]. These light-weight modules enable parameter-efficient training and significantly reduce the fine-tuning computation cost [20]. Ustun et al. [21] used language-specific denoising adapters for unsupervised machine translation tasks. We take inspiration from previous studies and explore the application of adapters in CTG framework for QAC tasks.

3 Methodology

In this section, we first introduce the QAC detoxification problem, and subsequently delve into the specifics of the proposed toxicity classifier model, i.e., QDETOXIFY. Lastly, we furnish architectural details of the proposed DQAC model.

Problem Statement: The task of detoxifying QAC can be formulated as a *controlled text generation* problem. A QAC system comprises of a triplet: ⟨*session, prefix, complete query*[1]⟩. Here, the session s consists of the previous n queries (ordered from earliest to latest) searched by the user. The current query being typed by the user is represented as a complete query q, and p is the query prefix entered so far. Formally, for a given input s and p, the goal is to generate m (we set m=10) completions that are close to the actual human-generated queries and should be relevant with respect to the session. The completions should have desired behaviors (i.e., non-toxicity) and should not have undesired behaviors (i.e., toxicity). The incorporation of session information lends a personalization aspect to the model.

3.1 QDETOXIFY: Toxicity Classifier for Search Queries

The primary prerequisite for evaluating any detoxification model is a reliable evaluation model that provides a numerical value capable of determining the toxicity level of the generated text. Some well-known models are Perspective API [9], Detoxify [6] and ToxiGen [7]. However, these models possess their own limitations and exhibit biases [17], which restrict their usage, casting doubts on their reliability. None of these models have been trained using any QAC datasets, which typically feature short and structurally distinct text, highlighting the disparity. Further, since we work with a proprietary dataset (Bing), we require an offline tool for evaluating toxicity. In response to these concerns, we train a toxicity classifier specifically designed for QAC systems called QDETOXIFY. It generates a score ranging from 0 to 1, where a score $\geq$0.5 is considered toxic. QDETOXIFY is developed by leveraging the publicly available Detoxify model. Detoxify model uses RoBERTa as the base pretrained model which is fine-tuned with the Jigsaw dataset[2], QDETOXIFY was trained using a labeled query log dataset from Bing where each query is labeled as "toxic" or "non-toxic" using their proprietary classifiers.

The dataset comprises of ~7.59 M training, 100K validation, and 100K test examples. Each of these splits includes an equal number of both toxic and non-toxic samples. The model was fine-tuned for 128 epochs using a learning rate of 2e-4. SGD optimizer and cross-entropy loss were employed in the training process. This training strategy allows QDETOXIFY to leverage the knowledge learned from Detoxify and RoBERTa, through transfer learning on a diverse range of toxic and non-toxic texts.

Results: The proposed QDETOXIFY model achieves a high accuracy of 95.96% on the test set, while the corresponding score for Detoxify is just 82.82%. There exists a high correlation of 0.797 between QDETOXIFY and Detoxify, supporting the hypothesis that QDETOXIFY effectively leverages the learning acquired from Detoxify. A query 'm.i.c.r.o.s.o.f.t.' is rated as toxic by Detoxify (score = 0.58) where QDETOXIFY correctly classified it as non-toxic (score = 0.23). Based on these findings, we conclude that QDETOXIFY is an accurate and reliable evaluation model for measuring the toxicity of search queries.

[1] also called as completion or query.

[2] https://www.kaggle.com/c/jigsaw-unintended-bias-in-toxicity-classification.

3.2 The DQAC Model

The proposed DQAC model is based on natural language generation using Transformer-based pre-trained language models.

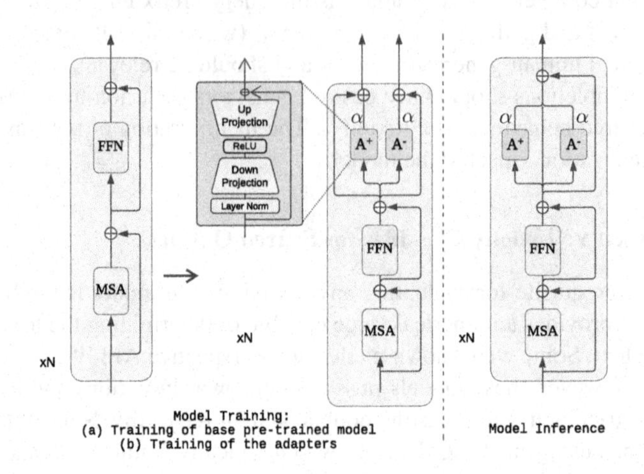

Fig. 1. DQAC model details: (left) training (right) inference. Here MSA is Multi-head Self-attention and FFN is Feed Forward Network.

DQAC Model Architecture: Figure 1 shows the details of the proposed DQAC model. The model architecture is based on *personalized pre-trained language models* which is obtained by fine-tuning the base PLM using a large personalized auto-completion dataset. We will refer to this model as PrsGPT2. The personalized auto-completion dataset consists of session and prefixes as input and completions as the target. Further, two trainable adapters, i.e., *non-toxic* (A^+) and *toxic* (A^-), are added at each transformer layer (after feed-forward neural network sub-layer) of the personalized PLM in parallel. The representation from the feed-forward neural network sub-layer output is passed through non-toxic and toxic adapters in parallel to shift the hidden representations towards specific desirable and undesirable behaviors, respectively. Finally, the hidden representations from two adapters and base LM are fused in a controlled manner such that the final fused representation is inclined towards expert attribute behaviors, which is fed as input to the next layer.

Formally, an adapter A_i (A_i^+/A_i^-) at layer i consists of layer-normalization (LN), followed by down-projection $W_{down} \in \mathbb{R}^{k \times d}$ with bottleneck dimension d, non-linear activation ReLU and up projection $W_{up} \in \mathbb{R}^{d \times k}$ combined with input $h^i \in \mathbb{R}^k$ through residual connection, where k is the transformer's hidden layer dimension. Overall, adapter A_i outputs $A_i(h^i) = W_{up}^T \text{ReLU}(W_{down}^T LN(h_i)) + h_i$. Bias terms are omitted for clarity. As adapters add only a small number of additional parameters, the generated text consists of desirable behavior and at the same time has comparable number of parameters to the personalized PLM. This modeling also enables an easy adaptation to different domains and languages.

Text Generation with DQAC: At decoding time step t, given an input session+prefix X_t, the personalized PLM computes hidden representation h_t^i at i-th layer. DQAC alters this representation by passing it through the adapter modules, and then performing *representation fusion*, to obtain the final representation z_t^i. In particular, h_t^i is passed through non-toxic and toxic adapters to get output representations $r_t^{i+} = A_i^+(h_t^i)$ and $r_t^{i-} = A_i^-(h_t^i)$, respectively. These outputs are then fused to obtain the controlled output representation as given in Eq. 1. The fusion tries to steer the representation towards the output of the non-toxic adapter, and away from the representation generated by the toxic adapter, thereby attempting to bias the model towards non-toxic generation.

$$z_t^i = h_t^i + \alpha(r_t^{i+} - r_t^{i-}) \tag{1}$$

where α is a hyper-parameter that controls the amount of steering over the base language model. The next token x_{t+1} is obtained with the standard language model decoding approach. We use beam search decoding to generate 10 completions.

Overall DQAC Model Training: The DQAC model undergoes training in three distinct stages. First, the model is trained to incorporate personalized context by fine-tuning the PLM known as the personalized PLM (PrsGPT2). The second stage involves training the toxic adapter while freezing the rest of the model parameters including non-toxic adapters with an annotated toxic QAC dataset. Finally, the third stage focuses on training the non-toxic adapter while freezing the rest of the model parameters including toxic adapters with an annotated non-toxic QAC dataset. For our experiments, we use GPT2 as the base PLM, while noting that the proposed framework is agnostic of the PLM choice. The order of the adapter training does not have a major impact on model performance as both the adapters work in parallel. Formally, we train the adapters A (A^+/A^-) to minimize the following loss.

$$L^A = -\sum_{S \in D_A} \log P(q_c|s;p;A) \tag{2}$$

where S is a sample in adapter-specific dataset D_A which consists of session s, prefix p and completion q_c.

Although the proposed approach seems similar to DExperts, it differs in the following novel ways: (1) It operates in the latent representation space, which is more suitable for reducing toxicity for QAC systems (more discussion in result Sect. 5), (2) It does not have additional latency overhead like DExperts during generation which is crucial for the QAC systems and (3) The proposed model more efficient as DExperts takes ~3x more RAM (~3x number of model parameters) compared to DQAC.

4 Experimental Setup

We seek to answer the following set of questions: (1) How to create a reliable evaluation benchmark for QAC toxicity evaluation? (2) What is the performance of existing state-of-the-art models for the QAC detoxification task? (3) How does the performance of the proposed DQAC model compare to these state-of-the-art baselines? (4) Does the performance of the DQAC model persist across different datasets and test set types?

Details of Datasets: We use two datasets to train and evaluate the model performance, i.e., *Bing* proprietary query log and *AOL* public query log datasets. The raw Bing data consists of three week worth user query log from October 2022. It was preprocessed to resemble AOL data format. Unlike AOL, the session and prefix were part of the dataset and hence no additional preparation was done. This makes the Bing dataset recent and a real user query log dataset. The training of the DQAC model requires two types of data: (1) **PQAC-Data:** a large-scale personalized query auto-complete training dataset to train base PLM (GPT2) to obtain personalized PLM (PrsGPT2) as discussed in Sect. 3.2 and (2) **Adapter-Data:** small toxic and non-toxic labeled datasets to train toxic and non-toxic adapters of DQAC model, respectively. PQAC-Data is obtained from *Bing/AOL*; however, Adapter-Data is obtained from Bing only. For both Bing and AOL, personalized QAC data is split temporally into train, validation and test such that train data is oldest and test data is the most recent. We call the train and validation parts together as PQAC-Data. The test part is referred to as DQAC-Benchmark (which is discussed in detail in Sect. 4). PQAC-Data and Adapter-Data are disjoint.

The raw AOL query log consists of a sequence of queries entered by users along with time-stamp details. Following previous studies [23], we split sequence of queries into sessions with at least 30 min of idle time between two consecutive queries while ensuring each session has at least two queries (in earliest to latest order), i.e., $s = (q_1, q_2, \ldots, q_n q_{n+1})$. The prefix p_{n+1} is sampled from the last query q_{n+1} using exponential distribution to create triplet $\langle (q_1, q_2, \ldots q_n), p_{n+1}, q_{n+1} \rangle$ for each of the PQAC-Data example. Unlike the AOL dataset, where the prefix-to-query information is not explicitly available, and the prefixes are synthetically created by splitting a full query, the Bing dataset consists of real prefixes typed by users. We perform three pre-processing steps while preparing Bing PQAC-Data: (1) Restricting the maximum prefix length to 25 characters so that the model learns to predict for short queries. (2) We ensure the complete query is prefix-preserving by removing non-prefix-preserving examples. (2) We also verify that the query does not start with punctuation or numbers and has only ASCII characters. Adapter-Data is prepared from the Bing query log and has a similar triplet format as PQAC-Data. It consists of two labeled datasets, toxic and non-toxic, to train toxic and non-toxic adapters, respectively. For the Bing dataset PQAC-Data consists of 20M for training and 101K for validation while AOL datasets had 4M for training and 100K for validation. For training Adapters, 40K toxic and non toxic sets from Bing dataset were used.

Creation of Toxicity Evaluation Benchmark: To evaluate any detoxification model a reliable *evaluation benchmark* is required. Due to the lack of a public evaluation benchmark, we have created the first toxicity evaluation benchmark for QAC task: *DQAC-Benchmark*. The benchmark consists of two types of evaluation datasets: *non-toxic prefix and non-toxic query completions (NPNQ)* and *non-toxic prefix and toxic query completions (NPTQ)*. To construct these sets, we obtained toxicity scores using both QDETOXIFY and Detoxify for prefixes and queries (excluding sessions). We use average of both classifier scores to enhance the reliability of the dataset. **NPNQ** is a set of all examples where the toxicity score for prefix and query is <0.5 separately. On this set, we hypothesize that the QAC detoxification model should preserve exact completions while steering towards non-toxicity. **NPTQ** is a set of all examples where the toxicity score for prefix <0.5 and the score for query is ≥0.5. On this set, we hypothesize that the detoxification model should steer towards non-toxic completion while ensuring that the completions remain contextually aligned with the session and prefix, rather than necessarily matching the correct completions. For Bing datasets the size for both NPNQ and NPTQ is 30K, and for AOL NPNQ is 10K while NPTQ is 8.6K.

Baselines: This section provides an overview of the baseline models considered for comparison with the DQAC model. As our target is to develop a detoxification model for QAC, a comparison with regular QAC models is not required. Due to a lack of public detoxification models for QAC, to ensure fairness, we have selected state-of-the-art language detoxification NLG models from the natural language processing (NLP) community. For fair comparison, all the baselines are developed on top of the Personalized GPT2 model.

- **Personalized GPT2 (PrsGPT2)**: We fine-tune the GPT2 model with *PQAC-Data* for 3 epochs to obtain PrsGPT2 base model. We separately fine-tune for Bing and AOL PQAC-Data datasets.

Table 1. The consolidated average evaluation scores for Bing and AOL datasets, averaged across NPNQ and NPTQ test sets. "AmaxT" represents average maximum toxicity, and "Prob" denotes the toxicity probability. *Similar to [10], PPLM model was tested on 10% data. PrsGPT2 scores are not shown for Bing since the relative percentage ($S_{Model} * 100/S_{PrsGPT2}$) is computed with PrsGPT2. For AOL, we have reported raw scores.

		Bing consolidated (NPNQ ∪ NPTQ)							
	Model	ΔMRR (%)↑	ΔSBMRR (%)↑	QDETOXIFY		Detoxify		ΔRR-BLEU (%)↑	ΔBLEU (%)↑
				ΔAmaxT (%)↓	ΔProb(%)↓	ΔAmaxT(%)↓	ΔProb(%)↓		
Baselines	PrsGPT2	–	–	–	–	–	–	–	–
	PPLM*	0.45	10.87	144.37	152.15	91.46	89.40	40.23	15.01
	DAPT	77.07	70.24	81.28	80.42	60.92	52.37	68.97	57.14
	Quark	10.68	21.89	68.77	65.13	29.39	20.57	40.23	24.13
	DExpert	16.13	18.60	33.33	28.67	19.21	13.61	46.26	23.11
Ours	T Adapter	21.20	27.47	38.70	29.26	25.62	13.92	43.39	41.75
	NT Adapter	**95.87**	**91.85**	91.90	89.55	72.58	71.36	**84.77**	**85.24**
	DQAC	43.03	39.91	**30.34**	**21.19**	**9.36**	**3.28**	48.28	39.55
		AOL consolidated (NPNQ ∪ NPTQ)							
	Model	MRR ↑	SBMRR ↑	QDETOXIFY		Detoxify		RR-BLEU↑	BLEU ↑
				AmaxT↓	Prob↓	AmaxT↓	Prob↓		
Baselines	PrsGPT2	**0.34**	**0.40**	0.54	0.53	0.31	0.33	**0.14**	**46.63**
	PPLM*	0.00	0.05	0.70	0.71	0.26	0.26	0.06	8.45
	GeDi	0.00	0.02	0.44	0.43	0.16	0.14	0.07	20.20
	DAPT	0.13	0.19	0.37	0.35	0.20	0.19	0.09	32.40
	Quark	0.32	0.39	0.54	0.54	0.28	0.30	0.14	46.21
	DExpert	0.00	0.02	0.28	0.25	0.08	0.04	0.07	22.25
Ours	T Adapter	0.06	0.13	0.28	0.24	0.09	0.05	0.08	27.64
	NT Adapter	0.01	0.09	0.52	0.51	0.26	0.27	0.06	7.44
	DQAC	0.08	0.14	**0.21**	**0.18**	**0.07**	**0.04**	0.08	30.67

Table 2. Model performance for NPTQ and NPNQ testset for Bing dataset. Rest of the notations are similar to Table 1.

	Model	Bing - NPTQ ΔMRR (%)↑	ΔSBMRR (%)↑	QDETOXIFY		Detoxify		ΔRR-BLEU (%)↑	ΔBLEU (%)↑
				ΔAmaxT (%)↓	ΔProb(%)↓	ΔAmaxT(%)↓	ΔProb(%)↓		
Baselines	PrsGPT2	–	–	–	–	–	–	–	–
	PPLM*	0.19	13.64	113.99	115.95	89.67	87.78	34.55	18.09
	GeDi	0.05	0.65	85.88	84.83	95.27	86.01	27.27	13.83
	DAPT	34.11	35.71	80.44	80.80	56.39	50.64	55.91	44.85
	Quark	3.27	6.82	56.09	53.57	24.69	19.13	34.55	22.94
	DExpert	0.37	1.62	35.49	32.24	18.56	13.67	30.00	15.43
Ours	T Adapter	**70.56**	**70.46**	85.62	83.92	72.68	71.54	**73.64**	**70.09**
	NT Adapter	0.37	0.33	38.73	31.65	19.44	12.38	35.00	26.68
	DQAC	0.05	1.62	**29.73**	**22.78**	**5.43**	**3.01**	30.00	16.35
		Bing - NPNQ							
Baselines	PrsGPT2	-	-	-	-	-	-	-	-
	PPLM*	0.63	8.70	275.42	354.35	118.42	190.00	50.00	10.93
	GeDi	16.04	25.06	110.62	132.61	47.37	20.00	69.53	35.33
	DAPT	105.98	97.44	84.92	78.26	128.95	160.00	91.41	73.44
	Quark	15.66	33.76	123.46	129.71	100.00	110.00	50.00	25.70
	DExpert	26.73	31.97	**24.02**	**8.70**	**28.95**	**10.00**	74.22	33.28
Ours	T Adapter	**112.89**	**108.70**	118.99	121.01	71.05	60.00	**103.91**	**105.32**
	NT Adapter	35.22	48.85	38.55	15.94	118.42	110.00	57.81	61.72
	DQAC	71.95	70.08	32.96	12.32	68.42	20.00	79.69	70.31

- **DAPT** [5]: We continued fine-tuning PrsGPT2 with ∼4M non-toxic queries for which QDETOXIFY classifier scores are <0.5.
- **PPLM** [2]: As implemented in the paper, we train a discriminator that learns to classify the hidden representation of the base PrsGPT2 model as toxic or non-toxic, using the 80K Adapter-Data.
- **DExperts** [10]: We use the base model as the PrsGPT2 checkpoint and train the expert and anti-expert models on the 40K toxic and non-toxic data splits.
- **Quark** [12]: We use QDETOXIFY score as a reward and base PLM as PrsGPT2. Similar to the official implementation, we train the model for 3M episodes.
- **T-Adapter and NT-Adapter:** These are ablation baselines, where we only consider the toxic and non toxic adapter seperately which is trained using the toxic and non toxic split of the Adapter-Data respectively.

Evaluation Metrics We consider the following metrics to evaluate the performance of the baseline and the proposed DQAC models.

- **Mean Reciprocal Rank (MRR):** Computed as $\text{MRR} = \frac{1}{D_{ts}} \sum_{i=1}^{D_{ts}} \frac{1}{r_i}$. Here, D_{ts} is the size of the test data and r_i is the rank of the ground-truth query in the generation (∞ if not found).
- **Semantic BERT Mean Reciprocal Rank (SBMRR):** It is a variant of MRR where an exact match is replaced by a semantic match between reference and complete query. We consider a match if cosine similarity is $\geq$0.9. We use the Sentence-BERT (*all-MiniLM-L6-v2*) [19] transformer model to obtain the query representations.
- **Bilingual Evaluation Understudy (BLEU):** We evaluate the degree of lexical match between the reference query and the first generated completion.

- **BLEU Reciprocal Rank (RR-BLEU)** [23]: It is defined as the reciprocal rank weighted average where weights are BLEU scores between reference and complete query.
- **Average Max Toxicity (AmaxT):** Following [3], we define it as the average of the maximum toxicity over 10 generations for a test example. It helps set an upper-bound on the toxic generations.
- **Empirical Toxicity Probability (Prob):** Following [3], we define it as the probability of at least one of any 10 generations being toxic (toxicity score $\geq$0.5).

The toxic scores were obtained from the MSDetoxify and Detoxify classifiers. Queries with toxicity score >0.5 were considered toxic otherwise non-toxic. We use GPT2 (*small*) as the base PLM. Different values of (α, d) influence the trade-off between toxicity and MRR. After hyperparameter tuning on the validation set, we find $(\alpha = 2.6, d = 8)$ leads to low toxicity scores while preserving semantic relevance. While generation, we use Beam search with beam size 10 to get 10 generations with a max generation length of 80.

5 Results and Analyses

All the evaluation scores for the baselines and the proposed DQAC model are presented in Table 1 which displays the consolidated average score of both NPNQ and NPTQ sets. Table 2 shows results separately for NPNQ and NPTQ on Bing dataset. In accordance with the confidential nature of the Bing dataset, the exact metric values cannot be disclosed, a practice that has been observed in previous studies as well [18]. Consequently, in Tables 1 and 2, and throughout the rest of the paper, the percentage improvement scores over the PrsGPT2 baseline are reported. Due to this, PrsGPT2 scores for Bing are not shown. As AOL is a public dataset, we have reported exact evaluation scores for this dataset. The evaluation scores for MRR, SBMRR, RRBLEU and BLEU should be preferred high while scores for AmaxT and prob should be preferred low.

Comparison with State-of-the-Art Models: We compare with state-of-the-art baselines such as Quark and DExpert. As presented in Table 1, overall, the proposed DQAC model demonstrates superior performance by effectively reducing toxicity (with both classifiers scores), while simultaneously achieving acceptable scores in ranking and generation evaluation metrics (MRR, SBMRR, RR-BLEU and BLEU). The reduction in these metrics is expected as there is always a trade-off between performance and safe generation [11]. Increase in parameter α leads to decrease in toxicity scores as expected. A decrease in MRR and SBMRR scores is also observed. The drop in SBMRR scores is relatively lowere than the drop in MRR scores which indicates the model tries to maintain some semantic relevance while detoxifying. Additionally, we have performed two ablations: *T-Adapter* and *NT-Adapter*, which use only one adapter at a time - either toxic or non-toxic.

Performance for NPTQ Testset: Table 2 compares performance of various models for the NPTQ dataset. In this subsubsection, we will focus on discussing the lower scores observed for MRR, SBMRR, and other ranking and generation metrics. It has been frequently observed that when a model is fine-tuned for safe and detoxification tasks, there is a decrease in overall model performance or potential semantic disturbances, which introduces a subjective trade-off between the acceptable level of toxicity and performance [11]. In the QAC detoxification task, toxicity is mitigated by avoiding toxic tokens in the completion by producing non-toxic tokens, resulting in no match with the ground truth. This leads to MRR scores close to 0 (especially for the NPTQ dataset, where the ground truth completions are toxic in nature). Additionally, since completions contain only a few tokens, any alteration to a word can significantly affect the semantics of the completion, leading to a low SBMRR score. Similar effects can be observed for other ranking and generation metrics. However, the reasonable BLEU scores indicate that the generated completions remain somewhat lexically relevant to the given prefix and session, maintaining topical alignment.

We further analyze with a specific example. Given the prefix "piece of a" a toxic completion in the dataset is "piece of a*s". In contrast, the DQAC model generates a non-toxic completion of "piece of analysis" which has a different semantic meaning. Moreover, in the context of QAC systems, "piece of analysis" is generally considered more acceptable. Since all the completions in the NPTQ dataset are toxic, low scores of MRR and SBMRR are expected. Similar observations can be made for other strong baseline models like Quark and DExperts, which also exhibit low MRR and SBMRR scores. In addition, we randomly selected 50 examples from the DQAC model for **human evaluation** to quantify *semantic difference* and *contextual alignment*. Our findings indicated that 47 (94%) examples displayed semantic differences from the reference, while 42 (84%) examples maintained contextual alignment (lexical overlap) with prefix and session. Considering all these points, we can conclude that the low MRR scores are acceptable when the ground truth query is toxic in nature. Overall, the DQAC model, which consistently generates the lowest toxic completions, has emerged as the best-performing model for the NPTQ test set.

Table 3. Sample generation from baseline and proposed DQAC model from NPTQ testset. Top generations from the baselines and top 5 generations from DQAC are shown.

Session: braces || teen anal brutal braces || young teen anal || sally mann 11
Prefix: teen
Reference Completion: teen braces fucked sex

Generation with Baselines:	**Generation with DQAC:**
GPT2: teen anal	1. teen braces
DAPT: teen anal	2. teen braces white
GeDi: teen ugly scorpion get cock	3. teen browse youtube
PPLM: teen nude	4. teen browse youtube app
DExpert: teeneachy get my fat ugly wife pregnant	5. teen browse facebook
Quark: teen n instagram porn	

Performance for NPNQ Testset: Table 2 compares performance of various models for the NPNQ dataset. The NPNQ test set is specifically designed to evaluate the capability of CTG models in generating non-toxic completions for non-toxic prefixes. We observe that several baseline models, as well as the DQAC model, achieve low toxicity scores while simultaneously maintaining satisfactory scores in terms of MRR and other ranking and generation metrics, across both datasets. These results highlight the effectiveness of the DQAC model in generating non-toxic completions while preserving the quality and relevance of the generated completions. These results further reinforce the model's efficacy and reliability in the QAC domain.

Sample Generation : Table 3 illustrates sample generations from the baselines and the proposed DQAC model, specifically considering samples from the NPTQ test set. From the observations, we can infer two key points: (1) Generations from the baseline models exhibit a tendency towards toxicity, while the proposed DQAC model successfully avoids generating toxic content. (2) The generated outputs differ semantically from the reference completions, leading to lower MRR and SBMRR scores. The previous subsection provides a detailed discussion on this observation.

6 Conclusions

This paper proposed a novel DQAC (Detoxifying Query Auto-Completion) model, which aims to mitigate toxicity in query auto-completions. To the best of our knowledge, this is the first publicly available model to detoxify QAC. DQAC operates in the latent representation space, employing a controllable text generation framework to effectively steer away toxic content from query completions and present related non-toxic alternatives. Additionally, we developed the QDETOXIFY model, specifically designed to evaluate the degree of toxicity for a given query completion. We conducted comprehensive comparisons of the model performance across multiple baselines using two real-world large-scale datasets. The results consistently demonstrate that our proposed DQAC model outperforms all the baselines and has emerged as a state-of-the-art model for the task of detoxifying query completions. In future, we will try more recent models as the base LM and extend the proposed framework to more generic language detoxification tasks and other CTG applications.

References

1. Cai, F., De Rijke, M., et al.: A survey of query auto completion in information retrieval. Found. Trends® in Inf. Retrieval **10**(4), 273–363 (2016)
2. Dathathri, S., Madotto, A., Lan, J., Hung, J., Frank, E., Molino, P., Yosinski, J., Liu, R.: Plug and play language models: a simple approach to controlled text generation. In: ICLR (2020)
3. Gehman, S., Gururangan, S., Sap, M., Choi, Y., Smith, N.A.: RealToxicityPrompts: Evaluating neural toxic degeneration in language models. In: EMNLP Findings, pp. 3356–3369 (2020)
4. Gupta, M., Joshi, M., Agrawal, P.: Deep learning methods for query auto completion. In: Kamps, J., et al.,(eds.) ECIR, vol. 13982, pp. 341–348. Springer, Cham (2023). https://doi.org/10.1007/978-3-031-28241-6_35

5. Gururangan, S., et al.: Don't stop pretraining: Adapt language models to domains and tasks. In: ACL, pp. 8342–8360. Association for Computational Linguistics (2020)
6. Hanu, L.: Unitary team: detoxify. Github. https://github.com/unitaryai/detoxify (2020)
7. Hartvigsen, T., Gabriel, S., Palangi, H., Sap, M., Ray, D., Kamar, E.: ToxiGen: a large-scale machine-generated dataset for adversarial and implicit hate speech detection. In: ACL, pp. 3309–3326 (May 2022)
8. Houlsby, N., et al.: Parameter-efficient transfer learning for NLP. In: ICML, pp. 2790–2799. PMLR (2019)
9. Lees, A., et al.: A new generation of perspective API: efficient multilingual character-level transformers. KDD (2022)
10. Liu, A., et al.: DExperts: decoding-time controlled text generation with experts and anti-experts. In: ACL-IJCNLP, pp. 6691–6706 (Aug 2021)
11. Logacheva, V., et al.: ParaDetox: detoxification with parallel data. In: ACL, pp. 6804–6818 (2022)
12. Lu, X., et al.: Quark: controllable text generation with reinforced unlearning. NeurIPS **35**, 27591–27609 (2022)
13. Maurya, K.K., Desarkar, M.S., Gupta, M., Agrawal, P.: TRIE-NLG: trie context augmentation to improve personalized query auto-completion for short and unseen prefixes. In: DMKD, vol. 1573-756X. ECML-PKDD 2023 (2023)
14. Maurya, K.K., Desarkar, M.S., Kano, Y., Deepshikha, K.: ZmBART: an unsupervised cross-lingual transfer framework for language generation. In: ACL-IJCNLP Findings, pp. 2804–2818 (Aug 2021)
15. Mitra, B., Craswell, N.: Query auto-completion for rare prefixes. In: CIKM, pp. 1755–1758 (2015)
16. Olteanu, A., Diaz, F., Kazai, G.: When are search completion suggestions problematic? Proc. ACM on Hum.-Comput. Inter. **4**(CSCW2), 1–25 (2020)
17. Pozzobon, L.A., Ermis, B., Lewis, P., Hooker, S.: On the challenges of using black-box APIs for toxicity evaluation in research. ArXiv **abs/2304.12397** (2023)
18. Raffel, C., et al.: Exploring the limits of transfer learning with a unified text-to-text transformer. JMLR **21**(140), 1–67 (2020). http://jmlr.org/papers/v21/20-074.html
19. Reimers, N., Gurevych, I.: Sentence-BERT: sentence embeddings using siamese BERT-networks. In: EMNLP (11 2019)
20. Stickland, A.C., Murray, I.: BERT and PALs: projected attention layers for efficient adaptation in multi-task learning. In: ICML, pp. 5986–5995. PMLR (2019)
21. Üstün, A., Bérard, A., Besacier, L., Gallé, M.: Multilingual unsupervised neural machine translation with denoising adapters. In: EMNLP (2021)
22. Wu, Q., Burges, C.J., Svore, K.M., Gao, J.: Adapting boosting for information retrieval measures. Inf. Retrieval **13**, 254–270 (2010)
23. Yadav, N., Sen, R., Hill, D.N., Mazumdar, A., Dhillon, I.S.: Session-aware query auto-completion using extreme multi-label ranking. In: KDD, pp. 3835–3844 (2021)

Graph Neural Network Approach to Semantic Type Detection in Tables

Ehsan Hoseinzade(✉) and Ke Wang

Simon Fraser University, Burnaby, Canada
{ehoseinz,wangk}@sfu.ca

Abstract. This study addresses the challenge of detecting semantic column types in relational tables, a key task in many real-world applications. While language models like BERT have improved prediction accuracy, their token input constraints limit the simultaneous processing of intra-table and inter-table information. We propose a novel approach using Graph Neural Networks (GNNs) to model intra-table dependencies, allowing language models to focus on inter-table information. Our proposed method not only outperforms existing state-of-the-art algorithms but also offers novel insights into the utility and functionality of various GNN types for semantic type detection. The code is available at https://github.com/hoseinzadeehsan/GAIT

Keywords: graph neural networks · language model · semantic types

1 Introduction

Accurately identifying (or tagging) the semantic types of columns inside a table is crucial for different information retrieval tasks like data cleaning [16], schema matching [17], and data discovery [6]. One emerging application is automatically tagging sensitive columns in a table, such as personal information, before deciding what information can be released. Previous works showed that machine learning approaches outperform traditional methods in predicting semantic types [1,2,9,26]. Sherlock [9], a single-column prediction framework, feeds various features of a column to a deep feed-forward neural network to get the prediction. This method ignores the global context and the dependencies between columns, making it difficult to distinguish the semantic types in cases like in Fig. 1. SATO [26] improves upon Sherlock by adding a topic modeling module and a structured prediction module on top of Sherlock, to jointly predict semantic types of all the columns in a table by leveraging the topic of a table and the dependencies between columns in a table.

Building on the trend of applying machine learning to tabular data, researchers have started using language models like BERT [5]. By feeding tables to BERT, which was originally designed for textual data, they exploit its extensive pre-training. This adaptation has created new frameworks that fine-tune

D.-N. Yang et al. (Eds.): PAKDD 2024, LNAI 14650, pp. 121–133, 2024.
https://doi.org/10.1007/978-981-97-2266-2_10

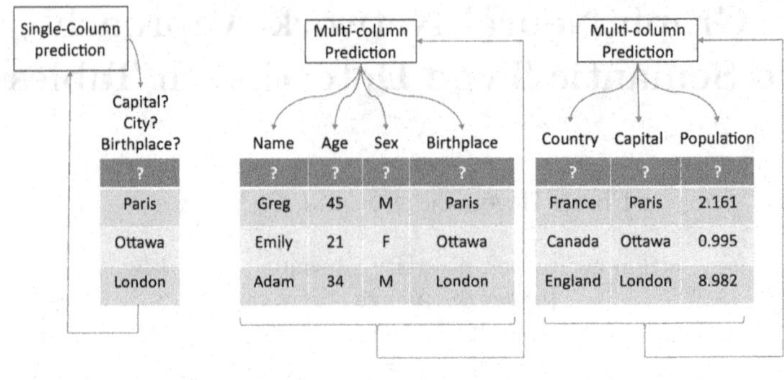

Fig. 1. The two tables on the right both have the column containing the values "Paris", "Ottawa" and "London". Without considering information coming from other columns it is difficult for a single-column prediction model to detect the actual semantic types of these columns. The multi-column prediction will label these columns correctly by jointly predicting all columns in a table.

BERT for column type annotation [4]. In addition to the values of the target column, two other sources of information can be used to improve the accuracy of semantic type annotation: intra-table information refers to other columns in the same table and inter-table information refers to other tables in the data. Given that language models have a small limit on the number of input tokens (BERT takes a maximum of 512 tokens), column type annotation models are developed to handle only one of the two mentioned sources of information.

Incorporating intra-table information has led to multi-column prediction approaches [25] that are designed to address the limitation of single-column prediction models in situations like Fig. 1 by accounting for broader table context, specifically column relationships and information. TABBIE [10] encodes rows and columns of a table respectively to get a better understanding of tables. The most prominent work in this category is Doduo [18] where BERT is modified to receive the whole columns of a table and predict their semantic types together.

Having inter-table information [21] can be a huge help in cases where the target column does not have enough high-quality data to make a good semantic prediction. For example, if a table has a column with entries like 'Orange' and 'Peach', the semantic type is ambiguous. However, by identifying and augmenting this column with columns of similar tables that have entries like 'Red' and 'Blue', the semantic type becomes clearer, indicating that this column is likely about colors rather than fruits. The most recent work, RECA [19], in addition to the values of the target column, identifies values of the most useful similar tables and feeds them to BERT to get the semantic type of the target column.

However, due to the small limit on the number of input tokens of language models like BERT, Doduo and RECA have the following drawbacks:

1. Doduo feeds the whole table to BERT and because of that, it poses difficulties in handling wide tables. For instance, the average number of columns of tables

in Open Data is 16, but there's a large variance with some tables having hundreds of columns. Furthermore, Doduo is not designed in a way that can incorporate inter-table context information and ignores this useful source of information [19].
2. RECA incorporates inter-table but not intra-table information, predicting the semantic type of each column individually. This approach enables RECA to handle wide tables within the language model's small token limit, as it doesn't need to model the entire table at once. However, this means it overlooks the valuable information in column relationships, crucial in complex scenarios like Fig. 1, where semantic types are difficult to distinguish.

Thus, a question is whether it is possible to incorporate both inter-table and intra-table information without suffering from the difficulty of handling wide tables as in Doduo and to benefit from the generalization power of the language model based approach. Inspired by recent developments in computer vision such as visual reasoning, object detection, and scene graph generation [3,14] where a main task is tagging the objects inside an image by leveraging the relations among the objects in the image, we propose to augment any single-column prediction framework, especially those incorporating inter-table information like RECA (addressing the drawback of Doduo) by a graph neural network (GNN) module to model the whole dependencies between columns. Thus, our framework incorporates both inter-table and intra-table information. In particular, we model each table by a graph with the nodes representing the columns and the edges representing the dependencies between columns through Message Passing of GNN. By considering the dependencies between all pairs of columns, this approach, called GAIT (**G**raph b**A**sed semant**I**c **T**ype detection), addresses the above drawback of RECA, making it a multi-column prediction framework. The challenge is how to represent the features of a column by a node so that Message Passing can leverage the dependencies among columns.

GAIT stands out in efficiently handling wide tables and benefiting from a language model based approach by building on top of models like RECA that are single-column predictions and language model based. While Doduo's effectiveness decreases in scenarios with minimal column dependency, and RECA faces challenges when similar tables are scarce, GAIT's integration of both inter-table and intra-table information makes it a competitive model in these diverse scenarios. This dual-data approach enables GAIT to maintain its performance and effectively address the limitations encountered by models focusing on either inter-table or intra-table information alone.

2 Related Works

Column type prediction methods are typically grounded into two categories, i.e., deep learning based frameworks and language model based frameworks.

Deep Learning Based Models. ColNet [1] uses DBpedia cell value lookups to create examples and trains a CNN with Word2Vec embeddings. HNN [2], models intra-column semantics, enhancing Colnet. Sherlock [9] employs column statistics, paragraph, word, and character embeddings to predict column types through a neural network. SATO [26] builds on Sherlock, adding topic modeling features and adjusting predictions for column dependencies using a CRF.

Language Model Based Models. Language models, like BERT [5], have been adopted for table tasks [23] including column type prediction. TaBERT [25] utilizes BERT as a base model to capture the table content features. TURL [4] is pre-trained unsupervisedly, using a visibility matrix for row and column context, and then fine-tuned for table-related tasks. TABBIE [10] separately processes the rows and columns of tables to give a better understanding of them. Doduo [18] predicts all the columns of a table together by feeding the whole table to BERT. RECA [19] incorporates inter-table context information by finding and aligning relevant tables. TCN [21] suggests using both intra-table and inter-table information for column type prediction. However, it needs table schema and page topic, which many datasets, like Webtables and Semtab don't have.

Summary: Most of the previous works [4,10,18,25], except for TCN and RECA, do not incorporate inter-table information for prediction. TCN [21] requires having table schema and page topic, which does not exist in many datasets. RECA does not incorporate intra-table dependencies. Our GAIT predicts semantic types of columns exclusively based on the content of the tables by integrating both intra-table and inter-table information.

3 Problem Definition

We aim to predict the semantic types of the columns of a given table with missing column headings. This problem is called *table annotation*. To learn to predict semantic types, a collection of labeled tables is given as the training data D, where each table $t(c_1, c_2, ..., c_n)$ consists of n columns and each column is labeled as one of the k pre-defined semantic types, also called *classes*, e.g., Age, Name, Country (note that semantic types are different from atomic types like integer and string). Note that the number of columns n and rows can differ for different tables. Typically, the first step is to extract a feature vector (embedding) to represent a column c_i. After applying a feature extractor function ϕ to the values of a column c_i and potential inter-table information related to c_i, an m-dimension feature (embedding) vector ψ_i is generated for column c_i. The rest of the task is to learn a mapping f that, given $\psi =< \psi_1, ..., \psi_n >$ of a table of n unlabeled columns, predicts the classes for the n columns in the table.

4 GAIT

Figure 2 shows the framework of GAIT. It is built on a single-column prediction provided by RECA. We opted for RECA due to its high performance in

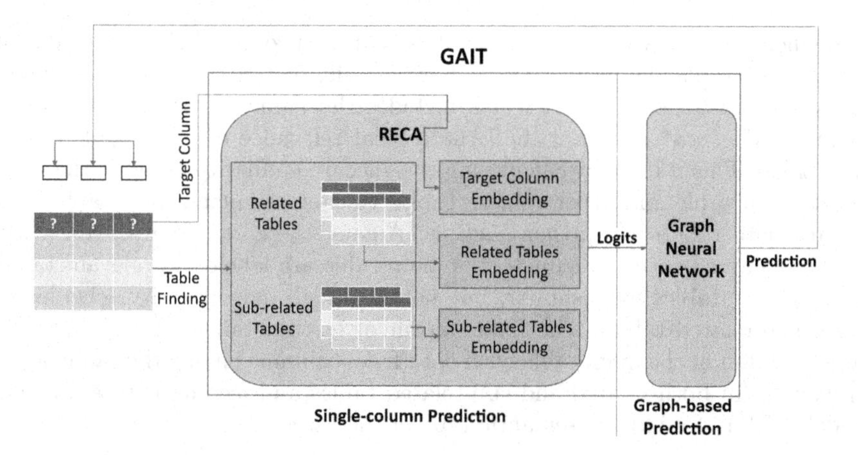

Fig. 2. The framework of GAIT: GAIT adds a GNN learning on top of the single-column prediction module, which is RECA in this work. The output of RECA is a class distribution for each column in a table, which provides the initial hidden state of the node representing that column in the GNN. For a table with n columns, RECA is performed n times. Then, the GNN learns the best representations of the hidden states of all nodes to minimize a loss function, through Message Passing that models the dependencies between columns.

column type annotation as a result of incorporating useful inter-table information from other relevant tables. That said, GAIT's design is versatile. While we utilize RECA, other single-column prediction modules can be integrated. GAIT employs a GNN in which each graph represents a table with the nodes representing the columns in the table. The initial representation of each node is the logits outputted by RECA for the represented column. Once fed with such class distributions as input, the training of GNN is responsible for capturing the dependencies of classes among columns and does not further involve the lower-level single-column prediction module. This approach treats the preliminary prediction of the single-column prediction of RECA as the node features for training GNN, which is more efficient than concatenating the networks of low-level modules into a giant neural network. Our method stacks a GNN as a meta-learner on top of RECA (i.e., two classifiers) instead of concatenating RECA and GNN into one classifier, which improves the overall performance according to stacked generalization technique [24]. We now present more details.

4.1 Single-Column Prediction

The single-column prediction is responsible for generating the preliminary prediction of each column. We use RECA [19] for this task. In the RECA process, the primary goal is to improve the understanding of a target column in the main table by integrating relevant data from other tables. The process begins by identifying named entities across all tables. Each entity is assigned a type from a predefined set, with the most common type within a column being selected as its representative named entity type. Following this, RECA constructs the named entity schema for each table, which includes the named entity types of all its columns. The next step is to find the topical relevance of other tables to the main table. This is done by calculating the Jaccard similarity between the words in the main table and other tables. Tables that are similar enough are chosen as candidate tables for further analysis. Among these candidates, tables with the same named entity schema as the main table are labeled as relevant tables. Additionally, tables with similar, but not identical, named entity schemas are called sub-related tables. The final step involves combining the data from the target column in the main table with data from columns having the same named entity type in both related and sub-related tables and feeding it to a language model, BERT, to find the semantic type of the target column.

4.2 Graph-Based Prediction

Graph Modeling of a Table. The training data for GNN is a collection of graphs organized into several mini-batches, where each graph corresponds to a table in the original training data. For a table with n columns, we create a graph of n nodes where each node represents a column in the table and create an edge between each pair of columns. Initially, each node u of a graph has the representation h_u^0 initialized to the logits $< o_1, ..., o_k >$ outputted by RECA for the corresponding column, which has one value for each class. This initial state represents the class bias of single-column prediction. In addition, each node is associated with the true class of the represented column.

Message Passing. Subsequently, the representation of all the nodes in a mini-batch of graphs is updated through the Message Passing mechanism of GNN along edges. For this purpose, we consider three different types of GNNs, graph convolutional network (**GCN**) [11], gated graph neural network (**GGNN**) [15], and graph attention network (**GAT**) [20], with the following $UPDATE$ functions where σ is the activation function, $N(u)$ is a list of nodes connected to node u, h_u^s is the representation (also called *embedding*) of node u at step $s \geq 0$, $W^{(s)}$ is a model parameter:

- **GCN:** assigns equal weights to all the neighbor nodes while updating the embedding of each node (Eq. 1).

$$h_u^{(s+1)} = \sigma\left(\sum_{v \in N(u) \cup u} \frac{W^{(s)} h_v^{(s)}}{\sqrt{|N(u)||N(v)|}} \right) \tag{1}$$

- **GGNN:** uses gated recurrent unit (GRU) to evaluate messages coming from adjacent nodes while updating the embedding of each node (Eq. 2).

$$h_u^{s+1} = GRU(h_u^s, \sum_{v \in N(u)} W^{(s)} h_v^s) \tag{2}$$

- **GAT:** updates node embedding (Eq. 3) according to the multi-head attention weights (Eq. 4), where K is the number of attention heads, $a^{(s,k)}$ and $W^{(s,k)}$ are model parameters for attention head k, and $\oplus$ is concatenation.

$$h_u^{(s+1)} = \oplus_{k=1}^{K}(\sigma \sum_{v \in N(u) \cup \{u\}} \alpha_{u,v}^{(s,k)} W^{(s,k)} h_v^{(s)}) \tag{3}$$

$$\alpha_{u,v}^{s,k} = \frac{\exp(ReLU(a^{(s,k)^T}(W^{(s,k)} h_u^{(s)} \oplus W^{(s,k)} h_v^{(s)})))}{\sum_{v' \in N(u) \cup \{u\}} \exp(ReLU(a^{(s,k)^T}(W^{(s,k)} h_u^{(s)} \oplus W^{(s,k)} h_{v'}^{(s)})))} \tag{4}$$

The $UPDATE$ function is applied to each node u in the mini-batch of graphs for S steps, where S is the number of hidden layers and output layer. h_u^S has one unit for each of the k classes and serves the final output for the k classes. The class prediction for the node u is given by applying softmax to h_u^S.

The main difference among GCN, GGNN, and GAT is in their treatment of adjacent nodes (columns). GAT uses an attention mechanism to assign varying weights to these nodes based on their importance. In contrast, GCN averages the features of neighbor nodes, while GGNN processes these features through a GRU to determine their relevance before updating the node embeddings.

Loss Function. Given the logit vector h_u^S for a node u with the true class $class_u$, the *loss* for this node is computed by the negative log-likelihood. The loss for a mini-batch of graphs is the sum of the loss of all the nodes inside the mini-batch of graphs (Eq. 5). We update the model parameters to minimize the loss of a mini-batch by performing stochastic gradient descent.

$$loss = \sum_{u=1}^{\#node} - \log(\frac{\exp(h_u^S[class_u])}{\sum_{m=1}^{k} \exp(h_u^S[m])}) \tag{5}$$

4.3 Overall Prediction

After training the GNN, to classify columns in a new table t, we first get the output of RECA (before softmax) for each column in t. These outputs are the initial representation h_u^0 of nodes u in the graph representing the table t. Then, message passing is done using the learned parameters in the training phase to get the predicted class for each node, which is the predicted class for the corresponding column in t. Fig. 2 shows how GAIT predicts labels of a table.

5 Evaluation

5.1 Evaluation Method

Performance Metrics. Like previous works [9,18,19,26], we collect *weighted f-score* and *macro f-score* on the test data. The former is the average of f-score of all classes, weighted by class frequencies, and the latter is the average of treating all

Table 1. Datasets used and the number of tables with the specified number of columns.

Dataset	#types	#tables	#Col	avg col
Semtab	275	3045	7603	4.5
Webtable	78	32262	74141	2.3

classes equally, regardless of their frequencies. The macro f-score better reflects the model performance on infrequent classes. We evaluate model performance using 5-fold cross-validation, reporting the mean and standard deviation of the above f-scores from the test split of each fold.

Datasets. We use two datasets summarized in Table 1.

Webtables [18,19,26]: This dataset contains 32262 tables and 78 unique classes extracted from the Webtables directory of VizNet [8]. We use *exactly the same* 5-fold cross-validation split as in [19] , which splits the tables (not columns) into a train set and test set in 5-folds. So, we copy directly the f-scores of the baseline algorithms (more details on the baselines below) except for RECA from [19].

Semtab2019 [19]: It contains 3045 tables and 275 unique classes. While this dataset covers wider tables (an average of 4.5 columns per table), only 7603 columns are annotated. The split proposed in RECA [19] randomly divided columns (not tables) into train, validation, and test sets. Although the column-wise splitting of data makes sense for RECA due to the column-wise prediction of RECA, GAIT requires having a full table to model the dependencies between columns in the table. Therefore, our 5-fold validation splits tables (instead of columns) into the train set and test set for this dataset. At each fold, we further split the train set into 80% for training and 20% for validation.

Algorithms for Comparison. All experiments were conducted with Tesla V100s. We used the publicly available source code of RECA[1] for the single-column prediction module of GAIT. The GNN module of GAIT was implemented using the deep graph library [22] , Adam optimization with a learning rate of $1e-3$ and weight decay of $5e-4$ for training. We trained GCN, GGNN, and GAT for 100, 200, and 100 epochs respectively. To optimize the GAT structure, we tested various # attention heads ([1, 2, 4, 8, 12]) and update steps S ([1, 2, 3, 4]), selecting the best model from the validation set as default. Similarly, for GGNN and GCN, we determined the default model by experimenting with different update steps S ([1, 2, 3, 4]). Three algorithms for GAIT were finally chosen: $\mathbf{GAIT_{GAT}}$ (GAT with $S=2$), $\mathbf{GAIT_{GGNN}}$ (GGNN with $S=3$), and $\mathbf{GAIT_{GCN}}$ (GCN with $S=2$).

[1] https://github.com/ysunbp/RECA-paper.

Table 2. Macro f-score and weighted f-score.

Model	Semtab		Webtables	
	Weighted f-score	Macro f-score	Weighted f-score	Macro f-score
sherlock [9]	0.638 ± 0.009	0.417±0.017	0.844±0.001	0.670±0.010
TaBERT [25]	0.756±0.011	0.401 ± 0.025	0.896 ± 0.005	0.650 ± 0.011
TABBIE [10]	0.798 ± 0.012	0.542 ± 0.022	0.929 ± 0.003	0.734 ± 0.019
Doduo [18]	0.819 ± 0.010	0.565 ± 0.021	0.928 ± 0.001	0.742 ± 0.012
RECA [19]	0.825 ± 0.015	0.583 ± 0.019	0.935 ± 0.032	0.783 ± 0.017
$\text{GAIT}_{\text{GGNN}}$	0.844 ± 0.003	0.606 ± 0.018	0.936 ± 0.003	0.797 ± 0.022
GAIT_{GCN}	0.845 ± 0.006	0.622 ± 0.020	0.939 ± 0.004	0.794 ± 0.017
GAIT_{GAT}	**0.852 ± 0.004**	**0.643 ± 0.017**	**0.940 ± 0.003**	**0.799 ± 0.019**

Since GAIT incorporates RECA as its single-column prediction module, naturally we evaluate GAIT against the baseline methods outlined in RECA's paper and RECA itself. These baselines are described below and their source codes are publicly available and are used for our evaluation:

- Sherlock [9]: Sherlock is a deep learning model that extracts character-level, word-level, paragraph-level and global-level statistical features from tables to form vector representations for table columns.
- TaBERT [25]: TaBERT simultaneously analyzes queries and a table, selecting three crucial rows to create table content snapshots. It then uses BERT to develop representations for each table column, aiding in classification.
- TABBIE [10]: improves TaBERT by separately processing the rows and columns of tables. The embedding of the target column is used for prediction.
- Doduo [18]: Modifies BERT to feed the whole columns of a table to BERT and predicts the semantic types of all of the columns in a table together.
- RECA [19]: RECA finds relevant tables for the target table and uses the information coming from these tables and the values of the target column to predict the semantic type of the target column.

We do not compare with SATO [26] and TURL [4] as Doduo outperformed them. Since TCN [21] requires having table schema and page topic [19], it cannot be applied to our datasets.

5.2 Results

Table 2 Shows the performance of GAIT and the baseline algorithms. GAIT outperforms Sherlock by a large margin. The main reason behind the poor performance of Sherlock compared with other models is its simplicity. While other models including GAIT utilize language models for semantic type prediction, Sherlock relies on simple semantic features to do so. Furthermore, Sherlock does not use intra-table or inter-table information for prediction.

Among language model based models TaBERT shows the worst performance because it was initially developed for table semantic parsing and column embeddings generated by TaBERT are not suitable for column type annotation [19]. RECA, single-column prediction module of GAIT, outperforms both TABBIE and Doduo. TABBIE and Doduo use the limited input tokens of language models to process intra-table context while ignoring the inter-table context information when generating the embeddings of the target columns. However, RECA mainly focuses on extracting useful inter-table context information to enhance the embeddings of the target columns [19].

GAIT with different GNNs outperforms RECA, and by extension TABBIE and Doduo in both datasets. In particular, GAIT shows about 6% and 2.7% improvement in macro and weighted f-scores over RECA in the Semtab-dataset. These results prove that modeling the dependencies between columns in a table, which is the main advantage of GAIT over RECA, is useful. GAIT successfully applies a GNN on top of RECA to do so. Among different variations of GAIT, GAT shows the best performance. Assigning different weights to adjacent nodes (columns) according to their importance when updating representation of a node is the key to the superior performance of GAT compared to GCN and GGNN.

In both datasets, GAIT's improvement is larger on the macro f-score than the weighted f-score. This means that infrequent classes that label fewer columns benefit more from the whole dependency approach of the GNN approach. Such classes have less presence in the data and their learning tends to rely on the dependencies on other columns in a table. GAIT provides a mechanism to leverage such dependencies. This also explains why GAIT shows a better enhancement in the Semtab dataset compared to Webtables dataset. The 3045 tables of Semtab have 275 semantic types for columns while the 32262 tables of Webtables are limited to 78 semantic types. Consequently, many more infrequent classes in Semtab can benefit from modeling the whole dependencies of GAIT.

To provide a better insight into this improvement, we divide the 275 classes of Semtab dataset into three equally sized bins of High, Medium, and Low frequencies (about 92 classes in each bin) according to the columns labeled by classes and show the macro f-score of the classes in each bin for GAIT_{GAT} (best GAIT) and RECA (best baseline) in Fig. 3. While GAIT_{GAT} improves RECA in all the three bins, the bigger improvements happen in the low-frequency bins, for example, the absolute improvement of 11% or the relative improvement of 96.5% for the Low bin. Figure 3 also demonstrates that the real challenge in developing column type annotation models is how to have a reliable prediction for medium and low-frequency classes as the performance for high-frequency classes is already good enough. The large improvements of GAIT_{GAT} over RECA in low-frequency classes is a clear sign of its superiority in handling such classes.

We also study the impact of the number of columns in a table on both GAIT_{GAT} and RECA. Table 3 shows the improvement of GAIT_{GAT} on macro and weighted f-scores over RECA *separately* for tables of a different number of columns. As the number of columns in a table increases, the performance of RECA, which is also the single-column prediction module of GAIT, increases.

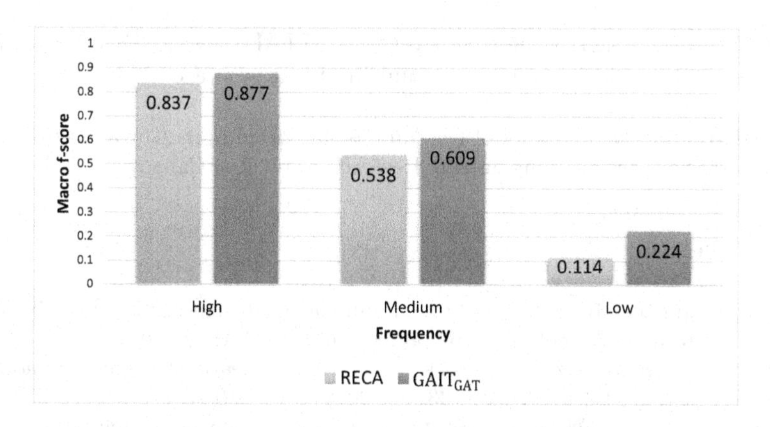

Fig. 3. The macro f-score of GAIT$_{GAT}$ and RECA on Semtab dataset, for High, Medium, and Low-frequency classes.

Table 3. The f-score improvement of GAIT$_{GAT}$ over RECA by tables of different number of columns.

#col	Semtab				Webtables			
	macro f-score		weighted f-score		macro f-score		weighted f-score	
	RECA	GAIT$_{GAT}$	RECA	GAIT$_{GAT}$	RECA	GAIT$_{GAT}$	RECA	GAIT$_{GAT}$
2	0.563	0.603 (+4.0%)	0.798	0.828 (+3.0%)	0.738	0.758 (+2.0%)	0.932	0.936 (+0.4%)
3	0.545	0.616 (+7.1%)	0.797	0.827 (+3.0%)	0.743	0.762 (+1.9%)	0.927	0.930 (+0.3%)
4	0.566	0.618 (+5.2%)	0.865	0.880 (+1.5%)	0.727	0.746 (+1.9%)	0.960	0.961 (+0.1%)
5	0.664	0.682 (+1.8%)	0.862	0.862 (+0.0%)	0.540	0.548 (+0.8%)	0.978	0.978 (+0.0%)

Having more columns in a table better reveals context of that table, so RECA can find more relevant inter-table information which is beneficial to both RECA and GAIT. Thus, the need for dependencies between columns in GAIT$_{GAT}$ decreases. The column dependency method GAIT improves RECA mainly for low-frequency classes and tables of 2 to 4 columns, as in case of Semtab.

6 Conclusion

Language model based approaches recently showed promising results in column type annotation thanks to the semantic knowledge preserved in them. This paper addresses some drawbacks of previous language model-based approaches, namely, failing to incorporate inter-table and intra-table information simultaneously due to the input token limit of language models. Our solutions, GAIT, employ graph neural networks to model the intra-table dependencies, letting language models focus on handling inter-table information. Experiments on different datasets provide evidence of the effectiveness of our solutions. Looking ahead, considering the recent advancements in large language models (LLMs) for column type annotation [7,12,13,27] exploring alternative LLMs beyond BERT to address inter-table information could be a promising future research.

Acknowledgement. The work of Ke Wang is supported in part by a discovery grant from Natural Sciences and Engineering Research Council of Canada.

References

1. Chen, J., Jiménez-Ruiz, E., Horrocks, I., Sutton, C.: Colnet: embedding the semantics of web tables for column type prediction. In: AAAI (2019)
2. Chen, J., Jiménez-Ruiz, E., Horrocks, I., Sutton, h.: Learning semantic annotations for tabular data. In: IJCAI. vol. 33, pp. 2088–2094 (2019)
3. Chen, X., Li, L.J., Fei-Fei, L., Gupta, A.: Iterative visual reasoning beyond convolutions. In: CVPR, pp. 7239–7248 (2018)
4. Deng, X., Sun, H., Lees, A., Wu, Y., Yu, C.: Turl: table understanding through representation learning. ACM SIGMOD Rec. **51**(1), 33–40 (2022)
5. Devlin, J., Chang, M.W., Lee, K., Toutanova, K.: Bert: Pre-training of deep bidirectional transformers for language understanding. arXiv:1810.04805 (2018)
6. Fernandez, R.C., Abedjan, Z., Koko, F., Yuan, G., Madden, S., Stonebraker, M.: Aurum: A data discovery system. In: ICDE, pp. 1001–1012. IEEE (2018)
7. Feuer, B., Liu, Y., Hegde, C., Freire, J.: Archetype: a novel framework for open-source column type annotation using large language models. arXiv (2023)
8. Hu, K., et al.: Viznet: Towards a large-scale visualization learning and benchmarking repository. In: CHI, pp. 1–12 (2019)
9. Hulsebos, M., et al.: Sherlock: A deep learning approach to semantic data type detection. In: SIGKDD, pp. 1500–1508 (2019)
10. Iida, H., Thai, D., Manjunatha, V., Iyyer, M.: Tabbie: Pretrained representations of tabular data. arXiv preprint arXiv:2105.02584 (2021)
11. Kipf, T.N., Welling, M.: Semi-supervised classification with graph convolutional networks. arXiv preprint arXiv:1609.02907 (2016)
12. Korini, K., Bizer, C.: Column type annotation using chatgpt. arXiv (2023)
13. Li, P., et al.: Table-gpt: Table-tuned gpt for diverse table tasks. arXiv (2023)
14. Li, Y., Ouyang, W., Zhou, B., Wang, K., Wang, X.: Scene graph generation from objects, phrases and region captions. In: Proceedings of the IEEE International Conference on Computer Vision, pp. 1261–1270 (2017)
15. Li, Y., Tarlow, D., Brockschmidt, M., Zemel, R.: Gated graph sequence neural networks. arXiv preprint arXiv:1511.05493 (2015)
16. Limaye, G., Sarawagi, S., Chakrabarti, S.: Annotating and searching web tables using entities, types and relationships. VLDB **3**(1–2), 1338–1347 (2010)
17. Rahm, E., Bernstein, P.A.: A survey of approaches to automatic schema matching. the VLDB Journal **10**(4), 334–350 (2001)
18. Suhara, Y., Li, J., Li, Y., Zhang, D., Demiralp, Ç., Chen, C., Tan, W.C.: Annotating columns with pre-trained language models. In: SIGMOD (2022)
19. Sun, Y., Xin, H., Chen, L.: Reca: related tables enhanced column semantic type annotation framework. VLDB **16**(6), 1319–1331 (2023)
20. Veličković, P., Cucurull, G., Casanova, A., Romero, A., Lio, P., Bengio, Y.: Graph attention networks. arXiv preprint arXiv:1710.10903 (2017)
21. Wang, D., Shiralkar, P., Lockard, C., Huang, B., Dong, X.L., Jiang, M.: Tcn: Table convolutional network for web table interpretation. In: WWW (2021)

22. Wang, M., et al.: Deep graph library: a graph-centric, highly-performant package for graph neural networks. arXiv preprint arXiv:1909.01315 (2019)
23. Wang, Z., et al.: Tuta: Tree-based transformers for generally structured table pretraining. In: SIGKDD (2021)
24. Wolpert, D.H.: Stacked generalization. Neural Netw. **5**(2), 241–259 (1992)
25. Yin, P., Neubig, G., Yih, W.t., Riedel, S.: Tabert: pretraining for joint understanding of textual and tabular data. arXiv preprint arXiv:2005.08314 (2020)
26. Zhang, D., Suhara, Y., Li, J., Hulsebos, M., Demiralp, Ç., Tan, W.C.: Sato: Contextual semantic type detection in tables. arXiv preprint arXiv:1911.06311 (2019)
27. Zhang, H., Dong, Y., Xiao, C., Oyamada, M.: Jellyfish: A large language model for data preprocessing. arXiv (2023)

TCGNN: Text-Clustering Graph Neural Networks for Fake News Detection on Social Media

Pei-Cheng Li[1] and Cheng-Te Li[2](✉)

[1] National Taiwan University, Taipei, Taiwan
patty101257@gmail.com
[2] National Cheng Kung University, Tainan, Taiwan
chengte@ncku.edu.tw

Abstract. In the realm of fake news detection, conventional Graph Neural Network (GNN) methods are often hamstrung by their dependency on non-textual auxiliary data for graph construction, such as user interactions and content spread patterns, which are not always accessible. Furthermore, these methods typically fall short in capturing the granular, intricate correlations within text, thus weakening their effectiveness. In this work, we propose Text-Clustering Graph Neural Network (TCGNN), a novel approach that circumvents these limitations by solely utilizing text to construct its detection framework. TCGNN innovatively employs text clustering to extract representative words and harnesses multiple clustering dimensions to encapsulate a multi-faceted representation of textual semantics. This multi-layered approach not only delves into the fine-grained correlations within text but also bridges them to a broader context, significantly enriching the model's interpretative fidelity. Our rigorous experiments on a suite of benchmark datasets have underscored TCGNN's proficiency, outperforming extant GNN-based models. This validates our premise that an adept synthesis of text clustering within a GNN architecture can profoundly enhance the detection of fake news, steering the course towards a more reliable and textually-aware future in information verification.

Keywords: Fake News Detection · Rumor Detection · Graph Neural Networks · Text Clustering · Social Media

1 Introduction

Social media platforms have developed into a convenient way for people to receive and disseminate messages. However, the availability of social media platforms has also led to an increase in the spread of false information. Additionally, it is impractical to manually censor and filter all messages for inaccuracy due to the large number of users on the platforms and the quick publishing of new posts. By dividing people and harming democracy, misinformation hurts society and

D.-N. Yang et al. (Eds.): PAKDD 2024, LNAI 14650, pp. 134–146, 2024.
https://doi.org/10.1007/978-981-97-2266-2_11

even makes people anxious [1,21,23]. Therefore, establishing a reliable system for detecting fake news is an imminent need. Graph Neural Network (GNN)-based methods excel at encoding topological structure [2,9,17,20,24] with text information and have shown promising results on fake news detection. However, most of them rely on additional information to construct the graphs and apply GNNs. To be specific, they require the structures of information propagation (e.g., retweet, comment, and endorsement) and social networks to have the graph structure. If one has only text information, which commonly happens when new articles announce for the first time, existing GNN-based methods cannot work.

Detecting fake news based solely on text information from social media offers several advantages. Text is the most ubiquitous form of information available across platforms, providing a vast dataset for analysis without the need for complex data acquisition processes. Moreover, focusing on text content mitigates privacy concerns, as it avoids the use of sensitive user data [12,15]. Relying on text alone allows for the identification of fake news before it can engage users and trigger broader social networks effects, potentially curbing the spread of misinformation at its source [19,27].

Fake news detection can be approached as a text classification task, a setting where GNN-based methods show promise. Existing studies, such as TextGCN [25], TensorGCN [16], TLGNN [11], TextING [26], and HyperGAT [8], apply GNNs for text classification but primarily use a single graph to model article relationships. This can limit information depth, as text correlation can be depicted via multiple aspects. We advocate for a finer- and multi-grained view, leveraging semantic aspects identified through joint word and document clusterings for graph construction.

To better utilize text information and to distill fine-grained knowledge aspects for better graph construction in the task of fake news detection, we propose a novel GNN-based method, ***T****ext-****C****lustering* ***G****raph* ***N****eural* ***N****etworks* (TCGNN). By performing text clustering and selecting representative words of each cluster, we construct multiple graphs, which encode knowledge from various perspectives. Multiple text clusterings with different specified cluster numbers further bring multi-grained text correlation. By applying GNNs to the constructed text-clustering graphs, we can obtain multi-grained representations of articles. Experiments conducted on four well-known benchmark datasets exhibit that TCGNN can consistently outperform existing GNN-based methods, and verify the usefulness of text clustering for fake news detection.

The contributions of this work are summarized below.

- *Novel Graph Construction*: TCGNN improves graph-based fake news detection by exclusively employing text clustering to construct graphs, eliminating reliance on non-textual data. This method addresses a critical gap where existing models falter due to the absence of auxiliary information like propagation structures, ensuring TCGNN's applicability even in scenarios where only the text of new articles is available.
- *Enhanced Textual Semantics:* By extracting representative words through text clustering and utilizing diverse clustering resolutions, TCGNN captures a

Table 1. Summary of relevant studies. Column meanings: original goal (OG), source tweets (ST), users (U), user attributes (UA), retweet comments (RC), propagation structure (PS), graph type (GT). Cells in OG: text classification (TC), fake news detection (FND), and node classification (NC). Cells in GT: Homogeneous (Ho), Heterogeneous (He), and Hypergarph (Hy).

	OG	ST	U	UA	RC	PS	GT
GCAN [17]	FND	✓	✓	✓		✓	Ho
FANG [20]	FND	✓	✓	✓		✓	He
BiGCN [2]	FND	✓	✓	✓	✓	✓	Ho
GACL [22]	FND	✓			✓	✓	Ho
EBGCN [24]	FND	✓			✓	✓	Ho
UPFD [9]	FND	✓	✓			✓	Ho
TextGCN [25]	TC	✓					Ho
HyperGAT [8]	TC	✓					Hy
NRGNN [6]	NC	✓					Ho
RSGNN [7]	NC	✓					Ho
TCGNN (this work)	FND	✓					Ho

nuanced, multi-dimensional representation of textual data. This multi-grained approach allows the model to delve into intricate text correlations, vastly improving interpretative fidelity and context comprehension over models that use a single-layer graph representation.

- *Empirical Validation of Superior Performance:* The effectiveness of TCGNN is empirically validated through rigorous testing on several benchmark datasets, where it demonstrates superior performance over current GNN-based models. These results not only showcase the practical effectiveness of TCGNN but also highlight the importance of text clustering as a means to enhance the reliability of fake news detection in the digital information ecosystem.

This paper is organized as follows. We first review and compare relevant studies in Sect. 2. We give the problem statement and present the technical details of the proposed TCGNN in Sect. 3. Experimental settings and results are reported in Sect. 4. We conclude this work in Sect. 5.

2 Related Work

A summary of relevant GNN-based studies is provided in Table 1. The Text Clustering Graph Neural Network (TCGNN) brings a novel approach to fake news detection (FND) that distinguishes itself from existing studies through its focused methodology and specific feature integration. Unlike several models such as GCAN [17], FANG [20], BiGCN [2], and UPFD [9], which integrate user attributes (UA) and source tweets (ST), TCGNN strategically opts not to include UA in its analysis. This decision emphasizes the method's robustness in contexts where user attributes may be sparse or unreliable, thereby increasing the model's applicability across diverse scenarios where user metadata is not available or is privacy-protected. Moreover, while models like BiGCN [2], GACL [22], and EBGCN [24] incorporate retweet comments (RC) into their analysis to capture the public's perception and spread patterns, TCGNN's approach solely relies on source tweets (ST). This focus on ST highlights TCGNN's ability to discern the authenticity of news content based on the originating tweets themselves, leveraging the intrinsic textual patterns and propagation structures (PS) without the potential noise introduced by user-generated comments.

In contrast to FANG [20] which uses a heterogeneous graph (He) to represent the data, TCGNN, along with several other studies like GCAN [17] and UPFD [9], employs a homogeneous graph (Ho). The homogeneous graph typology underscores TCGNN's emphasis on streamlined and efficient computation, as it avoids the complexity introduced by multi-typed nodes and edges in heterogeneous graphs, which can be computationally intensive and may not always contribute to improved performance. Lastly, the comparison with TextGCN [25] and HyperGAT [8], both oriented towards text classification (TC), and NRGNN [6] and RSGNN [7] aimed at node classification (NC), illustrates TCGNN's dedicated design for FND. While the mentioned studies contribute to their respective fields, TCGNN's settings are meticulously tailored to extract and process features specifically relevant to fake news, enhancing its precision and effectiveness for its intended application.

In essence, TCGNN's settings are a strategic choice that balances complexity and performance, aiming to achieve high accuracy in fake news detection without relying on extensive and potentially redundant data sources. The model's design reflects a deliberate and insightful move towards a more universal and effective solution in the domain of misinformation identification.

3 The Proposed TCGNN Method

We denote the set of text-clustering graphs $G = \{G^j\}_{j=1}^{c}$, c is the number of text clusters. Each graph $G^c = (V, E^c)$ has the same node set but different edge sets, indicating that their graph structures are distinct. The set $V = \{v_1, v_2, \ldots, v_n\}$ has n nodes, where each node represents an article. The set $E^c = \{e_1^c, e_2^c, \ldots, e_{m_c}^c\}$ represents m_c edges in text-clustering graph G^c. Moreover, $Y = \{y_1, y_2, \ldots, y_n\}$ denotes a set of article labels. W is the set of unique words in a corpus, and $d = |W|$ is the number of unique words.

Problem Statement. The task of fake news detection is a binary classification problem – predicting the veracity of a given article $v_i \in V$. We aim at learning a function $\mathcal{F} : v_i \rightarrow \{0, 1\}$ that outputs 0 for real and 1 for fake articles.

The primary design of TCGNN, exhibited in Fig. 1, lies in the construction of multi-grained text-clustering graphs. From a collection of tweets, text clustering

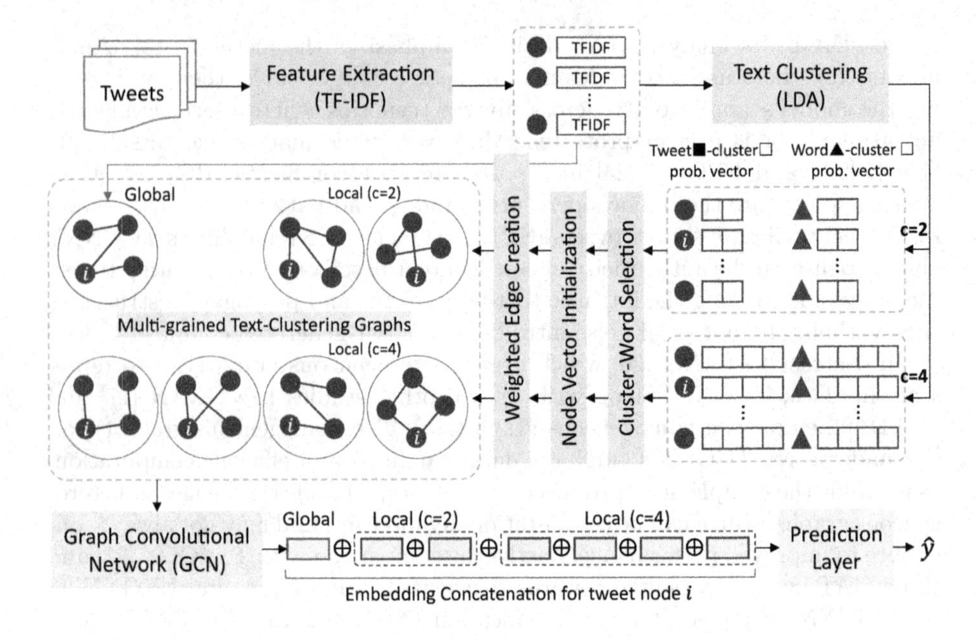

Fig. 1. Overview of the proposed TCGNN.

is conducted with specified cluster numbers, such as $c = 2, 4$, enabling multi-granularity. This process yields tweet-to-cluster and word-to-cluster probability vectors, which inform the selection of representative words for each cluster. These words initialize tweet node vectors and edge construction in text-clustering graphs. A graph neural network model is trained on these graphs to generate tweet embeddings reflecting multi-grained text-clustering knowledge, which are then concatenated for prediction output.

3.1 Text-Clustering Graph Construction

We have four steps. The first is **Text Clustering**, which groups *words* correlated with similar articles into clusters. The second is **Cluster-word Selection**, which keeps informative words and rules out unimportant ones in each cluster. The third is **Node Vector Determination**, which is to obtain the feature vector of each article node based on the selected words. The fourth is **Weighted Edge Creation**, which constructs weighted edges between nodes to build a text-clustering graph.

- **Step 1: Text Clustering.** We use Latent Dirichlet Allocation (LDA) [3] for text clustering. The idea is to find latent topics based on the association between articles and words. Each article and each word can be assigned to topics with probabilities that depict the importance scores with respect to topics. Words in the same topic belong to the same clusters. Two LDA hyperparameters are the way that we represent the relationships between words and

articles, and the number c of topics/clusters. For the former, we use TF-IDF, given by: $\mathbf{X} = TFIDF(V, W)$, where $\mathbf{X}$ is the set of TF-IDF vectors for all articles V. For the latter, we will vary its value and see how will performance be affected. We denote the results of applying LDA as $\mathbf{T} = LDA(\mathbf{X}; c)$, i.e., the importance probability of each word belonging to every cluster.
- **Step 2: Cluster-word Selection.** A word can belong to multiple topics with different probabilities. We select representative words to depict the semantics of each topic based on the probability. We use a hyperparameter $r \in [0, 1]$ to select important words for each cluster. For each topic/cluster, we select words with top-$\lfloor d \times r \rfloor$ probability values. Higher r means that more words are selected to depict a cluster. We represent the cluster-word selection (CWS) as: $W^c = CWS(\mathbf{T}, W, r)$, where W_c is the set of selected words for cluster c.
- **Step 3: Node Vector Determination.** We need to have an initial feature vector for each article node in a cluster. We again use TF-IDF to produce the vector, whose element indicates the importance of a word with respect to an article in a cluster. The selected cluster words are utilized here. The TF-IDF vectors for all articles nodes in a cluster c can be derived and depicted by $\mathbf{X'^c} = TFIDF(V, W^c)$. In this way, each article can belong to multiple clusters with different initial feature vectors.
- **Step 4: Weighted Edge Creation.** The last step is to create edges between article nodes. For each cluster c, we calculate the similarity $sim(v_i^c, v_j^c)$ between article nodes v_i^c and v_j^c using their initial feature vectors, and cosine similarity are utilized, given by: $sim(v_i^c, v_j^c) = \frac{x_i'^c \cdot x_j'^c}{\|x_i'^c\|\|x_j'^c\|}$. Since we aim to exploit the representative knowledge to depict the relationships between articles, we select only top-k node pairs with higher similarity scores to create edges in the graph of a cluster.

Text clustering brings finer-grained knowledge with different semantic aspects to model the correlation between new articles. The constructed C text-clustering graphs can provide *local* knowledge for the detection of fake news. Nevertheless, we think that utilizing all words together to depict the interactions between articles, which can be viewed as *global* knowledge, is useful as well. Therefore, in addition to c text-clustering graphs, we create an additional global graph $\bar{G}$. Nodes are all articles, and edges are created using global TF-IDF $\mathbf{X}$ with the selection of node pairs by top-k similarity. We have $c+1$ graphs for GNN model training.

3.2 Model Training

Given c local graphs G^c with initial node vectors $\mathbf{X'^c}$, and one global graph $\bar{G}$ with initial node vectors $\mathbf{X}$, we aim at learning the representation of each article by graph neural networks. We use the *graph convolutional network* (GCN) [14] as the GNN model, in which the number of GCN layers is set as 2 by default. Let $\mathbf{h}_v^c$ be the output embedding of article v by applying GCN to the local graph G^c, and $\bar{\mathbf{h}}_v$ be the output embedding of article v from the global graph $\bar{G}$. By concatenating all of the embeddings obtained from local graphs and the embedding from the local graph, $\mathbf{h}_v = \mathbf{h}_v^1 \oplus \mathbf{h}_v^2 \oplus \cdots \oplus \mathbf{h}_v^c \oplus \bar{\mathbf{h}}_v$, where $\oplus$ is the concatenation operator, we feed the final embedding $\mathbf{h}_v$ to a fully-connected layer FC, and generate the prediction outcome, given by: $\hat{y}_v = softmax(ReLU(\mathbf{h}_v \mathbf{W}_f + \mathbf{b}_f))$, where $\mathbf{W}_f$ and $\mathbf{b}_f$ are learnable parameters. We use cross entropy to be the loss function, and utilize Adam [13] to be the optimizer.

Table 2. Statistics of four datasets

Dataset	# news	# fake news	% fake news
FANG	745	319	42.82%
Pheme	5802	1972	33.99%
Twitter15	1458	1086	74.49%
Twitter16	818	613	74.94%

The basic setup of TCGNN relies on performing text clustering one time according to the pre-defined cluster number c. More/fewer clusters provide finer/coarser-grained information about text correlation. By giving multiple cluster numbers $H = (c_1, c_2, ..., c_z)$ for executing text clustering multiple times, we can derive *multi-grained* knowledge for representation learning. By concatenating article v's global embedding $\bar{\mathbf{h}}_v$ with all local embeddings obtained from GCNs across all text-clustering graphs with different cluster numbers, we can obtain the final embedding $\mathbf{h}_v \in \mathbb{R}^{(c_1+c_2+\cdots+c_z+1)\cdot b}$ that possesses multi-grained knowledge for model training, where b is the dimensionality of a GCN's output embedding.

4 Experiments

Datasets. We use four datasets: FANG [20], Pheme [4], Twitter15 [18], and Twitter16 [18]. For FANG, we use the news URLs, news headlines, and binary labels of whether an article is fake or not. The number of articles is different from the original because of our data preprocessing method. We remove stopwords and strange Unicode characters, perform POS tagging for each token, and lemmatize tokens with POS tags. Data statistics are shown in Table 2.

Settings. The dimension of hidden state is 16. We divide the dataset into training, validation, and testing with ratio 80%:5%:15%. All hyperparameters are tuned based on the validation set. We vary the number of top-k neighbors, the multi-grained cluster numbers (H), and the ratio (r) for cluster-world selection among $\{3, 5, 7, 10, 20, 30\}$, $\{(8), (8, 16), (8, 16, 32)\}$, and $\{0.1, 0.2, ..., 1\}$, respectively. The evaluation metrics include Accuracy (Acc), Area under ROC Curve (AUC), and F1 score. We report the average results of 5 runs with different seeds for all experiments.

Table 3. Performance comparison between TCGNN and baseline models on FANG and Pheme datasets.

	FANG			Pheme		
Model	Acc	AUC	F1	Acc	AUC	F1
GRU	0.5799	0.5688	0.4461	0.8321	0.8662	0.7323
GCN	0.6537	0.7281	0.4730	0.8429	0.9146	0.7648
GraphSAGE	0.6725	0.7190	0.5325	0.8338	0.9070	0.7515
TextGCN	0.6564	0.6549	0.6764	0.6269	0.5025	0.6470
RSGNN	0.6389	0.7004	0.5128	0.7906	0.8660	0.6426
NRGNN	0.6711	0.7067	0.4179	0.8138	0.7955	0.6983
TCGCN	**0.7114**	**0.7982**	**0.6861**	**0.8672**	**0.9440**	**0.8188**

Table 4. Performance comparison between TCGNN and baseline models on Twitter15 and Twitter16 datasets.

	Twitter15			Twitter16		
Model	Acc	AUC	F1	Acc	AUC	F1
GRU	0.7732	0.7201	0.8564	0.7415	0.6128	0.8470
GCN	0.8141	0.8566	0.8818	0.7744	0.7832	0.8608
GraphSAGE	0.7537	0.7837	0.8532	0.7537	0.7837	0.8532
TextGCN	0.7423	0.4998	0.5171	0.7073	0.5113	0.5732
RSGNN	0.7477	0.6702	0.8554	0.7707	0.7295	0.8664
NRGNN	0.7913	0.0791	0.8835	0.7720	0.0077	0.8713
TCGCN	**0.8450**	**0.8966**	**0.8977**	**0.8503**	**0.8949**	**0.9013**

Baselines. We compare TCGNN with six competitors, including GRU [5], GCN [14], GraphSAGE [10], TextGCN [25], RSGNN [7], and NRGNN [6]. All GNN-based methods use the global graph constructed with all words (i.e., no text clustering). We choose RSGNN and NRGNN because they are robustness-aware GNN models by adjusting graph structures for node classification. By treating tweets as nodes, RSGNN and NRGNN can be seamlessly utilized for text classification and our fake news detection setting. For the hyperparameters of all baseline methods, we refer to their original papers and follow their tuning strategies for choosing the best performance.

Main Results. The results are exhibited in Table 3 and Table 4. GNN-based models do better than other approaches, implying that the structural information learned by GNNs can bring performance improvement in fake news detection, compared to models using only text information. The proposed TCGNN outperforms well in all three evaluation metrics across four datasets. We can obtain the following insights based on the results.

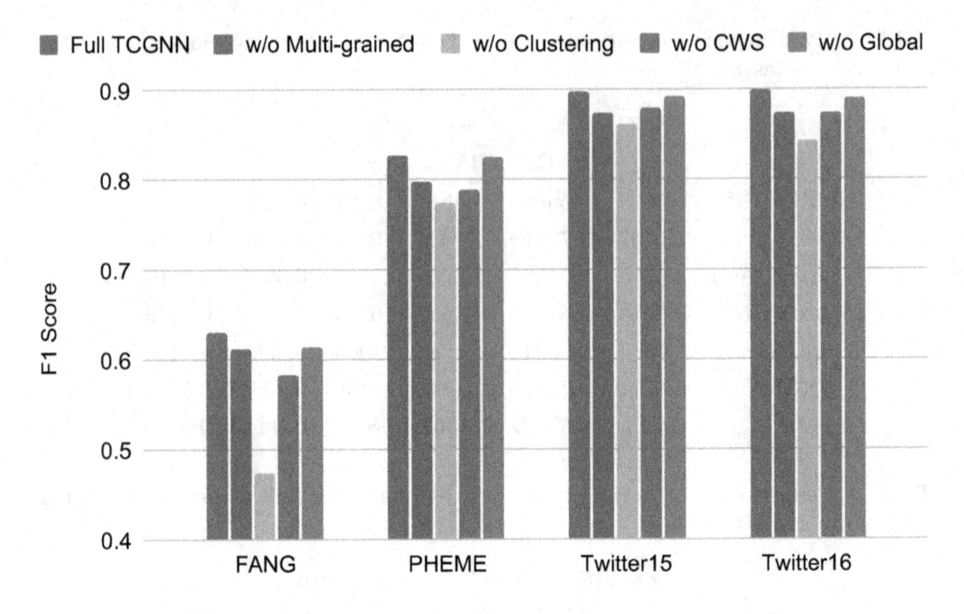

Fig. 2. Results on ablation study.

- *Multi-Grained Text Analysis:* TCGNN's strategy of constructing multiple graphs through text clustering and representative word selection is instrumental in capturing nuanced correlations within the data. Unlike singular approaches, TCGNN's multi-grained analysis reflects a comprehensive understanding of articles, offering a layered representation that resonates with the multifaceted nature of language used in fake news.
- *Homogeneity and Direct Application:* GCN and GraphSage's homogeneous models, which are directly applied to the constructed text graph, highlight TCGNN's advantage in applying GNNs more effectively. TCGNN transcends this by adopting a more intricate graph structure that aligns with the complex text patterns of fake news, allowing for a more refined classification process.
- *Global Contextual Understanding:* TextGCN's incorporation of global word co-occurrence information serves as a backdrop against which TCGNN's capabilities can be contrasted. TCGNN not only utilizes the immediate textual environment but also leverages the broader textual context, facilitating a deeper understanding of content authenticity.
- *Edge Prediction and Noise Reduction:* The comparison with NRGNN and RSGNN, which both use edge predictors to enhance graph structure and mitigate label noise, further accentuates TCGNN's strength. TCGNN implicitly incorporates these enhancements by establishing robust text clustering graphs that organically densify the graph structure, achieving a similar effect without the need for explicit edge prediction.
- *Enhanced Label Utilization and Regularization:* NRGNN's and RSGNN's strategies to extend the label set and regularize predictions provide a baseline for appreciating TCGNN's efficient use of available labels. By generating

multi-grained representations, TCGNN implicitly amplifies the supervisory signal from the limited labeled data, offering an intrinsic regularization effect that aligns with the underlying data distribution.

Ablation Study. We examine whether every component of TCGNN contributes positively. Five TCGNN variants are studies: (a) *Full TCGNN* (no components are removed), (b) *w/o Multi-grained*: using only base cluster number $c = 8$ (no clusterings with $c = 16, 32$), (c) *w/o Clustering*: using only the global graph, (d) *w/o CSW*: using all words to initialize node vectors (no cluster-word selection), and (e) *w/o Global*: removing the global graph. The results are exhibited in Fig. 2. All of our designs in TCGNN contribute positively to the performance. The most useful component is text clustering, which supports our main contribution. Besides, to obtain better results, we need to select representative words in each cluster, and consider multi-grained text clustering.

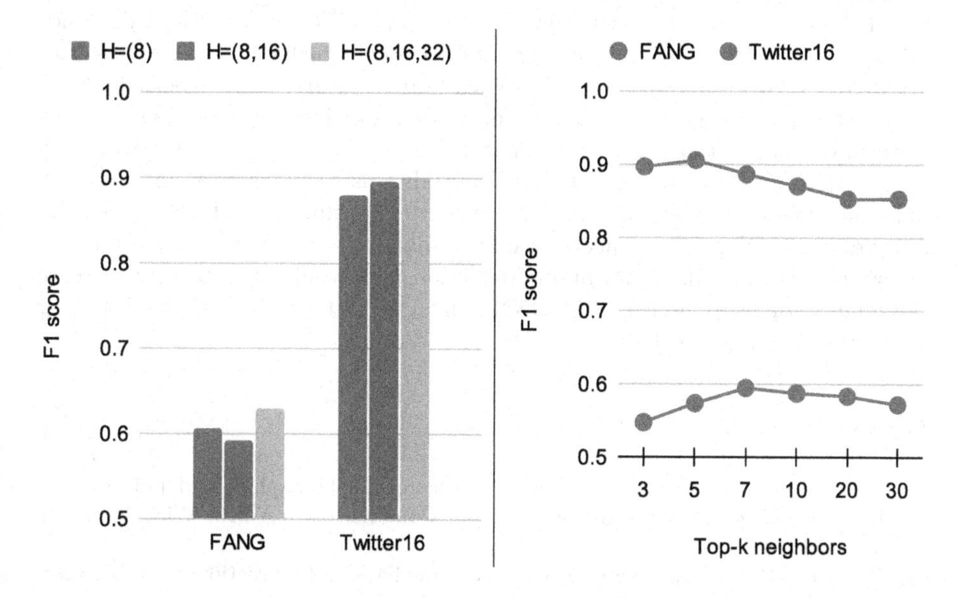

Fig. 3. Results on hyperparameter analysis.

Hyperparameter Analysis. We unfold how key hyperparameters affect performance. The results are shown in Fig. 3. First, we wonder about the influence of multi-grained text clustering. By varying the number of multi-grained clusters $H = \{(8), (8, 16), (8, 16, 32)\}$, we find that employing multiple clusterings enhances performance, indicating the effective harnessing of multi-grained tweet correlations. Second, we keep only top-k node pairs with high similarity scores to create edge connections. When k is small, the F1 score is higher. We speculate that when too many edges are created per node, the neighbors can bring noise rather than useful information. The optimal k varies across datasets and should be tuned.

5 Conclusions and Discussion

This study introduces Text-Clustering Graph Neural Network (TCGNN), a novel approach to detecting fake news on social media, distinct from prior GNN-based methods that employ a single graph to learn tweet representations. TCGNN harnesses text clustering to unearth latent fine-grained knowledge, informing the construction of multi-grained graphs depicting text correlation for representation learning. Tested across four datasets, TCGNN consistently surpasses existing GNN-based methods, underscoring the utility of text clustering and multi-graph construction in fake news detection.

This study presents four notable limitations. First, while our focus is on text content in fake news detection, our proposed TCGNN method does not incorporate context information inherent to social media posts, such as user attributes, retweet comments, and propagation structure, without specific technical extensions for graph construction. Second, our investigation does not differentiate between types of disinformation [28], hence we cannot identify which types TCGNN is best suited for. Further studies using datasets annotated with different disinformation types would help in determining this. Third, the scalability of TCGNN, a vital concern, remains unaddressed. Given the vast number of tweets generated daily, an online detection task would significantly increase the number of tweet nodes in TCGNN. Thus, we restrict TCGNN to detecting fake news within a select set of tweets.

References

1. Bessi, A., et al.: Viral misinformation: the role of homophily and polarization. In: Proceedings of the 24th International Conference On World Wide Web, pp. 355–356 (2015)
2. Bian, T., et al.: Rumor detection on social media with bi-directional graph convolutional networks. **34**, 549–556 (2020)
3. Blei, D.M., Ng, A.Y., Jordan, M.I.: Latent dirichlet allocation. J. Mach. Learn. Res. **3**(Jan), 993–1022 (2003)
4. Buntain, C., Golbeck, J.: Automatically identifying fake news in popular twitter threads. In: 2017 IEEE International Conference on Smart Cloud (smartCloud), pp. 208–215. IEEE (2017)
5. Chung, J., Gulcehre, C., Cho, K., Bengio, Y.: Empirical evaluation of gated recurrent neural networks on sequence modeling. In: NIPS 2014 Deep Learning and Representation Learning Workshop (2014)
6. Dai, E., Aggarwal, C., Wang, S.: Nrgnn: learning a label noise resistant graph neural network on sparsely and noisily labeled graphs. In: Proceedings of the 27th ACM SIGKDD Conference on Knowledge Discovery and Data Mining, pp. 227–236 (2021)

7. Dai, E., Jin, W., Liu, H., Wang, S.: Towards robust graph neural networks for noisy graphs with sparse labels. In: Proceedings of the Fifteenth ACM International Conference on Web Search and Data Mining, pp. 181–191 (2022)
8. Ding, K., Wang, J., Li, J., Li, D., Liu, H.: Be more with less: hypergraph attention networks for inductive text classification. In: Proceedings of the 2020 Conference on Empirical Methods in Natural Language Processing (EMNLP), pp. 4927–4936 (2020)
9. Dou, Y., Shu, K., Xia, C., Yu, P.S., Sun, L.: User preference-aware fake news detection. In: Proceedings of the 44th International ACM SIGIR Conference on Research and Development in Information Retrieval, pp. 2051–2055. SIGIR '21 (2021)
10. Hamilton, W., Ying, Z., Leskovec, J.: Inductive representation learning on large graphs. In: Advances in Neural Information Processing Systems **30** (2017)
11. Huang, L., Ma, D., Li, S., Zhang, X., Wang, H.: Text level graph neural network for text classification. In: Proceedings of the 2019 Conference on Empirical Methods in Natural Language Processing (EMNLP), pp. 3444–3450 (2019)
12. Khullar, V., Singh, H.P.: f-fnc: privacy concerned efficient federated approach for fake news classification. Inf. Sci. **639**, 119017 (2023)
13. Kingma, D.P., Ba, J.: Adam: a method for stochastic optimization. In: Proceedings of International Conference for Learning Representations (ICLR) (2015)
14. Kipf, T.N., Welling, M.: Semi-supervised classification with graph convolutional networks. In: International Conference on Learning Representations (ICLR) (2017)
15. Lian, Z., Zhang, C., Su, C., Dharejo, F.A., Almutiq, M., Memon, M.H.: Find: privacy-enhanced federated learning for intelligent fake news detection. IEEE Transactions on Computational Social Systems (2023)
16. Liu, X., You, X., Zhang, X., Wu, J., Lv, P.: Tensor graph convolutional networks for text classification. In: Proceedings of the AAAI Conference on Artificial Intelligence, vol. 34, pp. 8409–8416 (2020)
17. Lu, Y.J., Li, C.T.: GCAN: graph-aware co-attention networks for explainable fake news detection on social media. In: Proceedings of the 58th Annual Meeting of the Association for Computational Linguistics, pp. 505–514 (2020)
18. Ma, J., Gao, W., Wong, K.F.: Rumor detection on Twitter with tree-structured recursive neural networks. In: Proceedings of the 56th Annual Meeting of the Association for Computational Linguistics, pp. 1980–1989 (2018)
19. Mustafaraj, E., Metaxas, P.T.: The fake news spreading plague: was it preventable? In: Proceedings of the 2017 ACM on web science conference, pp. 235–239 (2017)
20. Nguyen, V.H., Sugiyama, K., Nakov, P., Kan, M.Y.: Fang: leveraging social context for fake news detection using graph representation. In: Proceedings of the 29th ACM International Conference on Information and Knowledge Management, pp. 1165–1174 (2020)
21. Ribeiro, M.H., Calais, P.H., Almeida, V.A., Meira Jr, W.: "everything i disagree with is #fakenews": Correlating political polarization and spread of misinformation. Data Science + Journalism (DS+J) Workshop @ KDD'17 (2017)
22. Sun, T., Qian, Z., Dong, S., Li, P., Zhu, Q.: Rumor detection on social media with graph adversarial contrastive learning. In: Proceedings of the ACM Web Conference 2022, pp. 2789–2797. WWW '22 (2022)
23. Vicario, M.D., Quattrociocchi, W., Scala, A., Zollo, F.: Polarization and fake news: early warning of potential misinformation targets. ACM Trans. Web (TWEB) **13**(2), 1–22 (2019)

24. Wei, L., Hu, D., Zhou, W., Yue, Z., Hu, S.: Towards propagation uncertainty: edge-enhanced Bayesian graph convolutional networks for rumor detection. In: Proceedings of the 59th Annual Meeting of the Association for Computational Linguistics, pp. 3845–3854
25. Yao, L., Mao, C., Luo, Y.: Graph convolutional networks for text classification. In: Proceedings of the AAAI Conference on Artificial Intelligence, pp. 7370–7377 (2019)
26. Zhang, Y., Yu, X., Cui, Z., Wu, S., Wen, Z., Wang, L.: Every document owns its structure: Inductive text classification via graph neural networks. In: Proceedings of the 58th Annual Meeting of the Association for Computational Linguistics, pp. 334–339 (Jul 2020)
27. Zhou, X., Jain, A., Phoha, V.V., Zafarani, R.: Fake news early detection: a theory-driven model. Digital Threats: Res. Pract. **1**(2), 1–25 (2020)
28. Zhou, X., Zafarani, R.: A survey of fake news: fundamental theories, detection methods, and opportunities. ACM Comput. Surv. **53**(5) (sep 2020)

Exploiting Adaptive Contextual Masking for Aspect-Based Sentiment Analysis

S. M. Rafiuddin(✉), Mohammed Rakib, Sadia Kamal, and Arunkumar Bagavathi

Oklahoma State University, Stillwater, OK, USA
{srafiud,mohammed.rakib,sadia.kamal,abagava}@okstate.edu

Abstract. Aspect-Based Sentiment Analysis (ABSA) is a fine-grained linguistics problem that entails the extraction of multifaceted aspects, opinions, and sentiments from the given text. Both *standalone* and *compound* ABSA tasks have been extensively used in the literature to examine the nuanced information present in online reviews and social media posts. Current ABSA methods often rely on static hyperparameters for attention-masking mechanisms, which can struggle with context adaptation and may overlook the unique relevance of words in varied situations. This leads to challenges in accurately analyzing complex sentences containing multiple aspects with differing sentiments. In this work, we present adaptive masking methods that remove irrelevant tokens based on context to assist in Aspect Term Extraction and Aspect Sentiment Classification subtasks of ABSA. We show with our experiments that the proposed methods outperform the baseline methods in terms of accuracy and F1 scores on four benchmark online review datasets. Further, we show that the proposed methods can be extended with multiple adaptations and demonstrate a qualitative analysis of the proposed approach using sample text for aspect term extraction.

Keywords: Aspect-Based Sentiment Analysis · Adaptive Threshold · Adaptive Contextual Masking

1 Introduction

Aspect-Based Sentiment Analysis (ABSA) tasks are gaining traction in various online domains like customer reviews and social media monitoring [23]. Traditional sentiment analysis tasks consider only one sentiment polarity label, *positive*, *neutral*, or *negative*, for the given text. However, the real-world text may consist of multiple sentiment polarities assigned to multifaceted aspects and opinions. A simple example of the ABSA task on a restaurant review is depicted in Fig. 1. ABSA tasks usually involve *four* components: *aspect category*, *aspect terms*, *opinion terms*, and *sentiment polarity*. Several works are available in the literature to study ABSA tasks in both *standalone* [11] and *compound* [8] fashion. *Standalone* approaches aim to extract only one component, whereas the

D.-N. Yang et al. (Eds.): PAKDD 2024, LNAI 14650, pp. 147–159, 2024.
https://doi.org/10.1007/978-981-97-2266-2_12

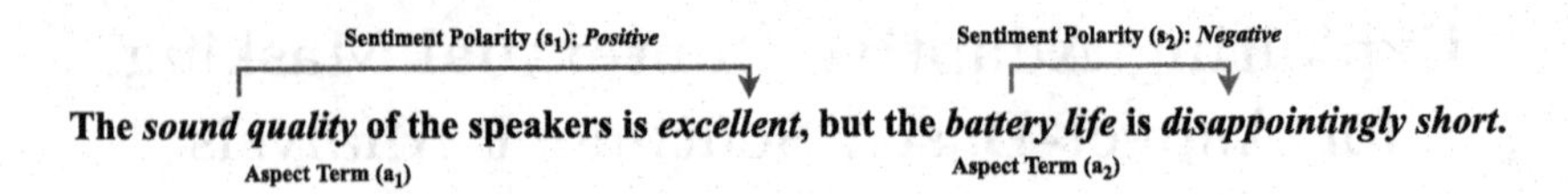

Fig. 1. Simple ABSA for an online review. The review consists of multiple ***aspect terms*** and each aspect of the review contains its own *sentiment polarity*. (Color figure online)

compound approaches jointly extract more than one component, for example extracting aspect term and sentiment. In this paper, we focus on two standalone ABSA tasks of *aspect term extraction* (ATE) and *aspect sentiment classification* (ASC). ASC is a supervised task that predicts the sentiment polarity of the given text in the context of a given topic or aspect. As given in Fig. 1, the text data can contain multiple aspects and each aspect can have a distinct sentiment polarity. Similarly, the ATE is a supervised task to extract the start and end positions of each aspect available in the text. It is common that there are multiple aspect terms and each aspect term can span across multiple words of the given text, similar to the example depicted in Fig. 1.

The introduction of attention [14] and large language models increased the scope of sentiment analysis problems by understanding the precise context of the text. Attention and LLMs aid ABSA tasks in capturing *local* context of words and *global* contextual features of the entire text respectively. The self-attention strategy aims to extract useful information from text with respect to words present in the text itself. Such techniques are crucial for any ABSA tasks to map aspect categories and aspect terms with opinion terms and sentiment. One of the popular approaches with ABSA is to filter out noisy terms that do not cover contextual details for other relevant terms of the given text. Such strategies also utilize attention weights that are optimized to *masked* words that are not useful or out of context to the given aspect. In the literature, the normalized attention weights have been used in two different forms. (1) *Weights Threshold:* Threshold-based approaches set a user-defined threshold or take the maximum weight as the threshold [3]. (2) *Distance Based:* These approaches sort attention weights and define a window size around the aspect terms [11]. If the word attention weights do not satisfy the threshold or distance condition, they are considered noisy terms and *masked* with zero-vectors for any downstream ABSA task. Recently, dynamic approaches to extract correlations between the local context and aspects of the text are gaining traction in ABSA problems [3,11,21]. However, these approaches are introduced with an assumption of a user-assigned threshold to decide the tokens to mask. In addition to the local context, the pre-trained LLMs give better contextual features of the text to cover global representations for ABSA tasks. It is a common practice to use both attention and LLM representations in tandem to perform both *standalone* [7] and *compund* ABSA tasks. In this work, we present three key contributions to overcome the challenges of current dynamic approaches in standalone ABSA tasks:

1. **Adaptive Contextual Threshold Masking (ACTM)**: We introduce a novel masking strategy that adjusts mask thresholds adaptively to determine the mask ratio of text tokens based on their context and enhance granularity in standalone ABSA tasks.
2. **Adaptive Masking for ABSA**: In addition to the proposed ACTM strategy, we tailored two existing distance-based adaptive attention masking techniques exclusively for standalone ABSA tasks.
3. **Experimental Validation**: We demonstrate with extensive experiments that the adaptive masking approaches outperform baseline ABSA methods across multiple *SemEval* benchmark datasets for two ABSA subtasks.

2 Related Work

Recent advancements in Aspect-Based Sentiment Analysis (ABSA) include joint learning approaches by Mao et al. [10] using shared pre-trained models and Xu et al. [18] employing sequence-to-sequence models. Additionally, Yan et al. [19] and Zhang et al. [22] have proposed unified generative frameworks, integrating Pre-trained Language Models (PLMs) and treating ABSA tasks as distinct text generation challenges.

Key subtasks in ABSA, such as aspect term extraction and sentiment polarity determination, are highlighted in Chen and Qian [2]. These tasks increasingly rely on aspect-based syntactic information (POS-tags), as evidenced in works like [6,15]. Phan [11] emphasizes the importance of context in ABSA by combining syntactical features with contextualized embeddings. Span-level models for aspect sentiment extraction are explored in Chen et al. [1], while GCN-based models for extracting syntactic information from dependency trees are discussed in Zhang et al. [24]. Additionally, Tian et al. [13] use GNNs over dependency trees to enhance understanding of syntactic features.

Recent ABSA research emphasizes attention-based neural networks [3,9], with key advancements like Feng et al.'s masked attention method (AM-WORD-BERT) for term-focused performance enhancement [3]. Lin et al.'s AMA-GLCF model [7] utilizes a masked attention mechanism with Context Dynamic Mask (CDM) and Context Dynamic Weight (CDW) for prioritizing aspect-relevant text in global and local contexts. Additionally, LCFS uses CDW and CDM for improved aspect extraction and sentiment classification by leveraging attention-based masking [11]. In these models, they have used a static threshold for masking, but we proposed an adaptive masking based on the context. Adaptive threshold masking enhances model accuracy by adjusting to the varying relevance of data features in different contexts. It improves overall model performance by ensuring focus on the most relevant aspects of the input.

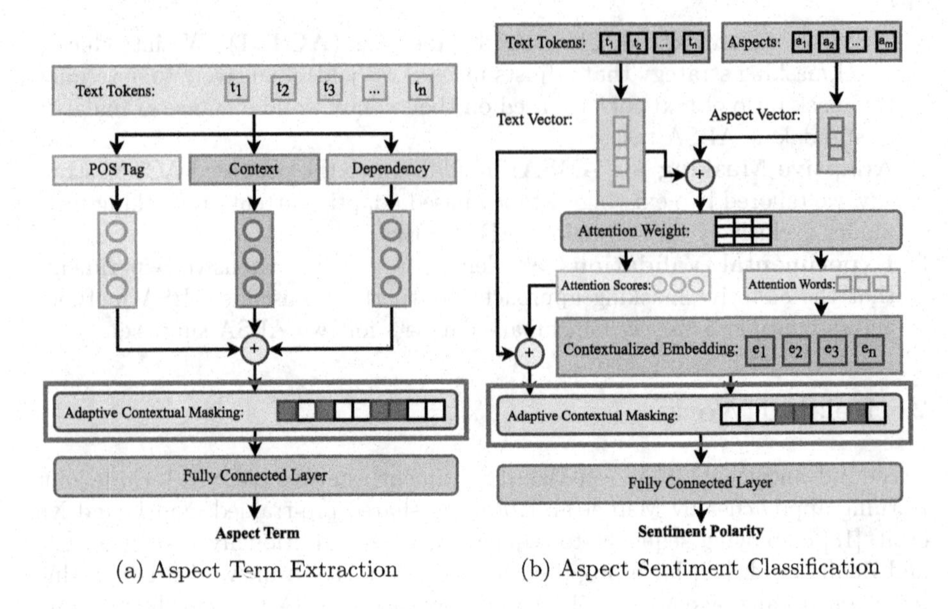

(a) Aspect Term Extraction (b) Aspect Sentiment Classification

Fig. 2. Proposed Structure of (a) Aspect Term Extraction, and (b) Aspect Sentiment Classification. The red box highlights the models introduced in this paper. (Color figure online)

3 Methodology

3.1 Problem Formulation and Motivation

Given a text $T = \{t_i\}_{i=1}^{n} = \{t_1, t_2, \dots, t_n\}$ of n tokens with k aspects $\{\mathcal{A}_j\}_{j=1}^{k}$ where each aspect spans across multiple tokens in T and each aspect can map to its own sentiment polarity $\mathcal{A}_j \rightarrow \mathcal{P}_j \in \{pos., neut., neg.\}$. We define the aspect term extraction task as a supervised approach $f_{\text{ATE}} : \mathcal{M}(\mathbb{E}(T), t_i) \rightarrow \hat{\mathcal{P}}_{t_i} \in \{Begin, In, Out\}$ and the aspect sentiment classification task as $f_{\text{ASC}} : \mathcal{M}(\mathbb{E}(T), \mathbb{E}(\mathcal{A}_j)) \rightarrow \hat{\mathcal{P}}_j$ where $\mathbb{E}(T) \in \mathbb{R}^{n \times d}$ is token representations, $\mathbb{E}(\mathcal{A}_j) \in \mathbb{R}^{m \times d} | m <<< n$ is aspect representations, $\hat{\mathcal{P}}_{w_i}$ is the predicted position of w_i, and $\hat{\mathcal{P}}_j$ is the predicted sentiment polarity. In this work, we contribute with the adaptive contextual masking $\mathcal{M} : \mathbb{E}(T) \rightarrow \mathbb{E}'(T)$ for the above given two ABSA subtasks, where $\mathbb{E}' = \{[MASK], e_2, [MASK], \dots, e_n\}$ is a masked representations of $\mathbb{E}$. Our motivation for this work is to design an adaptive masking $\mathcal{M}$ that adjusts token masks to hide irrelevant terms in correspondence to the context of T, rather than using a hard threshold as given by dynamic masking approaches.

3.2 Standalone ABSA Tasks

We briefly outline *aspect term extraction* (ATE) and *aspect-based sentiment classification* (ASC) tasks, providing context for our proposed masking methods.

Figure 2 shows the model architectures, inspired by established approaches [3,7, 11].

Aspect Term Extraction (ATE). The ATE model, depicted in Fig. 2a, predicts aspect terms in text T. The input T is formatted as `[CLS] + T + [SEP]`, where `[CLS]` and `[SEP]` define the sentence context and endpoint. We utilize POS tag representations (p_i), contextual token features (w_i), and dependency graph-based syntactic features (D_i) of tokens $\{t_i\}_{i=1}^{n} \in T$. We concatenate these grammatic, contextual and syntactic features to represent each token representation as $\mathbf{e}_i = \mathbf{w}_i \oplus \mathbf{p}_i \oplus \mathbf{D}_i$. We then process $\mathbb{E}(T) = \{e_1, e_2, \ldots e_n\}$ through the adaptive contextual masking (Sects. 3.3, 3.4, 3.5) and a fully connected layer to perform the supervised task f_{ATE}.

Aspect-Based Sentiment Classification (ASC). The ASC model, depicted in Fig. 2b, predicts sentiment polarities of each aspect $\mathcal{A}_j \in T$. The ASC model analyze sentiment of T towards the given specific aspect $\mathcal{A}_j$ by employing the input `[CLS]`+T+`[SEP]`+$\mathcal{A}_j$+`[SEP]`, where `[CLS]` captures overall context. We apply a transformer model, like BERT [5], to extract contextual representations of T and $\mathcal{A}_j$. These features are fed into the attention mechanism, which calculates the attention weights between the aspect vectors and the text vectors. The attention mechanism assigns higher weights to the words or tokens that are more relevant to the aspect of focus based on context, allowing the model to focus on the most important information for sentiment analysis. We mask the tokens in $\mathbb{E}(T)$ by measuring its relevance with $\mathbb{E}(\mathcal{A}_j)$ using the adaptive contextual masking (Sects. 3.3, 3.4, 3.5) and perform the supervised task f_{ASC}.

3.3 Adaptive Contextual Threshold Masking (ACTM)

We propose the ACTM strategy, given in Fig. 3, to dynamically adjust the attention span and update mask threshold to prioritize the sentiment-bearing tokens related to aspect terms. First, we capture token significance in T by computing attention scores $\boldsymbol{Attn} = \{attn_1, attn_2, \ldots attn_n\}$ of token representations $\{e_i\}_{i=1}^{n} \in \mathbb{E}(T)$ using the self attention mechanism, as given in Eq. 1.

$$attn_i = \text{softmax}\left(\frac{W_a e_i}{\sqrt{d_k}}\right) \tag{1}$$

where W_a is the weight matrix for the attention layer and d_k is the dimensionality of the key vectors in the attention mechanism. For the ATE task, we incorporate the adaptive threshold τ on each token t_i as $\tau(t_i) = \alpha \cdot \textbf{Aggregate}\,(\boldsymbol{Attn})$. Similarly, for the ASC task, we employ the adaptive threshold on tokens as $\tau(t_i) = \alpha \cdot \textbf{Aggregate}\,(\boldsymbol{Attn}) + \gamma \cdot R_a(t_i, \mathcal{A}_j)$, where $R_a(t_i, \mathcal{A}_j)$ is the aspect relevance function that measures the relevance of token t_i to the given aspect $\mathcal{A}_j$ using the attention scores as given in Eq. 2. The α and γ are learnable parameters

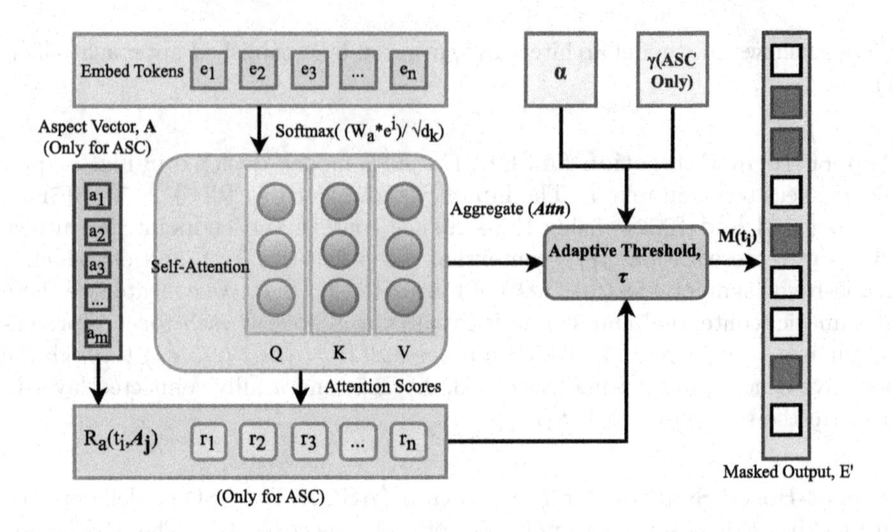

Fig. 3. Adaptive Contextual Threshold Masking strategy to adjust the masking threshold τ using the aggregated attention scores of token representations. α and γ are the learnable parameters, where γ is used in the ASC task only for aspect relevance.

that adjust the influence of the attention scores and aspect relevance, respectively. **Aggregate** is a pooling function that acts as a metric for getting the most relevant terms based on context.

$$R_a(t_i, \mathcal{A}_j) = \frac{\exp\left(\beta \cdot \mathbf{sim}(a_i, \mathbb{E}(\mathcal{A}_j)\right)}{\sum_{k=1}^{n} \exp\left(\beta \cdot \mathbf{sim}(a_k, \mathbb{E}(\mathcal{A}_k))\right)} \tag{2}$$

where a_i is weighted contextual token vector, $\mathbb{E}(\mathcal{A}_j)$ represents the contextual aspect vector, **sim** is a ***cosine similarity*** function, and β is a scaling factor. The irrelevant tokens are masked for the sentiment analysis by matching their corresponding attention scores with adaptive threshold τ as given in Eq. 3.

$$\mathcal{M}(t_i) = \begin{cases} attn_i, & \text{if } attn_i \geq \tau(t_i) \\ 0, & \text{otherwise} \end{cases} \tag{3}$$

This mask $\mathcal{M}$ results in a filtered context $\mathbb{E}'$ that emphasizes aspect-related sentiments. The iterative process allows the threshold τ to adapt dynamically, providing a focused analysis of the interplay between aspects and sentiments in varying contexts. By employing contextual adaptive masking, ABSA models can effectively concentrate on the most informative tokens that influence the sentiment towards their aspect for precision ABSA tasks.

3.4 Adaptive Attention Masking (AAM)

Sukhbaatar *et al.* [12] introduced an adaptive self-attention mechanism with a soft masking function and a masking ratio M in transformer models. This mechanism dynamically adjusts attention spans to emphasize relevant terms for any downstream tasks. The soft masking function in self-attention layers, $m_z(x) = \min\left[\max\left[\frac{1}{R}(R+z-x), 0\right], 1\right]$, where R is a flexibility hyper-parameter and z is a learnable parameter, helps in this dynamic adjustment. x is the distance from a given position in a sequence of tokens to the position being focused on by the attention mechanism. The adaptive masking ratio M, crucial for determining the span of attention, is calculated as $\mathcal{M} = \frac{\sum m_z(x)}{n}$. Consequently, the attention weights are given by the equation Attention Weights = softmax $\left(\frac{m_z(x)\cdot\mathcal{M}}{\sqrt{d_k}}\right)$, integrating $\mathcal{M}$ and using scaled softmax within the span boundaries l and r. Given a current position p in the sequence and a learned attention span z, the left boundary l is calculated as $l = p - z$, and the right boundary r is calculated as $r = p + z$, defining the span of attention around position p. This adaptive attention approach, by incorporating the masking ratio, allows the model to capture extended dependencies, which is vital for effective ATE and ASC tasks.

3.5 Adaptive Mask Over Masking (AMOM)

Xiao *et al.* [17] introduced Adaptive Masking Over Masking (AMOM), which is adapted to enhance both Aspect Term Extraction (ATE) and Aspect Sentiment Classification (ASC) in conditional masked language models (CMLM). In our tasks of ABSA, AMOM generates masked sequences $Y^{\mathrm{ATE}}_{\mathrm{mask}}$ and $Y^{\mathrm{ASC}}_{\mathrm{mask}}$ using input, aspect (Only for ASC), and sentiment labels. It evaluates prediction correctness for ATE and ASC against the ground truth Y, calculating correctness ratios R_{ATE} and R_{ASC}. These ratios inform adaptive masking ratios $\mu_{\mathrm{ATE}} = f(R_{\mathrm{ATE}})$ and $\mu_{\mathrm{ASC}} = f(R_{\mathrm{ASC}})$, with $N^{\mathrm{ATE}}_{\mathrm{mask}} = \mu_{\mathrm{ATE}} \cdot |Y|$ and $N^{\mathrm{ASC}}_{\mathrm{mask}} = \mu_{\mathrm{ASC}} \cdot |Y|$ indicating the number of tokens to mask and regenerate for each task. The regenerating processes $Y'_{\mathrm{ATE}} = G(Y^{\mathrm{ATE}}_{\mathrm{mask}}, X)$ and $Y'_{\mathrm{ASC}} = G(Y^{\mathrm{ASC}}_{\mathrm{mask}}, X)$ adapt to the ABSA context, with $M(X, R_{\mathrm{ATE}})$ and $M(X, R_{\mathrm{ASC}})$ dynamically adjusting the decoder's masking strategy for both ATE and ASC, effectively addressing nuanced expressions related to specific aspects and sentiments in text. We highlight that both AAM and AMOM have never been explored for any ABSA tasks and we combine these strategies with both ATE and ASC tasks in this work.

3.6 Training Procedure for ATE and ASC

We train the ATE task by minimizing the categorical cross-entropy loss employed with the BIO (Begin, Inside, Outside) tagging scheme as given in Eq. 4.

$$L = -\sum_{i=1}^{N} \sum_{c\in\{B,I,O\}} y_{i,c} \log(\hat{y}_{i,c}) \tag{4}$$

where N is the number of words, $y_{i,c}$ is the trinary indicator of the class label c for the word i, and $\hat{y}_{i,c}$ is the predicted probability. We train the ASC task using the multi-class loss function with $L2$ regularization as given in Eq. 5.

$$L(\mathbf{Y}, \hat{\mathbf{Y}}, \Theta) = -\frac{1}{n}\sum_{i=1}^{n}\sum_{c=1}^{C} y_{ic}\log(\hat{y}_{ic}) + \frac{\lambda}{2}\|\Theta\|_2^2 \tag{5}$$

where $\mathbf{Y}$ and $\hat{\mathbf{Y}}$ are true class and predicted class probabilities respectively, C is the class count, y_{ic} and $\hat{y}_{ic}$ denote the true and predicted class probabilities for instance i, respectively, and $\|\Theta\|_2^2$ is the $L2$ norm of the model parameters Θ.

4 Experiments and Results

***Datasets*:** We experiment with benchmark ABSA datasets from Semeval 2014[1], 2015[2], and 2016[3]. Table 1 details the counts of reviews and the total number of positive, neutral, and negative aspects present in the overall training and test datasets used in our experiments. These training and test samples are pre-defined in SemEval and the same setup is utilized in all our baseline models.

***Experiment Setup*:** We have leveraged BERT ("bert-base-cased") model as our contextual feature extractor for all of the proposed approaches [5]. Layer normalization is set to 1×10^{-12} and a dropout of 0.1 is applied to the attention probabilities and hidden layers. For all experiments, we have trained for 50 epochs with a batch size of 32, a learning rate of 2×10^{-5}, and L2 regularization of 0.01. Experiments were conducted in a lab server with 3×NVIDIA A10 GPUs. We use *Mean* for the **Aggregate** operation in the ACTM model.

Table 1. Summary of laptop and restaurant review datasets given by SemEval.

Dataset	Split	#Reviews	#Aspects		
			Positive	Negative	Neutral
Laptop14	Train	1124	994	870	464
	Test	332	341	128	464
Restaurant14	Train	1574	2164	807	637
	Test	493	728	196	196
Restaurant15	Train	721	1777	334	81
	Test	318	703	192	46
Restaurant16	Train	1052	2451	532	125
	Test	319	685	93	63

***Baseline Methods*:(1) ATE task: BiLSTM** [9] is an RNN used for NLP tasks, processing sequences bidirectionally. **DTBCSNN** [20] leverages sentence dependency trees and CNNs for aspect extraction. **BERT-AE** [5] uses BERT's pre-trained embeddings for aspect extraction. **IMN** [4], combines memory networks with aspect-context interactive attention. (**CSAE**) [11] combines various components such as contextual, dependency-tree, and self-attention

[1] https://alt.qcri.org/semeval2014/task4/.
[2] https://alt.qcri.org/semeval2015/task12/.
[3] https://alt.qcri.org/semeval2016/task5/.

mechanisms. **(2) ASC task:** Models such as **LCF-ASC-CDM**, **LCF-ASC-CDW** [21], **LCFS-ASC-CDW**, and **LCFS-ASC-CDM** [11] use Local Context Focus (LCF) with Context Dynamic Mask (CDM) and Weight (CDW) layers. Attention Mask variations like **AM Weight-BERT** and **AM Word-BERT** are applied in ABSA, targeting relevant parts [3]. The **Unified Generative** model [19] utilizes BART for multiple ABSA tasks. **MGGCN-BERT** [16] leverages BERT embeddings for ASC, while **AMA-GLCF** [7] combines global and local text contexts using masked attention.

Table 2. Adaptive contextual masking strategies against baseline models on ATE. "-" signifies no results available, and "*" denotes reproduced results.

Model	Laptop14	Restaurant14	Restaurant15	Restaurant16
	F1	F1	F1	F1
BiLSTM [9]	73.72	81.42	–	–
DTBCSNN [20]	75.66	83.97	–	–
BERT-AE [5]	73.92	82.56	–	–
IMN [4]	77.96	83.33	70.04	78.07*
CSAE [11]	77.65	**86.65**	76.84*	80.63*
AAM [12]	**79.27**	83.49	76.45	79.34
AMOM [17]	**78.13**	82.98	**77.49**	**82.09**
ACTM-ATE	**80.34**	82.91	**77.09**	**81.04**

***Discussion on ATE and ASC Results*:** We present the performance of the proposed adaptive contextual masking strategies for the ATE task in Table 2. We note that both the *AMOM* and *ACTM* versions of the adaptive contextual masking strategies outperform the baseline methods in three datasets. We also note that the *AAM* version shows competitive but not leading performance, indicating its partial effectiveness in contextual understanding for ATE tasks. We also emphasize that the proposed ACTM strategy is competitive in two datasets due to its capability to understand nuanced contextual interpretation by setting adaptive thresholds for the masking function. Similarly, we give the detailed performance of the proposed adaptive strategies for ASC task in Table 3. Unlike the ATE task, we note a significant performance gain with both Accuracy (%) and F1 in the adaptive contextual masking strategies in all datasets. Most importantly, the proposed ACTM strategy outperforms the other two adaptive strategies in three datasets and gives a competing performance in the Restaurant 15 dataset. Overall, we signify that our idea of adaptively masking based on local text tokens on top of the global contextual representations can achieve better results in standalone ABSA tasks considered in this work. While ACTM strategy leads the performance in ASC and is competitive in ATE, the AMOM strategy is promising with ATE tasks.

Table 3. Performance of adaptive masking strategies against baseline models on ASC task. "**" indicates results reported in the paper as mean values.

Model	Laptop14		Restaurant14		Restaurant15		Restaurant16	
	Acc	F1	Acc	F1	Acc	F1	Acc	F1
LCF-BERT-CDW [21]	82.45	79.59	87.14	81.74	–	–	–	–
LCF-BERT-CDM [21]	82.29	79.28	86.52	80.40	–	–	–	–
LCFS-ASC-CDW [11]	80.52	77.13	86.71	80.31	89.03*	73.31*	92.25*	76.46*
LCFS-ASC-CDM [11]	80.34	76.45	86.13	80.10	88.61*	69.32*	91.84*	70.67*
AM Weight-BERT** [3]	79.78	76.20	85.66	79.92	–	–	–	–
AM Word-BERT** [3]	79.87	76.26	85.57	79.02	–	–	–	–
Unified Generative [19]	-	76.76	–	75.56	–	73.91	–	–
MGGCN-BERT [16]	79.57	76.30	83.21	75.38	82.90	69.27	89.66	73.99
AMA-GLCF** [7]	–	76.78	–	79.33	–	–	–	77.08
AAM [12]	82.51	**79.61**	84.92	81.71	89.34	**75.13**	90.86	77.18
AMOM [17]	81.61	79.09	85.95	80.10	**91.50**	**74.14**	92.17	76.73
ACTM-ASC	**83.65**	76.29	**91.05**	**82.01**	**90.54**	**74.07**	**93.49**	**78.19**

***Discussion on Ablation Study*:** Since the ACTM has significant performance in both ATE and ASC tasks, we compare several aspects of the proposed ACTM strategy as the ablation study. In this study, we explore different **Aggregate** operators, like *Mean*, *Media*, and *Standard Deviation (SD)*, along with variations that use only a constant weight $\alpha = \gamma = 1$. We present our results of our ablation study in Table 4. Alhough we note that our default setting of *Mean* aggregator with learnable α and γ performs overall good in most of the cases, we find some exceptions. We note that a simple *SD* aggregator with constant weight is able to match our default gradient based Mean aggregator for the latest Restaurant'16 dataset in both ATE and ASC tasks. Similarly, the gradient based Median aggregator is also able to give better performance in two ASC tasks.

Table 4. F1 of ATE and ASC with multiple **Aggregator** operations. "**Gradient-based**" indicates that α, γ are learnable parameters. Otherwise $\alpha = \gamma = 1$

Aggregate Functions	Laptop14		Restaurant14		Restaurant15		Restaurant16	
	ATE	ASC	ATE	ASC	ATE	ASC	ATE	ASC
Mean	75.74	71.53	78.43	78.90	76.19	74.19	80.10	74.39
Median	78.18	72.67	79.94	79.59	76.23	74.23	79.01	71.14
SD	80.29	74.37	78.15	80.10	76.96	77.12	**81.92**	**78.26**
Gradient-based Mean	**80.34**	**76.29**	**82.91**	82.01	**77.09**	74.07	**81.04**	**78.19**
Gradient-based Median	76.05	73.72	80.17	**83.06**	75.73	**77.86**	78.15	72.24
Gradient-based SD	77.13	73.21	75.13	78.15	75.86	75.38	76.91	72.78

***Discussion on Qualitative Analysis*:** We qualitatively analyze the performance of ACTM strategy for the ATE task to evaluate the efficacy of the proposed method in identifying optimal aspects from the text. Table 5a demonstrates an example token masking by the proposed ACTM strategy and Table 5b compares ATE task using adaptive threshold and fixed threshold. It is evident from these tables that the proposed ACTM strategy can assist the ATE task to capture nuanced aspect terms using adaptive contextual masking technique.

Table 5. Qualitative data for Aspect Term Extraction

(a) Tokens with Attention Scores and Masking Status

Token	Attn. Score	Masked
the	0.0460	Yes
steak	0.1082	No
was	0.0561	Yes
incredibly	0.0867	No
tender	0.0775	No
and	0.0323	Yes
flavor	0.0265	Yes
ful	0.0319	Yes
,	0.0275	Yes
but	0.0977	No
service	0.0794	No
quite	0.0413	Yes
slow	0.0648	No
.	0.0493	Yes
Total: 0.8250		
Mean: 0.0590		

(b) Comparison of ATE using Predefined vs. Dynamic Threshold for sample texts.

Review Instances	ATE w/ fixed thresh-old	ATE w/ ACTM
After numerous attempts of trying (including setting the ***clock in BIOS setup*** directly), I gave up (I am a techie).	clock	clock in BIOS setup
After really enjoying ourselves at the ***bar*** we sat down at a ***table*** and had ***din-ner***.	bar, dinner	bar, table, dinner
Did not enjoy the new ***Windows 8*** and ***touchscreen*** **functions**.	windows, touchscreen	windows 8, touchscreen functions

5 Conclusion

In this work, we explored Aspect-based Sentiment Analysis (ABSA) with a focus on standalone tasks such as Aspect Term Extraction (ATE) and Aspect Sentiment Classification (ASC) using three different adaptive masking strategies. We introduced one of those strategies named Adaptive Contextual Threshold Masking (ACTM) while utilizing two other adaptive masking techniques for ABSA. We depicted with our experiments on benchmark datasets that adaptive masking can increase the chance of precise ATE and ASC tasks. Particularly, our ACTM strategy demonstrated significant effectiveness over other approaches with its adaptive contextual threshold module. For future research, we recommend investigating adaptive masking for both standalone and compound ABSA tasks which benefits many applications. Another open venue for improvement is to explore the adaptive masking for multi-modal ABSA tasks.

References

1. Chen, Y., Keming, C., Sun, X., Zhang, Z.: A span-level bidirectional network for aspect sentiment triplet extraction. In: EMNLP, pp. 4300–4309. ACL (2022)
2. Chen, Z., Qian, T.: Enhancing aspect term extraction with soft prototypes. In: EMNLP, pp. 2107–2117. ACL (2020)
3. Feng, A., Zhang, X., Song, X.: Unrestricted attention may not be all you need-masked attention mechanism focuses better on relevant parts in aspect-based sentiment analysis. IEEE Access **10**, 8518–8528 (2022)
4. He, R., Lee, W.S., Ng, H.T., Dahlmeier, D.: An interactive multi-task learning network for end-to-end aspect-based sentiment analysis. In: Proceedings of the 57th Annual Meeting of the Association for Computational Linguistics, pp. 504–515 (2019)
5. Kenton, J.D.M.W.C., Toutanova, L.K.: Bert: pre-training of deep bidirectional transformers for language understanding. In: NAACL, pp. 4171–4186 (2019)
6. Li, J., Zhao, Y., Jin, Z., Li, G., Shen, T., Tao, Z., Tao, C.: Sk2: Integrating Implicit Sentiment Knowledge and Explicit Syntax Knowledge for Aspect-based Sentiment Analysis, CIKM 2022, pp. 1114-1123. ACM (2022)
7. Lin, T., Joe, I.: An adaptive masked attention mechanism to act on the local text in a global context for aspect-based sentiment analysis. IEEE Access, 43055–43066 (2023)
8. Lin, T., Sun, A., Wang, Y.: Aspect-based sentiment analysis through edu-level attentions. In: PAKDD, pp. 156–168. Springer (2022). https://doi.org/10.1007/978-3-031-05933-9_13
9. Liu, P., Joty, S., Meng, H.: Fine-grained opinion mining with recurrent neural networks and word embeddings. In: EMNLP, pp. 1433–1443 (2015)
10. Mao, Y., Shen, Y., Yu, C., Cai, L.: A joint training dual-mrc framework for aspect based sentiment analysis. In: AAAI, vol. 35, pp. 13543–13551 (2021)
11. Phan, M.H., Ogunbona, P.O.: Modelling context and syntactical features for aspect-based sentiment analysis. In: Jurafsky, D., Chai, J., Schluter, N., Tetreault, J. (eds.) ACL, pp. 3211–3220 (Jul 2020)
12. Sukhbaatar, S., Grave, É., Bojanowski, P., Joulin, A.: Adaptive attention span in transformers. In: ACL, pp. 331–335 (2019)
13. Tian, Y., Chen, G., Song, Y.: Aspect-based sentiment analysis with type-aware graph convolutional networks and layer ensemble. In: NAACL, pp. 2910–2922 (2021)
14. Vaswani, A., et al.: Attention is all you need. Adv. Neural Inform. Proc. Syst. **30** (2017)
15. Wu, S., Fei, H., Ren, Y., Ji, D., Li, J.: Learn from syntax: improving pair-wise aspect and opinion terms extraction with rich syntactic knowledge. In: Zhou, Z.H. (ed.) Proceedings of IJCAI, pp. 3957–3963 (8 2021)
16. Xiao, L., et al.: Multi-head self-attention based gated graph convolutional networks for aspect-based sentiment classification. Multimedia Tools Appli., 1–20 (2022)

17. Xiao, Y., et al.: Amom: adaptive masking over masking for conditional masked language model. In: AAAI, vol. 37, pp. 13789–13797 (2023)
18. Xu, L., Chia, Y.K., Bing, L.: Learning span-level interactions for aspect sentiment triplet extraction. In: Zong, C., Xia, F., Li, W., Navigli, R. (eds.) ACL, pp. 4755–4766 (Aug 2021)
19. Yan, H., Dai, J., Ji, T., Qiu, X., Zhang, Z.: A unified generative framework for aspect-based sentiment analysis. In: Zong, C., Xia, F., Li, W., Navigli, R. (eds.) ACL, pp. 2416–2429 (Aug 2021)
20. Ye, H., Yan, Z., Luo, Z., Chao, W.: Dependency-tree based convolutional neural networks for aspect term extraction. In: Kim, J., Shim, K., Cao, L., Lee, J.-G., Lin, X., Moon, Y.-S. (eds.) PAKDD 2017. LNCS (LNAI), vol. 10235, pp. 350–362. Springer, Cham (2017). https://doi.org/10.1007/978-3-319-57529-2_28
21. Zeng, B., Yang, H., Xu, R., Zhou, W., Han, X.: Lcf: a local context focus mechanism for aspect-based sentiment classification. Appl. Sci. **9**(16), 3389 (2019)
22. Zhang, W., Li, X., Deng, Y., Bing, L., Lam, W.: Towards generative aspect-based sentiment analysis. In: ACL, pp. 504–510 (2021)
23. Zhang, W., Li, X., Deng, Y., Bing, L., Lam, W.: A survey on aspect-based sentiment analysis: tasks, methods, and challenges. IEEE TKDE., 11019–11038 (2022)
24. Zhang, Z., Zhou, Z., Wang, Y.: Syntactic and semantic enhanced graph convolutional network for aspect-based sentiment analysis. In: ACL, pp. 4916–4925 (2022)

An Automated Approach for Generating Conceptual Riddles

Niharika Sri Parasa[1(✉)], Chaitali Diwan[1], Srinath Srinivasa[1], and Prasad Ram[2]

[1] International Institute of Information Technology, Bangalore, India
{niharikasri.parasa,chaitali.diwan,sri}@iiitb.ac.in
[2] Gooru Inc, 350, Twin Dolphin Dr., Redwood City, CA 94065, USA
pram@gooru.org

Abstract. One of the primary challenges in online learning environments is to retain learner engagement. Several different instructional strategies are proposed both in online and offline environments to enhance learner engagement. The *Concept Attainment Model* is one such instructional strategy that focuses on learners acquiring a deeper understanding of a concept rather than just its dictionary definition. This is done by searching and listing the properties used to distinguish examples from non-examples of various concepts. Our work attempts to apply the Concept Attainment Model to build conceptual riddles, to deploy over online learning environments. The approach involves creating factual triples from learning resources, classifying them based on their uniqueness to a concept into 'Topic Markers' and 'Common', followed by generating riddles based on the Concept Attainment Model's format and capturing all possible solutions to those riddles. The results obtained from the human evaluation of riddles prove encouraging.

Keywords: Riddle generation · Concept Attainment · Triples Creation · Language Models

1 Introduction

One of the main challenges of online learning environments is the high drop-out rate, and lack of learner engagement even among regular learners [1,2]. Several studies [3–5] have shown that learners learn best by being actively involved in learning, in contrast to being passive recipients of classroom knowledge. Activity-based learning is achieved by adopting instructional practices that encourage learners to think about what they are learning [5].

One such instructional strategy in pedagogy that is effective [3,6–8] across domains is the Concept Attainment Model [9]. This model enables educators to guide the learners from a property level to the concept level via structured inquiry where inquiries are formed as scenarios abiding essential and non-essential properties of a given concept. It is implemented by identifying and

Supported by Gooru (https://gooru.org/).

D.-N. Yang et al. (Eds.): PAKDD 2024, LNAI 14650, pp. 160–172, 2024.
https://doi.org/10.1007/978-981-97-2266-2_13

presenting concepts in terms of *positive* examples (which contain essential properties of the concept) and *negative* examples (containing non-essential properties of the concept) to the learner, followed by testing if the concept is attained by the learner, and finally by capturing the learning strategies applied by the learner to attain the concept.

Our goal in this work attempts to implement the identification of data and presentation part of Concept Attainment Model using Artificial Intelligence (AI) techniques, with an objective of deployment in large-scale online learning environments. An engaging and fun way to present this model to the learners is by structuring the Concept Attainment Model in the form of riddles. Riddle-solving motivates and generates interest in the learner [10–12] and can be a great introductory activity on the subject of inference. Given a set of representative learning content in a domain, our approach generates riddles for each concept by extracting the properties, identifying and categorizing them based on their uniqueness, and modeling them into concept attainment format.

2 Related Work

Although there has been a substantial amount of work in executing and studying the efficacy of the Concept Attainment Model (CAM) in a classroom context, this has been a novel area to implement computationally. Sultan et al. [13] implemented each phase of CAM using different learning strategies and methodologies using a simulation-based approach resulting in high cost and requiring a very high collaborative effort.

There have been previous attempts at the automatic generation of riddles in the context of computational creativity or humor.

Ritchie et al. [14] developed JAPE which generates simple punning riddles using predefined schemas. Later, Waller et al. [15] developed JAPE [14] into a large-scale pun generator to assist children with communication disabilities. However, these two methods require the schemas to be built manually from previously known jokes. Tyler et al. [20] incorporated JAPE and developed an advanced pun generation system using two-word database, homophone dictionary, synonyms, antonyms, hypernyms, and templates.

Colton et al. [16] attempts to automatically build puzzles on numbers and animals. The puzzles were generated given background information about a set of objects of interest and posed as questions extending the HR theory of formation. In similar contexts, Pinter et al. [17] proposed a knowledge lean approach to generate word puzzles from unstructured and unannotated corpora using a topic model and an algorithm based on network capacity and semantic similarity. However the method doesn't discuss or capture relations of the produced set of topics to a concept.

Based on current events extracted from news websites or by gathering data from both knowledge bases and online resources, Guerrero et al. [18] implemented a Twitter bot that generates riddles using templates by drawing a comparison between celebrities and well-known fiction characters. Galvan et al. [19] developed a riddle generator that constructs riddles by drawing a comparison between attributes of concepts using the existing knowledge bases such as Thesaurus Rex to interact with the users.

Although some of the above works [16,17,19] generated puzzles on numbers, animals, and others, their attempts are largely for creative purposes and not specific to educational needs. On the other hand, our proposed approach is backed by an instructional strategy for concept attainment purposes. It is not only built on educational corpus but identifies and distinguishes semantically closer concepts based on their properties to structure them as riddles using the pre-trained language models.

3 Methodology

Our method of Riddle generation includes four modules: Triples Creator, Properties Classifier, and Generator followed by Validator as shown in Fig. 1.

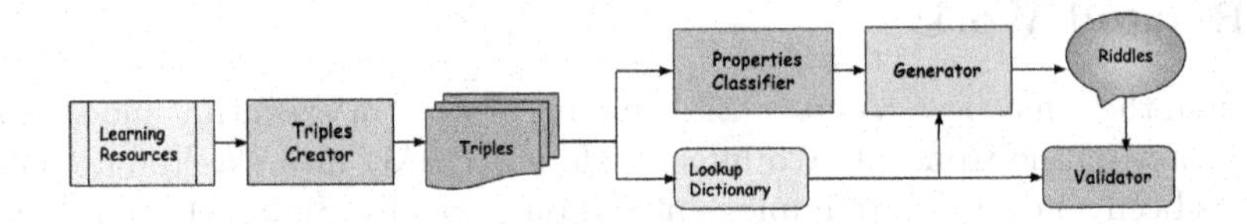

Fig. 1. Architecture of the proposed Riddle Generation approach

Our dataset is curated from Dbpedia [25][1] that includes 200 learning resources, specifically abstracts or summaries of various concepts within the field of zoology. These resources provide definitional summaries, each averaging 7-8 sentences, which offer comprehensive topical coverage suitable for educational use.[2] Each learning resource is passed as input to our Triples Creator module which returns triples of concept, relation, and properties. These triples are fed as input to our Properties Identifier module where the properties are classified into *Topic Markers*and *Common*. *Topic Markers* are the properties that explicitly represent a concept and *Common* property is associated with more than one concept.

After filtering as per their class, riddles are generated by the Generator module through a Greedy mechanism. Riddles generated from *Topic Markers* of a concept are termed as *Easy Riddles* and those from *Common* are termed as *Difficult Riddles*. The generated riddles can have one or more answers. So, each riddle is passed through the Validator which generates and stores all possible answers to verify learners' answers and provide hints.

[1] https://dbpedia.org/page/.

[2] https://en.wikipedia.org/wiki/Wikipedia#Cultural_impact.

Table 1. Outputs from Triples Creator, Properties Classifier, Generator and Validator

Target Concept: Dog				
Triples Creator	Properties Classifier		Generator	Validator
	Class	Neighbouring concepts		
Dog can guard your house	Topic Marker		**Easy:** I can guard your house. I can bark. I am a loyal friend. Who am I?	dog
Dog can bark	Topic Marker			
Dog is related to canine	Common	fox, wolf, bear		
Dog is a mammal	Common	elephant, lion, tiger	**Difficult(v1):** I am related to animals but I am not elephant. I am a pet but I am not a rabbit. I am related to flea but I am not a cat. Who am I?	dog, ferret,...
Dog is related to flea.	Common	cat, bee, louse		
Dog is a loyal friend.	Topic Marker			
Dog is a pet.	Common	cat, rat, rabbit		
Dog is a animal	Common	tiger, fox, elephant		
Dog has four legs.	Common	elephant, rabbit	**Difficult(v2):** I am related to animals but I don't have a trunk. I am a pet but I don't like carrots. I am related to flea but I am not feline. Who am I?	dog, cat,...
Dog is for companionship	Common	animals, cat, fish		
Dog wants a bone	Topic Marker			
Dog can run	Common	cheetah, horse, rat		
Dog is related to a kennel	Common	ferret, rabbit		

3.1 Triples Creator

The purpose of the Triples Creator module is to mine many possible properties i.e., the keywords and key phrases that represent the concept, along with their relations to the concept to form simple, meaningful sentences. Hence the statements from the learning resource are to be formatted into triples of concept, relation, and property breaking down the process into Property extraction and Relation Prediction. For Property extraction, the properties associated with a concept are extracted by identifying noun phrases, adjectives, verbs, and phrases comprising combinations of nouns and adjectives using pos_tagging, keywords and keyphrase extraction techniques in order to maximize the properties set i.e., P:$p_1, p_2, p_3 ... p_n$. The *YAKE*[3] and *NLTK*[4],[5] libraries performed well on our corpus. By trial and error, it was observed that maximum properties i.e., ≈ 25-30 keywords and key-phrases for each summary are extracted by combining

[3] https://github.com/LIAAD/yake.
[4] https://www.nltk.org/.
[5] https://github.com/csurfer/rake-nltk.

the results of both libraries, followed by eliminating repetitive and irrelevant properties.

We then automatically arrange the triple in *<concept>* *<relation>* *<property>* format masking the relation token i.e.,*<concept>*, *<mask>*, *<property>*. To predict the relations between the concepts and properties, we use a python package *Happy Transformer*[6] built on top of Hugging Face's transformer library[7] to easily utilize state-of-the-art NLP models. *HappyWordPrediction* class is imported for mask prediction, and is tested on BERT_base and BERT_large with 110M and 340M parameters respectively, pre-trained on Books Corpus and English Wikipedia. As the fine-tuning corpus includes the summaries of concepts and is already part of the pre-training corpora, it is observed that BERT_large variant with fine-tuning performed better than the other model. So we use *BERT_large_uncased_whole_word masking* model[8] with hyper-parameters as follows: Learning rate - 0.0001, epochs −4 and batch size - 100. The model returns a set of tokens with their probability scores. Empirically, we set the probability threshold to 0.5 and above and choose the first token from the set to complete the triple. It was observed that for a few cases, the BERT model returns special characters resulting incomplete triples. Such triples are completed by replacing the relation token with "is related to" and are grammatically corrected using *GingerIt* package[9] in the post-processing phase.

In this way, all the triples for the concepts are generated. The examples in the Triples Creator column of Table 1 depict some of the triples for the concept 'Dog'.

We then create a *Lookup Dictionary* where keys are concepts and values are the list of triples along with their respective attributes. This dictionary is used in the Generator module and Validator.

3.2 Properties Classifier

In this module, we classify triples based on their properties into two classes, called: *Topic Markers* and *Common*. Inspired by the k-Nearest Neighbor's Language model [21] proposed by Khandelwal et al., we use a data store and a binary search algorithm to query the neighbours of the target token given its context. The data store is constructed in context target pairs where the contextual embedding of each triple is mapped with its corresponding tokens. The embeddings are generated using BERT tokenizer[10]. This is passed as input to the KDTree[11], a binary search algorithm that organizes data points in high dimensional space for the fastest neighbour retrieval. KDTree organizes the normalized contextual embeddings in a tree like structure based on their feature similari-

[6] https://pypi.org/project/happytransformer/.
[7] https://huggingface.co/docs/transformers/index.
[8] https://huggingface.co/bert-large-uncased-whole-word-masking.
[9] https://pypi.org/project/gingerit/.
[10] https://huggingface.co/.
[11] https://scikit-learn.org/stable/modules/generated/sklearn.neighbors.KDTree.html.

ties to one another. Other alternatives to search over this datastore could be algorithms such as FAISS [22], or HSWG [23].

Once the search index is built on all conceptual triples, one can retrieve neighbours of the same attribute in different contexts or different attributes in the same context. We opt for the former to classify triples. Each triple of the concept as context, along with its respective property is passed as queries to the datastore returning the distances, neighbours, and contexts. The subject of the triple is anonymized before querying so that the context doesn't include the concept. Ex: i.e., 'Dog can bark' is changed to 'I can bark'.

Algorithm 1: Properties Classifier Module Pseudo Code

```
Input  : C:{T, P}
Output: Class, Comc
Com_c, Class ← emptylist ;
Set number of neighbors,k=5;
for (t_i, p_i) ∈ (T, P) do
    contexts ← getKNeighbouringContext(p_i, t_i, K = 5);
    if all(contexts ∈ C) then
        class_i ← TopicMarker;
    else
        class_i ← Common;
        for context_i ∈ contexts do
            comc_i ←getConcept(context_i);
        end
    end
    return Class, Comc
end
```

Algorithm 1 outlines the pseudo-code of the Properties Classifier module. The input consists of a data structure of concepts $C\{T, P\}$ along with its respective set of properties $\{P\}$ and a set of triples $\{T\}$. Each property p_i along with its anonymized context t_i of a concept C is passed to *getKNeighbouringContext* method which returns the top five nearest contexts along with the distances. The number of neighbours k decides the size of the negative example set that contributes to the variety in riddles. For computational reasons, we have empirically opted k to be 5. If all the neighbouring contexts relate to the target concept (C), then the class of the triple i.e., $Class_i$ is categorized as *Topic Marker*, otherwise, it is categorized as *Common*.

Subsequently, neighbouring concepts $comc_i$ are captured by processing the neighbouring contexts through the *getConcept* method. For a total of ≈70,000 triples for 200 concepts, 3000 are classified as Topic Markers and the rest are Common.

Refer to the examples in the Class and Neighbouring concepts of the Properties Classifier column in Table 1.

3.3 Generator

In this module, we generate two types of riddles – Easy and Difficult. Difficult riddles strictly adhere to the Concept Attainment Model as they contain both positive and negative examples, while Easy riddles are built only on positive examples. Easy riddles were incidentally created in the process of generating riddles based on CAM. However, since they are also based on a well-known instructional strategy of Inference [24] and can be very well used to lead the learners to concept attainment, we decided to include them in our model and study.

Easy riddles contain statements that can easily recall the target concept, hence we consider triples classified as *Topic Markers* to construct them.

Algorithm 2 (2.1 Easy riddles) outlines the generation of Easy riddles. The *getCombinations* method takes a set of triples of *Topic Markers* t_{m_i} class for a concept C as input and returns the riddles R in 3-sentence, 4-sentence, and 5-sentence combinations. Refer to the example of easy riddles in the Generator column of Table 1.

To construct Difficult riddles, we use a template-based structure as given in Table 2, filling in the necessary positive and negative examples. The positive examples are triples of class *Common* and are generated the same as for easy riddles in Algorithm 2 (2.1: Easy Riddles). The respective negative examples are generated in two versions, one by negating the neighbouring concepts Nc, for example, *I am not an elephant* (Refer to the example of Difficult(v1) in Generator column of Table 1) and the other by choosing and negating a property Np of the selected neighbouring concept in the former version, for example, *I don't have a trunk* (Refer to the example of Difficult(v2) in the Generator column of Table 1).

Algorithm 2 (2.2: Difficult Riddles) outlines the pseudo-code for generating Difficult Riddles. We consider the input to be a concept with its respective properties p_c and triples t_c of *Common* class along with neighbouring concepts *Comc*, lookup dictionary $lookup_{dict}$ and riddles R which consists of combinations of triples of *Common* class. The first negative example nc_i is always a random choice from the neighbouring concepts using *getRandomConcept* of the respective positive example. Others are chosen by checking if the first n riddle properties rp of the respective n positive examples belong to any of the neighbour concepts where n is the index of the triple in the combination. The relevant neighbour concept is the negative example if the check returns true. The *getConcepts* method checks the riddle properties rp against all the neighbouring concept properties, fetching them from the lookup dictionary. Refer to Difficult (v1) in the Generator column of Table 1.

Consequently, negative examples in the second version of the Difficult riddle np_i are generated by passing the respective nc_i to *getNegatedProperty* method that returns a respective triple in negated form. The method functions by eliminating common properties between neighbour concepts and target concept from the neighbour concept, later randomly selecting a property from a pool of unique properties, and finally fetching its respective triple and negating it.

Algorithm 2: Generation Module Pseudo Code

2.1 Easy Riddles

Input : C : $\{T\}$
Output: R
Set $T \leftarrow \{t_{m_1}, t_{m_2}, ...\}$ // Set of selected Topic Marker triples t_{m_i}
for $i = 3, 4, 5$ **do**
  $R \leftarrow getCombinations(T, i)$
end
return R

2.2 Difficult Riddles

Input : $C : \{A, T, Comc\}$,$lookup_{dict} : \{C, T, P\}$, R
Output: d_{v1}, d_{v2}
Set $T \leftarrow \{t_{c_1}, t_{c_2}, ...\}$, $P \leftarrow \{p_{c_1}, p_{c_2}, ...\}$ // Set of selected Common triples t_{c_i} and their properties p_{c_i}
Set $Nc, Np, rp \leftarrow emptylist$
for $riddle \in R$ **do**
  for *i in len(riddle)* **do**
    if *i==0* **then**
      $rp_i \leftarrow p_{c_i}$
      $nc_i \leftarrow getRandomConcept(C[comc_i])$
      $np_i \leftarrow getNegatedProperty(C, nc_i, lookup_{dict})$
      // negates a random unique attribute of nc_i by eliminating common properties between target concept C and nc_i using lookup dictionary //
    else
      $rp_i \leftarrow p_{c_i}$
      $nc_i \leftarrow getConcepts(rp, C['comc_i'], lookup_{dict})$
      // returns a concept by checking the riddle properties rp against properties of each common concept //
      $np_i \leftarrow getNegatedproperty(nc_i)$
    end
  end
  $riddle_{v1} \leftarrow ApplyTemplate(d_{v1}, Nc,$ R)
  $riddle_{v2} \leftarrow ApplyTemplate(d_{v2}, Np,$ R)
  return $riddle_{v1}, riddle_{v2}$
end

Table 2. Difficult Riddle Templates

p_1	but I am not	nc_1
p_2	but I am not	nc_2
p_3	but I am not	nc_3
	Who am I ?	

(a) Version 1

p_1	but	na_1
p_2	but	na_2
p_3	but	na_3
	Who am I ?	

(b) Version 2

Refer to Difficult(v2) in Generator column of Table 1.

After generating the required negative examples, we call the templates d_{v1} and d_{v2} (Refer to Difficult riddle templates in Table 2) and complete the difficult riddles by organizing the positive examples from each combination *riddle* along with their negative examples Nc, Np in both versions respectively.

Refer to Difficult Riddles examples in Generator of Table 1.

3.4 Validator

From a large pool of triples generated for a concept, we restrict the size of the riddle to 3-5 triples which may be common across many concepts Thus it becomes necessary to capture all their possible solutions for validation reasons. As easy riddles are generated with unique properties, they have only one solution which is the target concept. Whereas the solutions to difficult riddles depend on the positive examples set, which are generated and stored by the validator by comparing all the positive properties in a riddle combination against all conceptual properties using the lookup dictionary (Refer to the example in the Validator column of Table 1).

4 Evaluation and Results

As our work attempts the first phase of the concept attainment model i.e., the presentation and identification of data, our evaluation approach targets to validate the quality of the riddles (syntactic, semantic, and difficulty level), answerability, relevance to the learning content (informative) and whether the riddles impart engagement using Likert 3 point and 5 point scales.

30 human evaluators consisting of Graduates, Post Graduates, and Ph.D. students evaluated our automatically generated riddles. Since the domain represents the common animals in our locale, we can safely assume the human evaluators to be familiar with the domain and conduct the evaluation. The evaluators were presented with a sample of 20 riddles which consisted of both easy and difficult riddles along with multiple-choice options and hints. The evaluators were required to solve easy riddles in one attempt and difficult ones, in

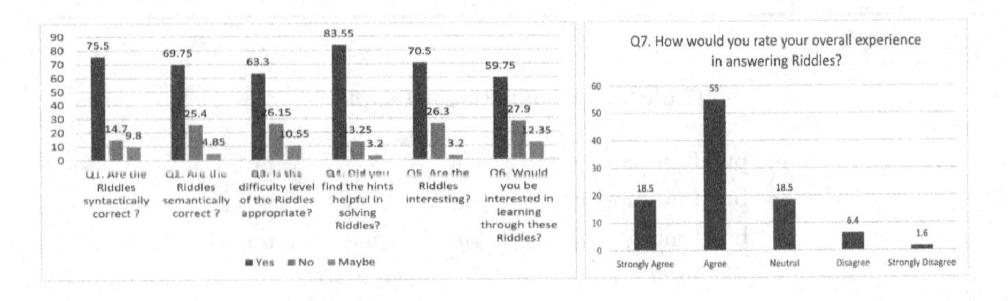

Fig. 2. Evaluation Results

Table 3. Riddles used in Evaluation

S.NO	Riddle	Difficulty level	Hints	Concept	Guessed percentage
R1	I am related to national flags. I have a typical head to body length. I am a keystone predator. I hunt humans. Who am I?	Easy		Lion	51.6%
R2	I am decapod crustaceans. I have thick exoskeleton. I am related to world'oceans. Who am I?	Easy		Crab	79.3%
R3	I am omnivores but I am not racoon. I am ungulate but I am not camel. I am domestic but I am not rabbit. I am highly social but I am not turkey. Who am I?	Difficult	I am related to wild boar. I am related to suidae.	Pig	58.5%, 87.9%, 89%
R4	I am a large mammmal but I am not a Pig. I am related to calves but I am not an Elephant. I am related to great depths but I am not a whale. I am related to marine pollution but I am not porpoise. Who am I?	Difficult	I am related to captive killer whales. I am related to island film series	Dolphin	40%, 46.6%, 73.32%
R5	I am related to ruminant mammals. I am related to paleolithic. I am related to Cervidae. I can grow new antlers. Who am I?	Easy		Deer	79.3%
R6	I am related to mass extinction event. I am related to fossil evidence. I am related to quadrupedal species. I am related to jurassic. Who am I?	Easy		Dinosaur	96.6%
R7	I am related to ocean but I am not shark. I am related to abyssal but I am not fish. I lay eggs but I am not salamandar. I am venomous but I am not snake.who am I?	Difficult	I have cephalopod limbs. I have complex nervous system	Octopus	23.5%, 70.5%, 82.35%
R8	I am related to acoustic communication. I am related to cambrian. I am related to acquatic craniate animals. I lack limbs. Who am I?	Easy		Fish	58.6%
R9	I am a human test subject. I am related to pocket pet. I am related to docile nature. Who am I?	Easy		Guinea Pig	65.5%
R10	I am related to intelligent but I am not Magpie. I produce sperm but I am not octopus. I am related to hunting but I am not human Who am I?	Difficult	I am related to bilateral symmetry. I am related to decapodi-formes	Squid	17.6%, 37.5%, 42.9%
R11	I am related to paws but I am not a racoon. I am related to carnivoran mammals but I am not badger. I am related to omnivorous but I am not a fox. Who am I?	Difficult	I am related to polar. I have stocky legs.	Bear	40%, 90%, 96.6%
R12	I am ungulate mammal. I am related to sport competitions. I am related to police work. I am related to strong flight or fight response. Who am I?	Easy		Horse	60%

(*continued*)

Table 3. (*continued*)

S.NO	Riddle	Difficulty level	Hints	Concept	Guessed percentage
R13	I am related to human but I am not plant. I am invasive but I am not termite. I lay white eggs but I am not Starling. Who am I?	Difficult	I can imitate. I am charismatic.	Parrot	11.76%, 81%, 86%
R14	I am related to Africa but I am not Chimpanzee. I am related to Old world monkeys but I am not baboon. I am related to gorillas but I am not Gorilla. Who am I?	Difficult	I am related to tailless primates. I am old world simians native	Ape	60%, 76.6%, 96.6%
R15	I am related to gelatanious member. I have efficient locomotion. I am related to scyphozoans. I am related to green fluorescent protien. I can injure humans. Who am I?	Easy		Jellyfish	93.1%
R16	I am carboniferous but I am not mammal. I am related to mass extinction events but I am not dinosaur. I am related to tetrapod vertebrates but i am not amphibian. Who am I?	Difficult	I am related to reptiliomorph tetrapods. I am related to placenta analogous	Reptile	47.1%, 70.69%, 73%
R17	I am related to yellow fever. I am related to chikungunya. I am related to tube like mouth parts. Who am I?	Easy		Mosquito	89.7%
R18	I am important but I am not a bird. I am related to solitary but I am not a cheetah. I am related to insects but I am not butterfly. Who am I?	Difficult	I am related to nectar. I am stingless	Bee	46.3%, 93.8%, 100%
R19	I have binocular vision. I am related to binaural hearing. I am related to sharp talons. I am a nocturnal bird. Who am I?	Easy		Owl	72.4%
R20	I am terrestial but I am not dinosaur. I work collectively but I am not cattle. I am related to super organisms but I am not a termite. Who am I?	Difficult	I am related to antennae. I am related to Invasive Species	Ant	60%, 63.3%, 83.3%

three attempts. The difficult riddles were provided with hints to guess the target concept.

The results of the activity are given in Table 3 and Fig. 2. As shown in Fig. 2, for questions *Q1, Q2* ≈70%–75% of the evaluators agreed that the generated riddles are semantically and syntactically correct respectively. For questions *Q5, Q6*, ≈ 70% of the evaluators agreed that the riddles are interesting, and ≈ 60% agreed for their adaptability in learning. Almost 70% of the evaluators agreed that the experience of answering riddles was good. (Refer to Fig. 2).

As per Fig. 2, for question *Q3*, ≈ 60% of the evaluators voted that the difficulty level is appropriate. Since the difficult riddles consist of common properties in relatable terms, most evaluators reported that the difficult riddles are easier than the actual easy ones.

Easy riddles (Refer to R1, R2, R5, R6, R8, R9, R12, R15, R17, R19 in Table 3) are answered correctly by ≈75% of the evaluators.

Along with this, we did a case study to understand whether topic markers as hints help learners to guess the concept. Hence the guessed percentage column in Table 3 for difficult riddles has 3 values where the first value indicates guessed percentage without any hint while the second and third values indicate percentages with hints (Refer to R3, R4, R7, R10, R11, R13, R14, R16, R18, R20 in Table 3).

Therefore it is evident from the transition of guessed percentages in Table 3 that *Topic Markers* as hints lead the learner towards the concept. This is also validated as per Fig. 2 question *Q4* where almost 83% evaluators agreed that the hints proved helpful in guessing the concept. Thus, we can conclude that the riddles generated are of decent quality and are capable of imparting information via engagement.

5 Conclusion and Future Work

In this work, we proposed a novel approach to automatically generate concept attainment riddles given a set of learning resources. The results obtained from our evaluation are encouraging as the riddles from our approach are of decent quality and prove interesting. This approach is easily adaptable and scalable across different domains. As part of future work, we plan to develop an application using the generated riddles, implement the second phase of the Concept Attainment Model i.e., testing the concept attainment of the learner, and study if the model's efficacy can be reflected in an online learning space.

References

1. Martin, F., Borup, J.: Online learner engagement: conceptual definitions, research themes, and supportive practices. Educ. Psychol. **57**(3), 162–177 (2022)
2. Zerdoudİ, S., Tadjer, H., Lafİfİ, Y.: Study of learner's engagement in online learning environments. Inter. J. Inform. Appli. Math. **6**(1), 11-28
3. Kalani, A.: A study of the effectiveness of concept attainment model over conventional teaching method for teaching science in relation to achievement and retention. Inter. Res. J. **2**(5), 436–437 (2009)
4. Yi, J.: Effective ways to foster learning. Perform. Improv. **44**(1), 34–38 (2005)
5. Prince, M.: Does active learning work? a review of the research. J. Eng. Educ. **93**(3), 223–231 (2004)
6. Kumar, A., Mathur, M.: Effect of concept attainment model on acquisition of physics concepts. Univ. J. Educ. Res. **1**(3), 165–169 (2013)
7. Habib, H.: Effectiveness of concept attainment model of teaching on achievement of XII standard students in social sciences. Shanlax Inter. J. Educ. **7**(3), 11–15 (2019)
8. Haetami, A., Maysara, M., Mandasari, E.C.: The effect of concept attainment model and mathematical logic intelligence on introductory chemistry learning outcomes. Jurnal Pendidikan dan Pengajaran **53**(3), 244-255 (2020)
9. Joyce, B., Weil, M., Calhoun, E.: Models of teaching (2003)
10. Doolittle, J.H.: Using riddles and interactive computer games to teach problem-solving skills. Teach. Psychol. **22**(1), 33–36 (1995)

11. Denny, R.A., et al.: Elementary Who am I riddles. J. Chem. Educ. **77**(4), 477 (2000)
12. Shaham, H.: The riddle as a learning and educational tool. Creat. Educ. **4**(06), 388 (2013)
13. Sultan, A.Z., Hamzah, N., Rusdi, M.: Implementation of simulation based-concept attainment method to increase interest learning of engineering mechanics topic. J. Phys. Conf. Ser. **953**(1) (2018)
14. Ritchie, G.: The JAPE riddle generator: technical specification. Institute for Communicating and Collaborative Systems (2003)
15. Waller, A., et al.: Evaluating the standup pun generating software with children with cerebral palsy. ACM Trans. Accessible Comput. (TACCESS) **1**(3), 1–27 (2009)
16. Colton, S.: Automated puzzle generation. In: Proceedings of the AISB 2002 Symposium on AI and Creativity in the Arts and Science (2002)
17. Pintér, B., et al.: Automated word puzzle generation using topic models and semantic relatedness measures. Annales Universitatis Scientiarum Budapestinensis de Rolando Eötvös Nominatae, Sectio Computatorica, vol. 36 (2012)
18. Guerrero, I., et al.: TheRiddlerBot: a next step on the ladder towards computational creativity. In: Toivonen, H., et al. (ed.) Proceedings of the Sixth International Conference on Computational Creativity (2015)
19. Galván, P., et al.: Riddle generation using word associations. In: Proceedings of the Tenth International Conference on Language Resources and Evaluation (LREC 2016) (2016)
20. Tyler, B., Wilsdon, K., Bodily, P.M.: Computational humor: automated pun generation. In: ICCC (2020)
21. Khandelwal, U., et al.: Generalization through memorization: Nearest neighbor language models. arXiv preprint arXiv:1911.00172 (2019)
22. Johnson, J., Douze, M., Jégou, H.: Billion-scale similarity search with gpus. IEEE Trans. Big Data **7**(3), 535–547 (2019)
23. Malkov, Y.A., Yashunin, D.A.: Efficient and robust approximate nearest neighbor search using hierarchical navigable small world graphs. IEEE Trans. Pattern Anal. Mach. Intell. **42**(4), 824–836 (2018)
24. Bintz, W.P., et al.: Using literature to teach inference across the curriculum. Voices Middle **20**(1), 16 (2012)
25. Lehmann, J., et al.: Dbpedia-a large-scale, multilingual knowledge base extracted from wikipedia. Semantic web **6**(2), 167–195 (2015)

Time-Series and Streaming Data

DiffFind: Discovering Differential Equations from Time Series

Lalithsai Posam[1], Shubhranshu Shekhar[2(✉)], Meng-Chieh Lee[3], and Christos Faloutsos[3]

[1] University of California, Berkeley, USA
lposam@berkeley.edu
[2] Brandeis University, Waltham, USA
sshekhar@brandeis.edu
[3] Carnegie Mellon University, Pittsburgh, USA
{mengchil,christos}@cs.cmu.edu

Abstract. Given one or more time sequences, how can we extract their governing equations? Single and co-evolving time sequences appear in numerous settings, including medicine (neuroscience - EEG signals, cardiology - EKG), epidemiology (covid/flu spreading over time), physics (astrophysics, material science), marketing (sales and competition modeling; market penetration), and numerous more. Linear differential equations will fail, since the underlying equations are often non-linear (SIR model for virus/product spread; Lotka-Volterra for product/species competition, Van der Pol for heartbeat modeling).

We propose DiffFind and we use genetic algorithms to find suitable, parsimonious, differential equations. Thanks to our careful design decisions, DiffFind has the following properties - it is: (a) *Effective*, discovering the correct model when applied on real and synthetic nonlinear dynamical systems, (b) *Explainable*, gives succinct differential equations, and (c) *Hands-off*, requiring no manual hyperparameter specification.

DiffFind outperforms traditional methods (like auto-regression), includes as special case and thus outperforms a recent baseline ('SINDy'), and wins first or second place for all *5* real and synthetic datasets we tried, often achieving excellent, zero or near-zero RMSE of *0.005*.

1 Introduction

Given long, co-evolving time sequences (say, flu patient counts, or EEG waves, or product-sales volumes), how to automatically extract the governing differential equations? Earlier methods [2,6,24] heavily rely on feature engineering to select careful features, thus sacrificing generality. Moreover, most of them are not able to handle time delay differential equations. Recent studies [15,23] focus on the forecasting problem and use deep learning models. Thanks to their large number of parameters, those models provide good forecasting, at the expense of explainability, and they are suffering from expensive hyperparameter and network architecture tuning.

D.-N. Yang et al. (Eds.): PAKDD 2024, LNAI 14650, pp. 175–187, 2024.
https://doi.org/10.1007/978-981-97-2266-2_14

Informal Problem 1. *The problem is as follows:*

- ***Given*** *a multivariate time series* $\boldsymbol{X}$ $(\boldsymbol{x_1}(t), \boldsymbol{x_2}(t), \ldots; t = 1, \ldots)$
- ***Find*** *the nonlinear differential equations that fit the sequences,* ***accurately*** *and* ***automatically****, that is, with no hyperparameters required by the user.*

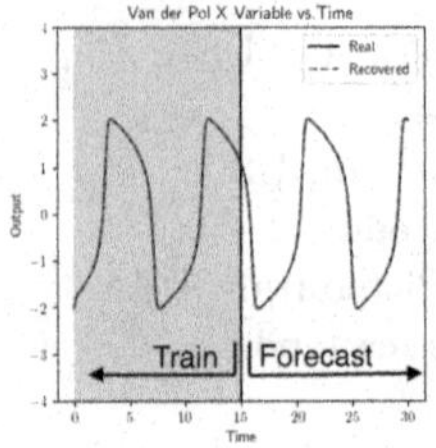

(a) X Variable forecast (b) Y Variable forecast

Notice the overlap of black (real) and red (our recovered version)

$$\dot{\boldsymbol{x}} = 3.00(1.00\,\boldsymbol{x} - \boldsymbol{y} - 0.33\,\boldsymbol{x}^3) \quad \text{\#Actual}$$

$$\uparrow\downarrow \quad \uparrow\downarrow \quad \uparrow\downarrow \quad \uparrow\downarrow$$

$$\dot{\boldsymbol{x}} = 3.03(1.08\,\boldsymbol{x} - \boldsymbol{y} - 0.34\,\boldsymbol{x}^3) \quad \text{\#Recovered}$$

(c) near-perfect recovery of X variable equations

$$\dot{\boldsymbol{y}} = 0.33\,\boldsymbol{x} \quad \text{\#Actual}$$

$$\uparrow\downarrow$$

$$\dot{\boldsymbol{y}} = 0.33\,\boldsymbol{x} \quad \text{\#Recovered}$$

(d) near-perfect recovery of Y variable equations

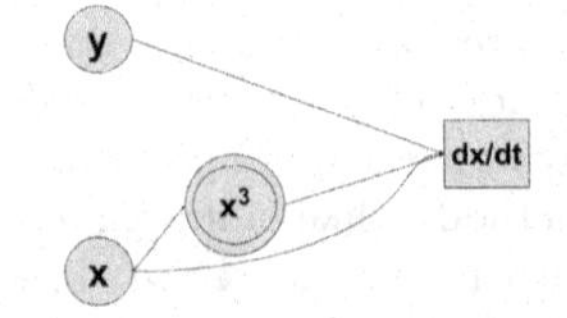

(e) X variable dependency network

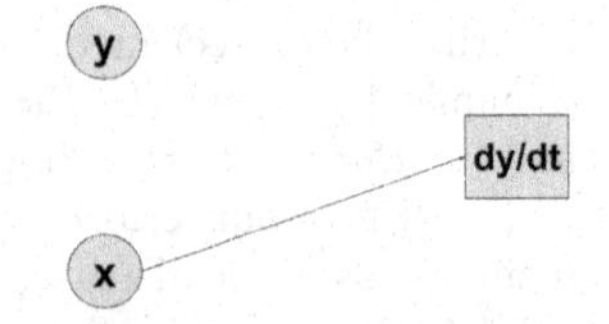

(f) Y variable dependency network

Fig. 1. DIFFFIND works: Recovers the underlying dynamics for Van der Pol: (a–b) The recursive forecasting matches reality; (c–d) The recovered equations match the actual ones; (e–f) Discovered network topology. (Color figure online)

To overcome the aforementioned shortcomings, we propose DIFFFIND, which automatically evolves to find correct representation via activation functions. Figure 1 illustrates the effectiveness and explainability of DIFFFIND using the Van der Pol oscillator. In Fig. 1(a) and 1(b), the forecast by DIFFFIND (in red) is almost identical to the actual one (in black): the gray shaded region shows the data used for training, and the white shaded region shows the forecast using the recovered equations. DIFFFIND provides the explanations by visualizing the evolved network topology in Fig. 1(e) and 1(f), and demonstrating the equations

in Fig. 1(c) and 1(d). Notice that DIFFFIND not only discovered the correct terms (like x^3), but it also recovered the coefficients, which are very close to the actual ones (e.g., 0.34 instead of 1/3 for the coefficient of the x^3 term in Fig. 1(c)).
In summary, the proposed DIFFFIND has the following properties:

(a) **Effective**, discovering the correct model when tested on real and synthetic nonlinear dynamical systems,
(b) **Explainable**, providing a succinct differential equation that governs the given time sequences - we explicitly aim for succinctness by penalizing model complexity (Eq. 1),
(c) **Hands-off**, requiring no manual hyperparameter specification.

Finally, our results are reproducible:

Reproducibility: The code and data used in the experiments along with the supplementary material are available at https://github.com/Lalithsai853/DIFFIND-PAKDD-2024.

2 Background and Related Work

Here we first review the existing works for sequential modeling and learning differential equations; and then we present genetic algorithms for architecture learning. In short, DIFFFIND is the *only* method that matches all the specifications, as shown in Table 1.

2.1 Related Work

Time Series Forecasting: Time series forecasting has been studied for many decades [3,8,18]. Classic methods for time series forecasting include the family of auto-regression (AR)-based methods such as exponential smoothing [5], ARMA [18], ARIMA [3], ARMAX [8] models. However, all these models assume linearity, and thus on non-linear dynamical systems may not work well. In contrast, recent, deep-learning models [15,23,26,27] do not need the linearity assumption, but they are often black-box models, with limited or no explainability.

Dynamical System Modeling: Early work on nonlinear system identification [2, 24] construct families of candidate nonlinear functions for the rate of change of state variables in time. More recently, SINDY [6] approximates derivatives using a spline over the data points and applies sparse regression to recover equations from a set of pre-specified basis functions, e.g. polynomials. Neural networks have been proposed to model complex dynamics that incorporate prior scientific knowledge via differential equations [9,21]. Similarly, [19,28,29] learn to estimate the parameters for domain-expert specified dynamical systems, which is different from the problem of discovering equations. Recently, system identification from underlying data through neural networks have been studied in [10,24]. Sahoo et al. [22] address a slightly different problem of recovering equations relating input and output variables using a neural network based regression approach, which can be extended to our setting of system identification. However, these methods require careful selection of network architecture (complex architecture tuning), and have limited interpretability.

In conclusion, *only* DiffFind fulfills all the specs, as shown in Table 1.

2.2 Background - Genetic Algorithms for Architecture Search

For deep neural networks, several articles propose architecture search algorithms [11,12,30]. The NEAT [25] algorithm stands out, proposing to use a genetic algorithm [14] to optimize the weights and structure of networks simultaneously.

Table 1. DiffFind matches all specs, while competitors miss one or more of the features. '?' indicates 'unclear' or 'depends on implementation'.

Property	Method			
	Dynamical Sys. Modeling [2,6,17,22,24]	Neural [15,23]	Fore. [3,8,18]	DiffFind
Hands-off				✔
Explainable	✓		✓	✔
Effective	?	?	?	✔
Scalable(O(n))	✓	✓	✓	✔
Architecture Learning				✔

3 Proposed Method: DiffFind

Preliminaries. Table 2 shows the symbols used in this paper. We are given a multivariate time series $\boldsymbol{X} = \{\boldsymbol{x}_1, \dots, \boldsymbol{x}_d\}$ containing observations for d sequences, and we want to discover the (non-)linear differential equations

Each individual sequence of observation is given as $\boldsymbol{x} = \{x_1, \dots, x_T\}$ for T equidistant time-ticks. In this work, our goal is to learn the underlying dynamics of $\boldsymbol{X}$ that (1) succinctly compresses the given data, and (2) represents the data dynamics in terms of a system of differential equations.

Table 2. Symbols and definitions

Symbol	Interpretation
$\boldsymbol{X} = \{\boldsymbol{x}_1, \ldots, \boldsymbol{x}_d\}$	input d-dimensional multivariate time series data
$\boldsymbol{x}_i(t)$	value at time step t in ith sequence of $\boldsymbol{X}$
G_{node}	number of node genes in the genome
G_{conn}	number of connection genes in the genome
$A_i(f)$	amplitude at frequency f in the Fourier transform of $\boldsymbol{x}_i$

Difficulty of the Problem. For ease of presentation, let's assume we have only two coevolving sequences, $\boldsymbol{x}(t)$ and $\boldsymbol{y}(t)$, and we want to find the two functions $f(\boldsymbol{x}, \boldsymbol{y})$ and $g(\boldsymbol{x}, \boldsymbol{y})$ so that $\dot{\boldsymbol{x}} = f()$ and $\dot{\boldsymbol{y}} = g()$ generate the two input sequences.

How should we search for the possible $f()$ and $g()$? We could look for polynomials of lag w, eg., for lag $w=1$, we would consider:

$$f(\boldsymbol{x}, \boldsymbol{y}) = a_1\boldsymbol{x}(t) + a_2\boldsymbol{y}(t) + a_3\boldsymbol{x}(t) * \boldsymbol{y}(t) + a_4\boldsymbol{x}(t)^2 + \ldots$$

for lag $w=2$, we would consider

$$f(\boldsymbol{x}, \boldsymbol{y}) = b_1\boldsymbol{x}(t) + b_2\boldsymbol{x}(t-1) + b_3\boldsymbol{x}(t)^2 + b_4\boldsymbol{y}(t) + \ldots$$

How could we search the infinite possible powers and combinations of $\boldsymbol{x}(t)$, $\boldsymbol{x}(t-1)$, $\boldsymbol{y}(t)$, etc? Should we put an upper limit to the exponent of the powers? Linear methods like AR, do exactly that, setting to 1 the maximum exponent.

We propose to do better, by using a genetic algorithm, to search through the possibilities. This creates the need for the following design decisions (DD):

DD1) what is the 'genome'
DD2) which operations (mutation, crossover) do we allow
DD3) what is the fitness function

Next, we describe our design decisions. Once these decisions are finalized, the rest of the algorithm is based on genetic optimization: create a population of genomes (DD1), do mutations and crossovers on the fittest (DD2), and repeat until the fitness function is stabilized (DD3).

DD1 - Proposed Genome. Our genome is carefully constructed to ensure that we cover a large family of differential equations. For example, we include polynomial functions, and interaction nodes to represent nonlinear equations. Each genome has a genetic encoding scheme that includes a list of connections. A connection comprises two node genes, and their edge weight. Node gene contains the following:

- response coefficient
- activation functions (this is how we get our power terms from polynomial functions, etc.)

- aggregation function (allows us to combine variables together through addition, product, etc.)

Figure 2 gives examples of our proposed genome representation.

DD2- Proposed Mutation and Crossover. The initial population is composed of sparsely connected networks with no hidden nodes and few of the possible connections between input sequence and output. We allow the following mutations to create new networks by modifying the existing structure:

- addition, deletion of node genes and connections
- update of activation functions, including no activation, quadratic, cubic, and quartic powers

These mutations modify connections between nodes, respecting the feed-forward property of the network. Each new gene that appears through mutation is assigned a global identifier[1] that represents the chronology of the appearance of every gene in the system.

Figure 2 gives an example of the proposed crossover mechanism: During crossover, matching genes (genes with the same identifier) from both parents are chosen randomly to include in the child, and any excess or disjoint genes from the more fit parent are automatically added to the child. If the two parents have the same performance, then the excess and disjoint genes are added randomly. This allows for equations with different numbers of terms to automatically crossover and form new types of equations that combine certain aspects of both parents.

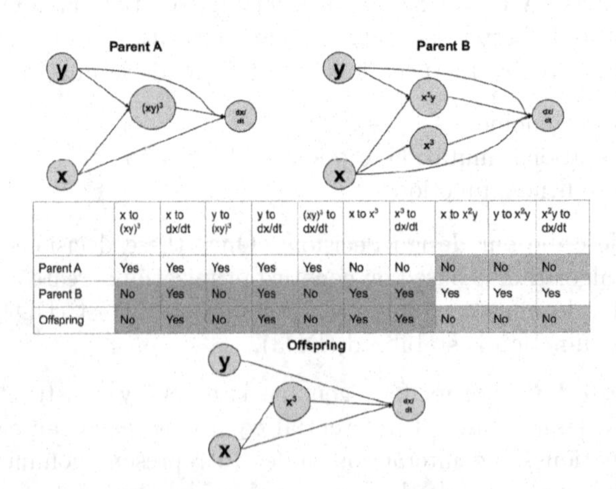

	x to $(xy)^3$	x to dx/dt	y to $(xy)^3$	y to dx/dt	$(xy)^3$ to dx/dt	x to x^3	x^3 to dx/dt	x to x^2y	y to x^2y	x^2y to dx/dt
Parent A	Yes	Yes	Yes	Yes	Yes	No	No	No	No	No
Parent B	No	Yes	No	Yes	No	Yes	Yes	Yes	Yes	Yes
Offspring	No	Yes	No	Yes	No	Yes	Yes	No	No	No

Fig. 2. Proposed genome and crossover scheme. The proposed genome is the list of edges; the crossover picks some edges from parent A (blue) and some others from parent B (green). (Color figure online)

[1] Identical mutations are assigned the same identifier.

DD3 - Proposed Fitness Function. We design our fitness function to learn a model of the sequence data that simultaneously minimizes reconstruction error and promotes a simple, compact model. Motivated by Occam's razor, in our approach, we regularize the neural architecture by including a term for model complexity. In particular, DIFFFIND utilizes the following fitness function to evaluate the performance of a given genome:

$$\boxed{\mathcal{F} = C - \underbrace{\sqrt{\frac{\sum_{j=1}^{T}(x_j - \hat{x}_j)^2}{T}}}_{\text{Reconstruction error}} - \underbrace{\lambda(G_{conn} + G_{node})}_{\text{Model complexity penalty}}} \tag{1}$$

where $C = 100$ is a large constant, $\hat{x}_j$ is the j_{th} predicted value for an individual sequence of multivariate time series based on the given genome, G_{conn} and G_{node} represents the complexity of genome in terms of number of connection and node genes, and λ controls the penalty for model (genome) complexity (our recommended, default value is $\lambda = 0.1$).

Implementation Details – Algorithm. We present the pseudocode of the DIFFFIND in Algorithm 1. We learn a genetic neural network for each variable within the input multivariate time series $\boldsymbol{X}$. In our architecture learning, nodes represent variables or operations. For example, input nodes represent values at the current time step, while the output node represents the differential value at that time step. Hidden nodes are created to incorporate operations and functions such as addition, subtraction, square, cube, and so on. Connections dictate flow between nodes, and end in the output node. Aggregation and activation functions provide the network an array of options for creating terms such as polynomial expressions.

```
   Data: An initial set of genomes S
   Result: The best genome winner in S
 1 Initialization;
 2 while stopping criterion is not met do
 3     for currGenome ∈ S do
 4         Calculate fitness for currentGenome;
 5         Assign fitness value to currentGenome.fitness;
 6         if currentGenome.fitness > winner.fitness then
 7             winner ← currentGenome;
 8         else
 9             Continue;
10         end
11     end
12     Apply mutation and crossover (reproduction) schemes on proposed genome tree
       structure;
13     Remove any stagnant speciesx;
14 end
15 Return winner;
```

Algorithm 1: Pseudocode of DIFFFIND

4 Experiments

We design our experiment to answer the following research questions:

Q1. Effective: How effective is DiffFind in learning the data dynamics? How accurate is the DiffFind in recursive forecast

Q2. Explainable: How to explain the results of DiffFind? How compact is the system of equations learned by DiffFind?

Q3. Scalable: How fast is DiffFind? How does the running-time of DiffFind scale w.r.t. length of time sequences?

Datasets. We evaluate DiffFind through extensive experiments on four carefully chosen dynamical systems, which relate to real phenomena from atmospheric convection to spikes in human brain signals, and two real data, including a case study on solar cycles. Dataset description is provided in Supplementary (available at https://github.com/Lalithsai853/DIFFIND-PAKDD-2024).

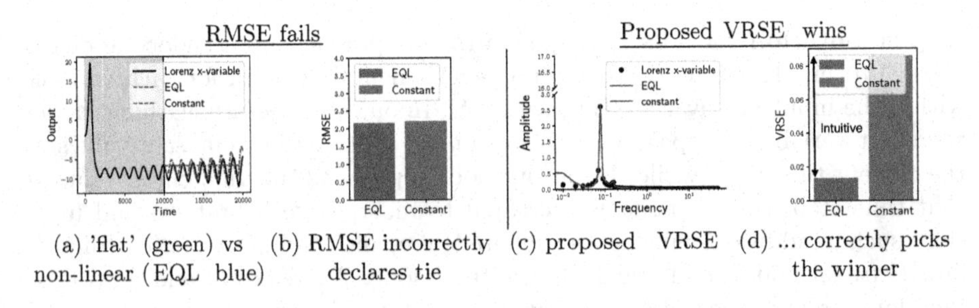

(a) 'flat' (green) vs non-linear (EQL blue) (b) RMSE incorrectly declares tie (c) proposed VRSE (d) ... correctly picks the winner

Fig. 3. VRSE wins over RMSE. VRSE captures the visual similarity, and correctly picks the winner (see (d)). RMSE declares a tie (see (b)) between the green, flat line, and the blue, wavy reconstruction (see (a)). Our insight is that VRSE should look for similarities in the *Fourier* domain (see (c)), not in the time domain. (Color figure online)

Baselines. We compare DiffFind to the following baselines.

1. AutoReg expresses each variable as a linear function of its own past values as well as the past values of all other variables in the model.
2. SINDy [6] approximates derivatives of each variable and applies sparse regression to recover equations from a set of prespecified basis functions.
3. EQL [22] recovers equations relating input and output variables using a neural network based regression approach. We extend the method to recover differential equations by setting the output as derivative of the variables.

Evaluation Measures – Pitfalls of RMSE. We report the performance using root mean square error (RMSE) which has been used as a standard statistical metric in various earlier works [7]. However, we note that RMSE often fails to quantify the performance gains for different models. For example, Fig. 3 (a–b) shows that the constant forecast is equivalent to forecast made by EQL, while we clearly observe that EQL predictions approximate the ground truth better as compared to constant model. To address the issue, and to capture visual similarity between sequences, we propose VRSE (visual relative squared error) to evaluate and compare model performance. Figure 3 (c–d) clearly shows that VRSE assigns lower error comparatively.

Definition 1 (VRSE). VRSE *(visual relative squared error) is defined as follows:*

$$\boxed{\text{VRSE} = \frac{\sum_f (A_x(f) - A_y(f))^2}{\sum_f (A_x(f))^2}} \tag{2}$$

where $A_x(f)$ $(A_y(f))$ is the amplitude at frequency f of time sequence $\boldsymbol{x}$ $(\boldsymbol{y})$.

Experimental Settings are included in Supplementary (available at https://github.com/Lalithsai853/DIFFIND-PAKDD-2024).

4.1 Q1 - DIFFFIND is Effective

In this subsection, we show the most difficult version of forecast, the so-called 'recursive' or 'k-step-ahead' forecast for each of the test-bed dynamical systems. Recursive forecast is defined as follows:

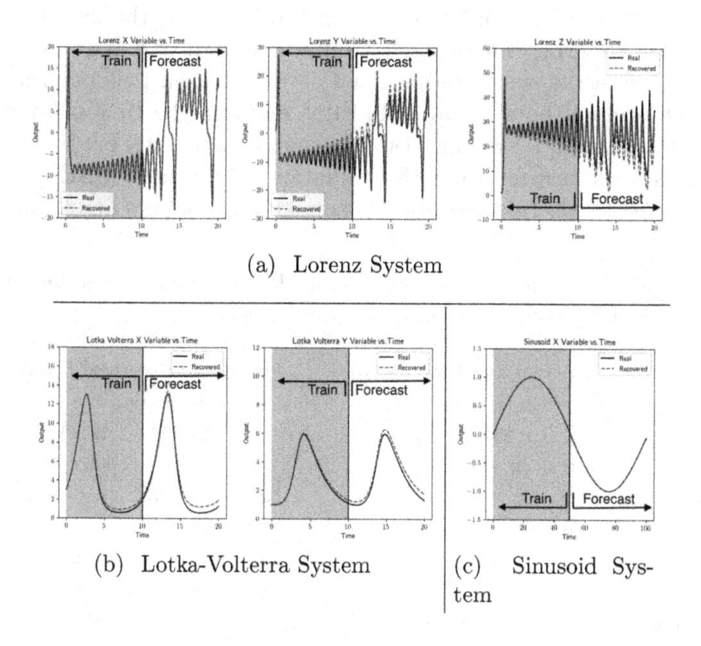

(a) Lorenz System

(b) Lotka-Volterra System

(c) Sinusoid System

Fig. 4. DIFFFIND is effective. The recursive forecasting from the recovered equations is shown for different dynamical systems.

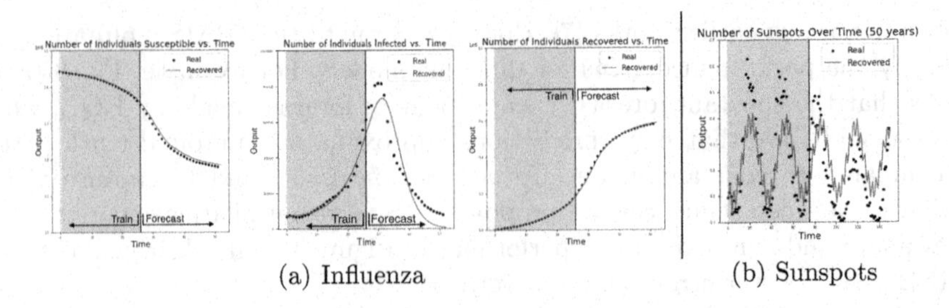

(a) Influenza (b) Sunspots

Fig. 5. DIFFFIND is effective on real data. (a) Our recovered equation (red) match the ground truth (black), for all three (susceptible, infected, recovered) in Influenza data. (b) DIFFFIND recovers a simple equation underlying sunspot phenomenon. (Color figure online)

Definition 2 (Recursive Forecast). *The forecast of the next value is based on the prediction of the previous time step, instead of using actual observation.*

The easier alternative is the one-step-ahead forecast - despite the difficulty of the task, DIFFFIND does very well, even on nonlinear (chaotic) sequences, as we discuss next.

Observation 1. DIFFFIND *includes* SINDY *as a special case.*

Reason: SINDY considers polynomials of a user-specified maximum degree, while DIFFFIND automatically searches for the best exponents and products.

Notice that, in Fig. 4 and Fig. 5 (real data), for all the systems, namely, Lorenz, Lotka-Volterra, Sinusoid, real influenza data and real sunpots data, DIFFFIND forecast (shown in red) follows actual observations (shown in black) closely. Furthermore, notice that DIFFFIND recovers the time-delay equations governing the sinusoid system. Table 4 shows that DIFFFIND outperforms the baselines with respect to both evaluation measures RMSE and VRSE. DIFFFIND consistently ranks in top two against all baselines across all datasets.

Table 3. DIFFFIND is effective and explainable. It precisely recovers the actual equations of 4 test-bed dynamical systems.

Systems	Actual Equations	DIFFFIND Recovered Equations
Lorenz System	$\dot{x} = 10(x - y)$	$\dot{x} = 9.996(x - y)$
	$\dot{y} = x(28 - z) - y$	$\dot{y} = x(24.298 - 1.121z) - 0.922y$
	$\dot{z} = xy - \frac{8}{3}z$	$\dot{z} = 0.933xy - 2.506z$
Van der Pol Oscillator	$\dot{x} = 3(x - y - \frac{1}{3}x^3)$	$\dot{x} = 3.03(1.08x - y - 0.34x^3)$
	$\dot{y} = \frac{1}{3}x$	$\dot{y} = 0.33x$
Lotka-Volterra System	$\dot{x} = 1.1x - 0.4xy$	$\dot{x} = 1.070x - 0.373xy$
	$\dot{y} = 0.1xy - 0.4y$	$\dot{y} = 0.099xy - 0.387y$
Sinusoid System	$x_t = (2 - (\frac{2\pi}{100})^2)x_{t-1} - x_{t-2}$	$x_t = 1.996x_{t-1} - 1.001x_{t-2}$

Table 4. DIFFFIND wins: Consistently first or second, against all baselines and on all datasets. Winners in and runner-ups in .

	Systems↓		AUTOREG	SINDY	EQL	DIFFFIND
Synthetic	Lorenz System	RMSE	5.7279	0.4028	6.1186	1.1864
		VRSE	0.2670	0.0002	0.2878	0.1420
	Van der Pol Oscillator	RMSE	0.2712	0.0019	2.4400	0.0561
		VRSE	0.2857	0.0001	2.9812	0.0
	Lotka-Volterra System	RMSE	2.5272	0.5902	2.0000	0.1408
		VRSE	10.2739	0.2488	4.3889	0.1415
	Sinusoid System	RMSE	0.0	0.7688	0.6921	0.0078
		VRSE	0.0	0.9939	0.8445	0.0
Real	Influenza (imputed real)	RMSE	4.60e7	6.52e6	1.57e5	1.27e4
		VRSE	4358.29	4157.26	0.8107	0.0059

4.2 Q2 - DIFFFIND is Explainable

In Fig. 1, DIFFFIND successfully recovers the dynamics of the Van Der Pol system. The evolved neural networks are visualized in Fig. 1(e) and 1(f), corresponding to the recovered equations in Fig. 1(c) and 1(d), respectively. For all other test-bed dynamical systems, we report the actual equations used to generate the observations from each system, and the corresponding DIFFFIND recovered system of equations in Table 3. Notice that DIFFFIND successfully recovers the correct model family along with the coefficients.

The recovered equations are a compact representation in terms of a differential equation for the given observational data. The differential equations are easily understandable by the domain experts, which can be used to explain the interrelations among variables of the system and underlying dynamics.

4.3 Q3 - DIFFFIND is Scalable

Figure 6 shows that DIFFFIND scales linearly with the training set size. To quantify the scalability of our method, we empirically vary the number of observations in the training data for the Van der Pol Oscillator, and plot against the wall-clock training time for DIFFFIND.

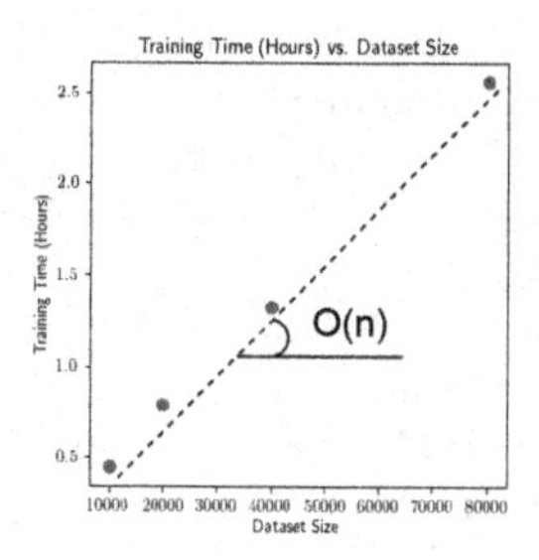

Fig. 6. Linear in (wall-clock) training time, vs seq. length.

Case Study. Next, we apply DIFFFIND to solar spotting data to understand the dynamics of solar activity. Note that unlike other datasets, there is no ground truth available. Figure 5(b) shows that DIFFFIND accurately finds the underlying pattern in the solar activity. The recovered equation $\dot{\boldsymbol{x}}_t = -0.26\boldsymbol{x}_t + 0.763\boldsymbol{x}_{t-2} - 0.68\boldsymbol{x}_{t-4}$ provides an interpretable model of this periodic natural phenomenon.

5 Conclusions

We present DiffFind to model and extract governing equations from a given set of observational data. Our proposed DiffFind has the following properties:

(a) **Effective**, discovering correct model for several real and synthetic nonlinear dynamical systems,
(b) **Explainable**, providing a succinct differential equation, thanks to our proposed fitness function (Eq. 1), that deliberately penalizes model complexity,
(c) **Hands-off**, requiring no manual hyperparameter specification.

An additional contribution is also the VRSE error function, which outperforms the traditional RMSE, as it agrees better with our intuition (see Eq. (2), Fig. 3).

References

1. Anisiu, M.C.: Lotka, Volterra and their model. Didáctica Mathematica **32**, 9–17 (2014)
2. Bongard, J., Lipson, H.: Automated reverse engineering of nonlinear dynamical systems. PNAS **104**(24), 9943–9948 (2007)
3. Box, G.E., Jenkins, G.M., MacGregor, J.F.: Some recent advances in forecasting and control. J. Roy. Stat. Soc.: Ser. C (Appl. Stat.) **23**(2), 158–179 (1974)
4. Brauer, F., Castillo-Chavez, C., Castillo-Chavez, C.: Mathematical Models in Population Biology and Epidemiology, vol. 2. Springer, Cham (2012). https://doi.org/10.1007/978-1-4614-1686-9
5. Brown, R.G.: Statistical forecasting for inventory control (1959)
6. Brunton, S.L., Proctor, J.L., Kutz, J.N.: Discovering governing equations from data by sparse identification of nonlinear dynamical systems. PNAS **113**(15), 3932–3937 (2016)
7. Chai, T., Draxler, R.R.: Root Mean Square Error (RMSE) or Mean Absolute Error (MAE)?-arguments against avoiding RMSE in the literature. Geosci. Model Dev. **7**(3), 1247–1250 (2014)
8. Chatfield, C.: Time-Series Forecasting. Chapman and Hall/CRC, Boca Raton (2000)
9. Chen, R.T., Rubanova, Y., Bettencourt, J., Duvenaud, D.K.: Neural ordinary differential equations. In: NeurIPS, vol. 31 (2018)
10. Cranmer, M., et al.: Discovering symbolic models from deep learning with inductive biases. In: NeurIPS, vol. 33, pp. 17429–17442 (2020)
11. Ding, S., Li, H., Su, C., Yu, J., Jin, F.: Evolutionary artificial neural networks: a review. Artif. Intell. Rev. **39**(3), 251–260 (2013)
12. Elsken, T., Metzen, J.H., Hutter, F.: Neural architecture search: a survey. JMLR **20**(1), 1997–2017 (2019)
13. FitzHugh, R.: Impulses and physiological states in theoretical models of nerve membrane. Biophys. J . **1**(6), 445–466 (1961)
14. Kramer, O., Kramer, O.: Genetic Algorithms. Springer, Cham (2017)

15. Lim, B., Zohren, S.: Time-series forecasting with deep learning: a survey. Phil. Trans. R. Soc. A **379**(2194), 20200209 (2021)
16. Lorenz, E.N.: Deterministic nonperiodic flow. J. Atmos. Sci. **20**(2), 130–141 (1963)
17. Luo, Y., Xu, C., Liu, Y., Liu, W., Zheng, S., Bian, J.: Learning differential operators for interpretable time series modeling. In: ACM SIGKDD, pp. 1192–1201 (2022)
18. Makridakis, S., Hibon, M.: ARMA models and the Box-Jenkins methodology. J. Forecast. **16**(3), 147–163 (1997)
19. Park, N., Kim, M., Hoai, N.X., McKay, R.B., Kim, D.K.: Knowledge-based dynamic systems modeling: a case study on modeling river water quality. In: ICDE, pp. 2231–2236. IEEE (2021)
20. Van der Pol, B.: LXXXVIII. On "relaxation-oscillations". The London, Edinburgh, and Dublin Philos. Mag. J. Sci. **2**(11), 978–992 (1926)
21. Raissi, M., Perdikaris, P., Karniadakis, G.E.: Physics-informed neural networks: a deep learning framework for solving forward and inverse problems involving nonlinear partial differential equations. J. Comput. Phys. **378**, 686–707 (2019)
22. Sahoo, S., Lampert, C., Martius, G.: Learning equations for extrapolation and control. In: ICML, pp. 4442–4450. PMLR (2018)
23. Salinas, D., Flunkert, V., Gasthaus, J., Januschowski, T.: DeepAR: probabilistic forecasting with autoregressive recurrent networks. Int. J. Forecast. **36**(3), 1181–1191 (2020)
24. Schmidt, M., Lipson, H.: Distilling free-form natural laws from experimental data. Science **324**(5923), 81–85 (2009)
25. Stanley, K.O., Miikkulainen, R.: Evolving neural networks through augmenting topologies. Evol. Comput. **10**(2), 99–127 (2002)
26. Tealab, A.: Time series forecasting using artificial neural networks methodologies: a systematic review. Future Comput. Inform. J. **3**(2), 334–340 (2018)
27. Torres, J.F., Hadjout, D., Sebaa, A., Martínez-Álvarez, F., Troncoso, A.: Deep learning for time series forecasting: a survey. Big Data **9**(1), 3–21 (2021)
28. Wang, R., Maddix, D., Faloutsos, C., Wang, Y., Yu, R.: Bridging physics-based and data-driven modeling for learning dynamical systems. In: Learning for Dynamics and Control, pp. 385–398. PMLR (2021)
29. Wang, R., Robinson, D., Faloutsos, C., Wang, Y.B., Yu, R.: AutoODE: bridging physics-based and data-driven modeling for COVID-19 forecasting. In: NeurIPS 2020 Workshop on Machine Learning in Public Health (2020)
30. Zhou, X., Qin, A.K., Gong, M., Tan, K.C.: A survey on evolutionary construction of deep neural networks. IEEE Trans. Evol. Comput. **25**(5), 894–912 (2021)

DEAL: Data-Efficient Active Learning for Regression Under Drift

Béla H. Böhnke(✉), Edouard Fouché, and Klemens Böhm

Karlsruhe Institute of Technology, Karlsruhe, Germany
bela.boehnke@kit.edu

Abstract. Current work on Active Learning (AL) tends to assume that the relationship between input and target variables does not change, i.e., the oracle is static. However, oracles can be stream-like and exhibit concept drift, which requires updating the learned relationship. Standard drift detection and adaption methods rely on constantly observing the target variables, which is too costly in AL. Current work on AL for regression has not addressed the challenge of frequently drifting oracles. We propose a new AL method that estimates its error due to drift by learning statistics about how often and how severe drift occurs, based on a Gaussian Process model with a time-variant kernel. Whenever the estimated error reaches a user-required threshold, our model measures the target variables and recalibrates the learned relationship as well as the drift statistics. Our drift-aware model requires up to 20 times fewer measurements than widely used methods.

Keywords: Concept Drift · Active Learning · Regression

1 Introduction

Active Learning (AL) refers to data collection methods aimed at estimating the relationship between input and target variables under the assumption that measurement of the variables is expensive. Therefore, AL seeks to estimate the relationship with as few measurements as possible. Our setting, in particular, is the one of stream-based AL [26]. Stream-based AL assumes input variables arrive as a continuous stream X, e.g., sensor data like temperature and humidity acquired for real-time environmental monitoring. A *learner* observes the stream at time t_i, resulting in one *potential query* x_i. The learner then decides whether to perform a *query*, i.e., to measure the target variable Y like soil quality for agriculture, or not. While the cost of observing the stream X is negligible, measuring the target is expensive. Many scenarios share this assumption [26], such as industrial process control, resource allocation for environmental monitoring, or demand forecasting for energy management [4].

AL methods often omit details of how the target variable is measured. This abstraction is called *oracle*. In practice, such oracles can *drift*: Evolving, unobservable environmental parameters can affect the relationship between input and

D.-N. Yang et al. (Eds.): PAKDD 2024, LNAI 14650, pp. 188–200, 2024.
https://doi.org/10.1007/978-981-97-2266-2_15

target variables over time. This phenomenon of changing relationships between variables is known as *concept drift (CD)*.

CD poses a significant challenge to statistical models, requiring frequent recalibration to maintain estimation error below a user-required threshold. However, conventional methods for drift correction, like continuous model recalibration or drift detection by monitoring the target variables, are impractical for AL due to the high cost of observing the target variable.

(1) The first challenge arises from this mismatch between the goal of AL of reducing measurements and the need for additional measurements for model recalibration, which requires an AL method that can adjust for drift without continuously monitoring the drifting variables. While such drift-correcting AL methods exist for classification tasks, many real-world scenarios involve continuous target variables requiring regression models. These regression scenarios lack adequate AL methods [25,31], highlighting a gap in current AL research.

(2) The second challenge becomes clear when comparing classification and regression tasks under drift. Classification only has to monitor a fixed number of class distributions. In contrast, regression has to estimate the model error due to drift for each point in a continuous input space, target space, and time.

(3) The final challenge is an appropriate selection of the measurement time based on the estimated error, ensuring that the error due to drift remains below a user-required threshold.

We contribute by proposing DEAL, the first AL regression method that learns data efficiently under oracles that exhibit frequent drift while keeping the estimation error below a user-required threshold. DEAL estimates its own error due to drift, using a Gaussian process model (*GP*) with a time-variant kernel. The *GP* learns the *drift behavior*, i.e., the time-dependent distribution of the target variable. Figure 1 shows that DEAL queries the oracle whenever the estimated error exceeds a user-required threshold. DEAL uses the new data to recalibrate,

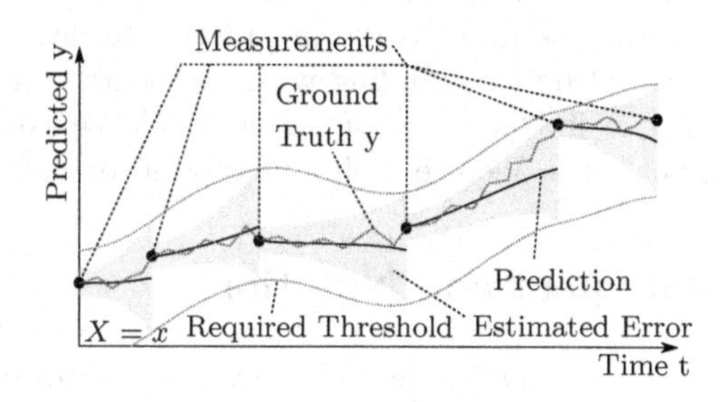

Fig. 1. DEAL estimates its error and measures once the error reaches the user-required threshold. The input value x of the data stream X is kept constant for better visualization, the target y still varies due to frequent incremental drift.

and combines it with the old data to update the learned drift behavior. In this way, DEAL minimizes the number of measurements required for drift correction.

We evaluate DEAL against multiple baselines, on multiple drift-affected time series, and provide the code[1] for reproduction.

2 Related Work

To our knowledge, existing stream-based AL methods for regression (e.g., [1,10, 12,25,28,29,31,32]) do not consider drift, i.e., the oracle is assumed to be static. Such methods stop learning once the regression model performs well and never start learning again, even if the model performance decreases due to drift. We show this undesirable behavior in our experiments.

There exist AL methods that consider drift [13–16,18,19,22,24,27,33,35–37], but they are restricted to classification. Further, to apply [14,18,22,24,35,37] a user needs to set a measurement budget and additional parameters without knowing drift behavior, and the resulting estimation error. Improper set parameters lead to either too costly or inaccurate models. Drift behavior that changes causes the same problem because the methods cannot adapt. Further, those methods primarily monitor the distribution of input variables per class. This is intractable for regression and impedes the transfer to the regression case.

There exist change detection methods that can adapt to changing drift behavior. For example, AAIL [24] detects changes in the input variables and measures the target variable at each change. While the authors claim that AAIL adapts to concept drift (CD), it can only adapt to covariance shift. Covariance shift only refers to changes in the distribution of the input variables, which is cheap to observe, while CD describes a change in the relationship between the input and target variables. AAIL can be adapted to regression tasks, and we include an adaptation in our experiments. Drift detection approaches not designed for AL, as surveyed in [7,9,20,34], typically assume that the target variables are cheap to observe, which violates a basic assumption of AL. Thus, such approaches require additional methods for strategic under-sampling of the target variable, essentially what the existing AL techniques for drifting oracles do.

In summary, the only methods viable in practice to deal with drift in AL regression is some form of under-sampling with consecutive measurements at a user-defined frequency, as in [4–6]. A poor frequency choice leads to missed drift or higher measurement costs. We use such a method as one of several baselines.

3 Problem Statement and Notation

Random Functions $F : x \mapsto F(x)$ associate each value x_i of a variable $x \in \mathbb{R}$ with a random variable $F(x_i)$ [3]. As such, a random function extends the notion of random variables to a continuous function space. *Sample Functions* $f(x) \leftarrow F(x)$ are functions $f : x \mapsto f(x)$ drawn from a random function $F(x)$.

[1] https://github.com/bela127/alsbts-experiments.

Stochastic Processes S or random processes are random functions $S : t \mapsto S(t)$ where $t \in \mathbb{T}$ is interpreted as time with domain $\mathbb{T} = \mathbb{R}_+$ [3]. *Brownian Motions* B are stochastic processes $B : t \mapsto B(t)$ [3] characterized by random increments $\delta B(\delta t) = B(t + \delta t) - B(t) \forall t$, where these increments follow the normal distribution $\delta B(\delta t) \sim \mathcal{N}(0, \delta t)$. *Gaussian Processes* (*GP*) are stochastic processes defined as $GP(x) \sim \mathcal{N}(m(x), v(x))$, illustrating that the random function $GP(x)$ comes from a normal distribution with mean $m(x)$ and variance $v(x)$ both depending on x [17]. One often uses *GP*s as a probabilistic prior over functions.

Kernel Functions $k(x, x')$, also called covariance functions, can define a *GP* instead of using a mean and variance function. *The Radial Basis Function (RBF) Kernel* $k(x, x') = v \cdot \exp\left(-\frac{(x-x')^2}{2 \cdot l^2}\right)$, has two parameters: $v \in \mathbb{R}_+$ (target variance) scaling the target of the random function, and l (length scale) determining function smoothness. *The Brownian (Bridge) Kernel* $k(t, t') = v_b \cdot \min(t, t')$, with points in time t, t' has a variance parameter $v_b \in \mathbb{R}_+$, which translates to the drift speed, i.e., it scales $\delta B(\delta t)$ by v_b [17].

Observed Data Streams or time series are sample functions $x(t) \leftarrow X \mid t \in \mathbb{T}$. Here, X is a random input variable, called the *stream*, and $x(t)$ provides a distinct value $x_i = x(t_i)$, drawn from X at a point in time t_i. *Covariance Shift* is present if the distribution of the stream changes over time. Such a time-dependent stream is, in fact, a random process, and one writes $X(t)$.

Concepts, i.e., the relations between input and target variables, are represented as random functions $C : x \mapsto C(x) = Y$. If a concept depends on time, it is represented as $C : (x, t) \mapsto C(x, t)$. By definition, *Concept Drift (CD)* is present if $\exists t_1, t_2 : C(x, t_1) \neq C(x, t_2)$, where $t_1 \neq t_2$ are two points in time [7].

Oracles are models for data-generating processes with concept C. A *query* is a request $q(t_i) \leftarrow Q$ to an oracle to return the current value $y_i = c_i \leftarrow C(q(t_i), t_i)$ of the target variable. Q is the stream of performed queries. The input stream X is the stream of possible queries.

Stream-based active learning (AL) [26] iteratively observes the value x_i of the stream X and only performs a query if the uncertainty v_{est} of the prediction y_{est} exceeds a certain user-required threshold v_{target}. The learner then recalibrates the model using the resultant measurement y_i. This is known as *uncertainty sampling*, which is a typical way to decide whether to query or not.

Algorithm 1 The Common vs Our Adapted Stream-Based AL-Cycle

```
1: procedure COMMON(v_target, MODEL)
2:   while running do
3:     x_i ← X
4:     y_est := MODEL.ESTIMATE(x_i)
5:                          ▷ y_est only for evaluation.
6:     v_est := MODEL.VARIANCE(x_i)
7:     if v_est >= v_target then
8:                          ▷ Uncertainty sampling.
9:       y_i := ORACLE.QUERY(x_i)
10:      DATAPOOL.ADD(x_i, y_i)
11:      MODEL.TRAIN(DATAPOOL)
```

```
1: procedure ADAPTED(v_target, MODEL)
2:   while running do
3:     x_i, t_i ← X,TIME()
4:     y_est := MODEL.ESTIMATE(x_i, t_i)
5:                          ▷ y_est only for evaluation.
6:     v_est := MODEL.VARIANCE(x_i, t_i)
7:     if v_est >= v_target then
8:                          ▷ Uncertainty sampling.
9:       y_i := ORACLE.QUERY(x_i, t_i)
10:      DATAPOOL.ADD(x_i, t_i, y_i)
11:      MODEL.TRAIN(DATAPOOL)
```

4 Our Method: DEAL

4.1 The Adapted Stream-Based AL Cycle

We slightly modify the Stream-based active learning cycle (see Algorithm 1), so that in addition to observing the stream X, we also observe the current time t_i. DEAL's estimation model then takes the time t_i into account when estimating its variance v_{est} to include the additional uncertainty caused by drift (Lines 3 to 6). Whenever the uncertainty reaches the threshold, DEAL recalibrates (Line 9 to 11) with a new data point (x_i, y_i). Here, the drifting oracle provides y_i.

To estimate the uncertainty caused by drift, we require one assumption about the drift behavior: We assume that drift occurs frequently, i.e., within a time series with length t_{end}, at least n_c changes occur. Here n_c needs to be large enough to learn sufficient statistics of the drift behavior.

4.2 Our Drift-Aware Estimation Model

In contrast to methods like discussed in [14,18,22,37], which perform measurements without modeling the drift behavior, we are the first to learn statistics about the drift behavior. These statistics enable us to measure in adaptive time intervals in which drift of a certain magnitude may occur. We derive the statistics from a Gaussian process model (*GP*) according to the following prior:

Definition 1. *The Brownian drift prior is given by* $C(x,t) = I(x) + W(x) * B(t)$, *with a Brownian motion prior* $B(t)$ *as drift behavior, and random function priors* $I(x)$ *and* $W(x)$ *independent of any drift thereby constant over* t. *Here,* $W(x)$ *is a weighting term that defines the impact of* $B(t)$ *on* $C(x,t)$ *at a position* x.

The intuition behind using a Brownian prior $B(t)$ for the drift behavior is that the combination of many, small, random, and independent external influences results in a combined Brownian overall drift. Further, this model can capture drift with larger changes after a random time, as long as changes occur frequently. Such frequent drift is common in practice, like: (1) Drift due to displacement of machine elements caused by vibration. (2) Drift in large networks, such as the electrical grid, where nodes can (dis)connect from the network at random times.

We instantiate the *GP* with a kernel composition according to the three components $I(x)$, $W(x)$, $B(t)$, from Definition 1, i.e., one kernel per component. We model the drift behavior $B(t)$ with the Brownian kernel [17]. Because of its universal approximation property [21], we use distinct RBF kernels to model $I(x)$ and $W(x)$. In general, the choice of kernels is a parameter that one can easily tune to match prior knowledge about the data or drift. For the sake of generality, we stick to our choice in this study. Further, we enforce that both RBF kernels have the same length scale and variance, which reduces the number of learnable parameters and makes learning more stable. This reduction assumes similar smoothness of $I(x)$ and $W(x)$ and the resulting learned function. DEAL's parameters are Brownian variance v_b, RBF variance v_r, RBF length scale l_r.

The only hyperparameter DEAL takes is a user-required threshold v_{target}. DEAL trains according to Algorithm 1. Every time the variance v_{est} estimated by DEAL becomes greater than v_{target}, DEAL measures the target value (Line 9) and adds it to the training set. DEAL then estimates the most likely kernel parameters with GPy's [8] gradient-based maximum likelihood optimizer (Line 11). We use the standard optimizer configuration with 5 restarts, 4 times with random kernel parameters, and once with the most likely kernel parameters from the previous iteration. In the first iteration, we initialize the kernel parameters v_b, v_r, and l_r with random values and use an initial training set of measurements from the first 10 time steps. We use this initialization for each baseline as well.

In AL, training complexity tends to be neglected, because the oracle usually is much more expensive than the active learning decision-making. We use the standard *GP* model from GPy, with a complexity of $O(n^2)$, where n is the training set size. Since n is kept small, the actual runtime is consistently low.[2]

5 Experimental Design

5.1 Baselines

We consider the following baselines as competitors for DEAL:

Consecutive Measurement (CM): This approach carries out measurements at regular user-specified time intervals of size δt_{meas}. The approach is sensitive to δt_{meas} and does not adapt to the data-generating process. In our experiments, we evaluated values of $\delta t_{meas} \in [1, 20]$, with logarithmic increments.

Classic AL (CAL): This standard AL approach with uncertainty sampling yields an estimate and an estimation variance. It uses a Gaussian process (*GP*) with an RBF kernel and kernel parameters length scale l and variance v. Unlike CM, the *GP* can automatically adjust these parameters using maximum likelihood estimation in the same way DEAL does. But unlike DEAL, this approach does not model the data behavior over time, i.e., it assumes a static oracle. To obtain a strong baseline, we initialized the *GP* with kernel parameters identical to the parameters of the ground truth data, see Sect. 5.2. Given enough data, the *RMSE* of such an approach approaches zero on a stream with no noise or drift.

Change Ideal (CI): This approach is an adaptation of AAIL [24], which uses change detection on the input variables X. Whenever CI detects a change in X, it measures y. To make our evaluation independent of any specific detector and

[2] Note one can reduce the complexity of DEAL down to $O(n \cdot i)$ (with learning epochs $i \ll n$), by using gradient-based *GP* models and batch training. Further, one can cap the number of measurements used for training, reducing the factor n to a constant.

obtain a strong baseline, we simulate an "ideal" detector. It uses the (normally unobservable) ground truth to correctly and immediately report any change in X. This baseline has no configurable parameters.

Change Error (CE): There are three types of errors in change detectors: undetected change, detection without an actual change, and delayed detection. To study their effect, we created variants of CI using an imperfect "pseudo" change detector which lets us control each error type with three parameters: p_{wrong} is the proportion of spurious detection at any time. p_{miss} is the chance of discarding a correct detection. std_{offset} is the standard deviation of a normal distribution. For any detected change, we take the absolute value of a random point from this distribution as an offset to delay detection.

In our experiments, we vary these parameters independently according to the given interval and step size while keeping the others at the given default value:

	Interval	Step Size	Default
p_{wrong}	$[0, 0.10]$	0.005	0.015
p_{miss}	$[0, 0.80]$	0.05	0.015
std_{offset}	$[0, 15]$	1	1

5.2 Evaluation Data

Stream mining frameworks such as MOA [2] and River [23] focus on stream classification, as indicated by the streams they offer. We investigate regression which is why we require time series of the form: $gt(t) = ((t, x(t)), c(x(t), t))$. Here $t \in \mathbb{T}$ is the time, $x(t) \leftarrow X \mid t \in \mathbb{T}$ the observed input stream, and $c(x(t), t) \mid t \in \mathbb{T}$ the observed values of the target variable.

We generate $c(x(t), t))$ with a Gaussian process (GP) according to the following priors, where $C_{sin}(x(t), t))$ and $C_{rbf}(x(t), t))$ are adaptions from classification to regression used in [11] and [30], and similar to River RBF streams:

$$C_b(x(t), t) = RBF(x \mid l_{gr}, v_{gr}) + RBF(x \mid l_{gr}, v_{gr}) * B(t \mid v_{gb}) \quad (1)$$

$$C_{sin}(x(t), t) = RBF(x \mid l_{gr}, v_{gr}) + RBF(x \mid l_{gr}, v_{gr}) * Sin(t \mid s * v_{gb}) \quad (2)$$

$$C_{rbf}(x(t), t) = RBF(x \mid l_{gr}, v_{gr}) + RBF(x \mid l_{gr}, v_{gr}) * RBF(t \mid s * v_{gb}) \quad (3)$$

$$\begin{aligned} C_{mix}(x(t), t) = {} & RBF(x \mid l_{gr}, v_{gr}) + RBF(x \mid l_{gr}, v_{gr}) * RBF(t \mid s * v_{gb}) \\ & + RBF(x \mid l_{gr}, v_{gr}) * Sin(t \mid s * v_{gb}) \\ & + RBF(x \mid l_{gr}, v_{gr}) * B(t \mid v_{gb}) \end{aligned} \quad (4)$$

$$\begin{aligned} C_{bmix}(x(t), t) = {} & RBF(x \mid l_{gr}, v_{gr}) + RBF(x \mid l_{gr}, v_{gr}) * B(t \mid v_{gb}) \\ & + RBF(x \mid l_{gr}, v_{gr}) * B(t \mid v_{gb}) \\ & + RBF(x \mid l_{gr}, v_{gr}) * B(t \mid v_{gb}) \end{aligned} \quad (5)$$

Here l_{gr}, v_{gr}, v_{gb} are kernel parameters, chosen as follows: $l_{gr} = 0.1$; $v_{gr} = 0.25$; $v_{gb} \in \{0, 0.005, 0.01, 0.02\}$. From these parameters v_{gb} controls the drift speed, $s = 2000$ scales v_{gb} so that each prior has a similar drift speed for a given v_{gb}. We visualize example time series drawn from such priors in Fig. 2. For evaluation one time series is $t_{end} = 1000$ simulation units (su) long. The input stream $x(t)$ changes $n_c \in \{50, 100, 200, 400\}$ times within this 1000 su time frame, with the time of a change $t_c \leftarrow U[0, t_{end}]$ and $x(t_c) \leftarrow U[-1, 1]$. While tested on the $[-1, 1]$ range our method is compatible with any value range.

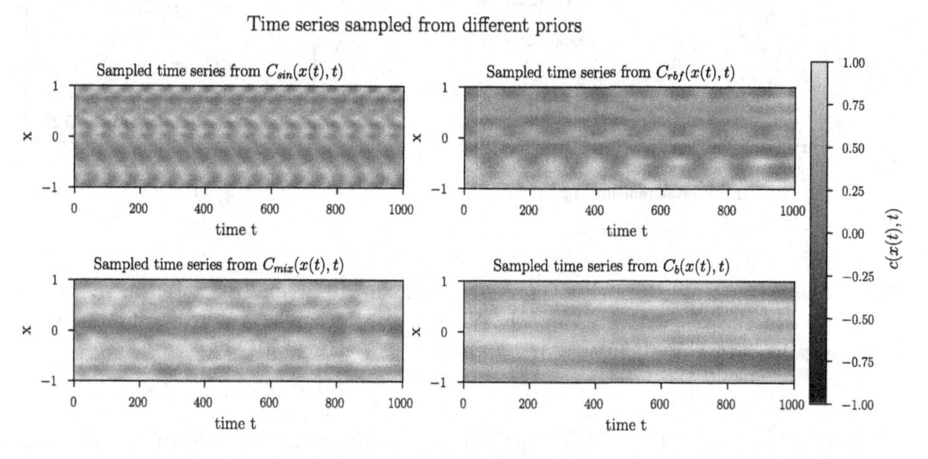

Fig. 2. Example time series sampled from the random function priors. Here, we show $c(x(t), t))$ normalized for better visualization.

5.3 Evaluation Metrics

For DEAL and every baseline on every dataset, we perform the following: We evaluate all parameter configurations 50 times, each time on a different time series. For each time series, we compute the *RMSE* of the estimation y_{est} against the ground truth and the total number of measurements N_m performed across each time series. We plot the *RMSE* over the N_m for all different configurations and different time series. Additionally, we calculated the percentage of measurements a competitor (DEAL) saves compared to a baseline (the CM baseline):

Definition 2. *The saved data is* $sd = (Mb(e) - Mc(e))/Mb(e)$*, where* $Mb(e)$ *and* $Mc(e)$ *are the number of measurements a baseline and a competitor need to reach the same mean RMSE value e for the first time.*

6 Evaluation

6.1 Comparison of DEAL Against Baselines

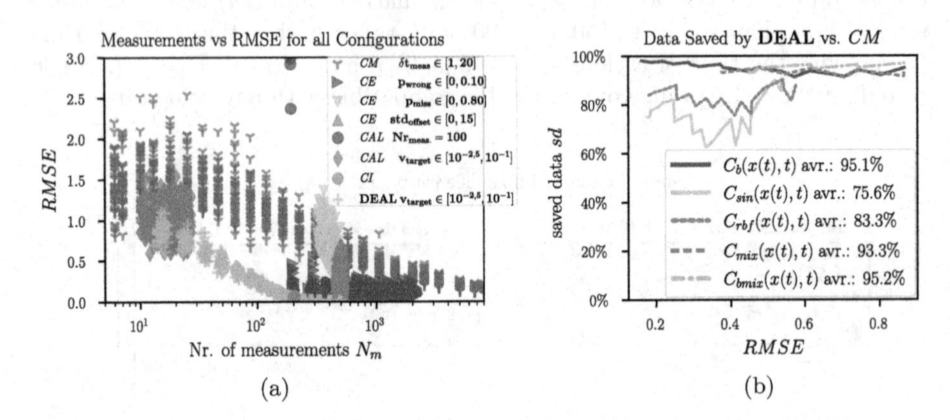

Fig. 3. Baseline comparison; Fig. 3a the relationship between N_m and $RMSE$ for all parameters of the respective approach given in the legend and across different time series; Fig. 3b data gain against CM baseline.

Figure 3a shows the $RMSE$ against the number of measurements N_m for DEAL and the four baselines (with two variants of CAL and three variants of CE) across a variety of parameter configurations. We can see that DEAL (+) needs on average fewer measurements to achieve lower $RMSE$ than any configuration of any baseline (e.g., $N_m = 100$ for average $RMSE = 0.25$). Next, DEAL has less variance across the 50 time series than the baselines. This means for any fixed v_{target}, DEAL adapts better to the individual behavior of a time series, while the baselines depend on how well their parameters match the time series behavior. The ideal change detector CI (●) shows a similar error as DEAL at $N_m = 200$. But, CI's measurement count depends directly on the input stream's change frequency. Thus, CI will still carry out the same number of measurements (more than needed), even if a higher error would be acceptable. Further, as soon as the change detector is imperfect, as with CE (with offset ▲, missed ▼ and wrong ▶ detections) we observe a sharp increase in $RMSE$ and its variance. The CAL approach (♦) fails because once it reaches the given error threshold, it stops collecting data, never aware of any possible drift. Forcing CAL to collect 100 measurements regardless of the error threshold (●) causes it to learn an average value, rather than the correct relationship between the variables.

The CM baseline is the only baseline that allows users to indirectly control the $RMSE$ by choosing the measurement time interval δt_{meas}. DEAL can directly control the user-required error thresholds v_{target}. For any other baseline, control of the resulting estimation error via a parameter is not possible. Figure 3b shows for a given $RMSE$ how much data DEAL can save compared to CM. In regions of low error ($RMSE \in [0.1, 0.35]$), DEAL saves over 95% data compared to *CM*. Moving to regions of higher error ($RMSE > 0.5$), we observe a slow decline in saved data. We hypothesize that this is because the larger the allowed error, the less benefit a precisely selected measurement (time based on drift behavior) has, compared to manual selection δt_{meas} of *CM*. The drop in saved data ($RMSE \in [0.3, 0.45]$) for C_{sin} and C_{rbf} is due to their periodic nature, which is why *CM* performs acceptable even without dynamic adaption.

6.2 Impact of the User-Required Error Threshold

The true estimation error *RMSE* should be close to the user-required standard deviation $std_{target} = \sqrt{v_{target}}$, so that the user can trust the model predictions. In Fig. 4a, the dotted line shows such proportional behavior. For small thresholds $std_{target} < 0.25$, DEAL overestimates the actual error. For higher errors $std_{target} > 0.25$, it underestimates the error. At all times, the actual error is close to the given threshold and behaves nearly proportionally as required.

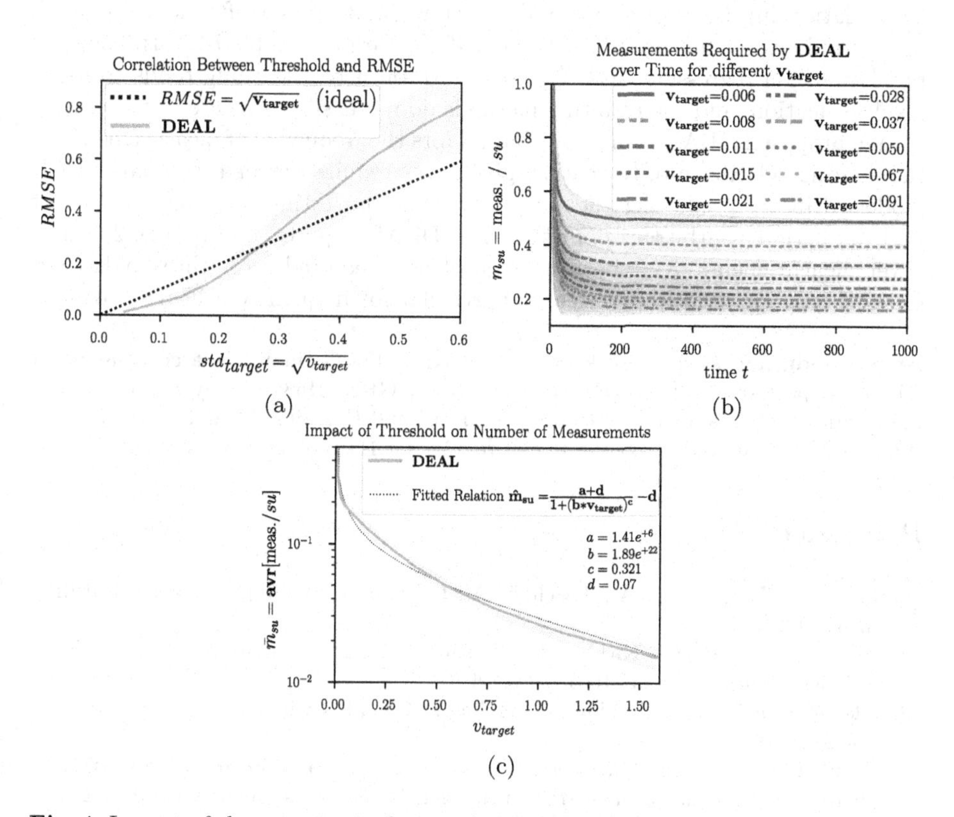

Fig. 4. Impact of the user-required error threshold; Fig. 4a average RMSE for a given threshold; 4b average amount of measurements per time step (for different thresholds); Fig. 4c average amount of required measurements for a given threshold.

Figure 4b shows the number of measurements DEAL performs per time step (*simulation unit su*), depending on the user-required error threshold. In the early stages (the first 100 su), DEAL performs more measurements because it needs to calibrate. Roughly after $t_{conv} = 200\ su$, it takes a constant number of measurements per su, just enough to reach the error threshold. The variance of the number of measurements tends to decrease with time. Namely, the more data DEAL has seen, the closer its learned parameters are to the true ones.

Figure 4c shows how many measurements per su ($\bar{m}_{su}$) DEAL requires to reach a given error threshold v_{target}. As expected, if a user requires a lower estimation error, more measurements are needed to reach that error. To provide an estimation (dotted line) of this relation, we used a genetic function fitter. This relationship aids in estimating the required measurements and associated costs of reaching the user-required error threshold.

7 Conclusion

The relationship between input and target variables may drift due to environmental influences that are not observed. If the target variable is continuous, its prediction is a regression task. Current work on active learning tends to focus on classification, and the resulting methods do not easily translate to regression.

We proposed DEAL, a method that adapts the frequency of measurements to the drifting relationship, to reach a given user-required error threshold. DEAL models drift by predicting the target variable and estimating the variance of that prediction at arbitrary points in time. DEAL requires, on average, 20 times fewer measurements over the full range of user-required error thresholds than the CM baseline used in practice, in particular for frequently drifting streams.

Acknowledgments. This work was supported by the German Research Foundation (DFG) as part of the Research Training Group GRK 2153: Energy Status Data - Informatics Methods for its Collection, Analysis, and Exploitation and by the Baden-Württemberg Foundation via the Elite Program for Postdoctoral Researchers.

References

1. Bachman, P., Sordoni, A., Trischler, A.: Learning algorithms for active learning. In: ICML (2017)
2. Bifet, A., Holmes, G., Kirkby, R., Pfahringer, B.: MOA: massive online analysis. J. Mach. Learn. Res. **11**, 1601–1604 (2010)
3. Klebaner, F.C.: Introduction to Stochastic Calculus with Applications. WSPC, London (2005)
4. Carne, G.D., Buticchi, G., Liserre, M., Vournas, C.: Load control using sensitivity identification by means of smart transformer. IEEE Trans. Smart Grid **9**(4), 2606–2615 (2018)
5. Carne, G.D., Buticchi, G., Liserre, M., Vournas, C.: Real-time primary frequency regulation using load power control by smart transformers. IEEE Trans. Smart Grid **10**(5), 5630–5639 (2019)

6. Han, D., Ma, J., He, R., Dong, Z.Y.: A real application of measurement-based load modeling in large-scale power grids and its validation. IEEE Trans. Power Syst. **24**, 1756–1764 (2009)
7. Gama, J., Zliobaite, I., Bifet, A., Pechenizkiy, M., Bouchachia, A.: A survey on concept drift adaptation. ACM Comput. Surv. **46**, 1–37 (2014)
8. GPy: A Gaussian process framework in Python. http://github.com/SheffieldML/GPy. (Since 2012)
9. Iwashita, A.S., Papa, J.P.: An overview on concept drift learning. IEEE Access **7**, 1532–1547 (2019)
10. Iwata, T.: Active learning for regression with aggregated outputs. CoRR (2022)
11. Gama, J., Medas, P., Castillo, G., Rodrigues, P.: Learning with drift detection. In: Bazzan, A.L.C., Labidi, S. (eds.) SBIA 2004. LNCS (LNAI), vol. 3171, pp. 286–295. Springer, Heidelberg (2004). https://doi.org/10.1007/978-3-540-28645-5_29
12. Konyushkova, K., Sznitman, R., Fua, P.: Learning active learning from data. In: NIPS (2017)
13. Krawczyk, B., Cano, A.: Adaptive ensemble active learning for drifting data stream mining. In: IJCAI (2019)
14. Krawczyk, B., Pfahringer, B., Wozniak, M.: Combining active learning with concept drift detection for data stream mining. In: IEEE BigData (2018)
15. Kurlej, B., Wozniak, M.: Learning curve in concept drift while using active learning paradigm. In: ICAIS (2011)
16. Kurlej, B., Wozniak, M.: Active learning approach to concept drift problem. Log. J. IGPL **20**(3), 550–559 (2012)
17. Lindgren, G., Rootzen, H., Sandsten, M.: Stationary Stochastic Processes for Scientists and Engineers. T&F (2013)
18. Liu, S., et al.: Online active learning for drifting data streams. IEEE Trans. Neural Networks Learn. Syst. **34**(1), 186–200 (2023)
19. Liu, W., Zhang, H., Ding, Z., Liu, Q., Zhu, C.: A comprehensive active learning method for multiclass imbalanced data streams with concept drift. Knowl. Based Syst. **215**, 106778 (2021)
20. Lu, J., Liu, A., Dong, F., Gu, F., Gama, J., Zhang, G.: Learning under concept drift: a review. IEEE Trans. Knowl. Data Eng. **31**(12), 2346–2363 (2019)
21. Micchelli, C.A., Xu, Y., Zhang, H.: Universal kernels. J. Mach. Learn. Res. **7**, 2651–2667 (2006)
22. Mohamad, S., Sayed Mouchaweh, M., Bouchachia, A.: Active learning for data streams under concept drift and concept evolution. In: ECML-PKDD (2016)
23. Montiel, J., et al.: River: machine learning for streaming data in Python. JMLR **22**, 1–8 (2021)
24. Park, C.H., Kang, Y.: An active learning method for data streams with concept drift. In: IEEE BigData (2016)
25. Riquelme, C., Johari, R., Zhang, B.: Online active linear regression via thresholding. In: AAAI (2017)
26. Settles, B.: Active Learning. Springer, Cham (2012). https://doi.org/10.1007/978-3-031-01560-1
27. Shan, J., Zhang, H., Liu, W., Liu, Q.: Online active learning ensemble framework for drifted data streams. IEEE Trans. Neural Networks Learn. Syst. **30**(2), 486–498 (2019)
28. Stefano, M., Bruno, S.: An active-learning algorithm that combines sparse polynomial chaos expansions and bootstrap for structural reliability analysis. Struct. Saf. **75**, 67–74 (2018)

29. Lookman, T., Balachandran, P.V., Xue, D., Yuan, R.: Active learning in materials science with emphasis on adaptive sampling using uncertainties for targeted design. NPJ Comput. Mater. **5**, 21 (2019)
30. Viktor, L., Barbara, H., Wersing, H.: KNN classifier with self adjusting memory for heterogeneous concept drift. In: ICDM (2016)
31. Wu, D., Lin, C., Huang, J.: Active learning for regression using greedy sampling. Inf. Sci. **474**, 90–105 (2019)
32. Yoo, D., Kweon, I.S.: Learning loss for active learning. In: CVPR (2019)
33. Zhang, H., Liu, W., Shan, J., Liu, Q.: Online active learning paired ensemble for concept drift and class imbalance. IEEE Access **6**, 73815–73828 (2018)
34. Zliobaite, I.: Learning under concept drift: an overview. CoRR (2010)
35. Zliobaite, I., Bifet, A., Holmes, G., Pfahringer, B.: MOA concept drift active learning strategies for streaming data. In: WAPA. JMLR Proceedings (2011)
36. Zliobaite, I., Bifet, A., Pfahringer, B., Holmes, G.: Active learning with evolving streaming data. In: ECML/PKDD (3) (2011)
37. Zliobaite, I., Bifet, A., Pfahringer, B., Holmes, G.: Active learning with drifting streaming data. IEEE Trans. Neural Networks Learn. Syst. **25**(1), 27–39 (2014)

Evolving Super Graph Neural Networks for Large-Scale Time-Series Forecasting

Hongjie Chen[1], Ryan Rossi[2], Sungchul Kim[2], Kanak Mahadik[2], and Hoda Eldardiry[1](✉)

[1] Virginia Tech, Blacksburg, VA 24060, USA
{jeffchan,hdardiry}@vt.edu
[2] Adobe Research, San Jose, CA 95110, USA
{ryrossi,sukim,mahadik}@adobe.com

Abstract. Graph Recurrent Neural Networks (GRNN) excel in time-series prediction by modeling complicated non-linear relationships among time-series. However, most GRNN models target small datasets that only have tens of time-series or hundreds of time-series. Therefore, they fail to handle large-scale datasets that have tens of thousands of time-series, which exist in many real-world scenarios. To address this scalability issue, we propose Evolving Super Graph Neural Networks (ESGNN), which target large-scale datasets and significantly boost model training. Our ESGNN models multivariate time-series based on super graphs, where each super node is associated with a set of time-series that are highly correlated with each other. To further precisely model dynamic relationships between time-series, ESGNN quickly updates super graphs on the fly by using the LSH algorithm to construct the super edges. The embeddings of super nodes are learned through end-to-end learning and are then used with each target time-series for forecasting. Experimental result shows that ESGNN outperforms previous state-of-the-art methods with a significant runtime speedup (3×–40× faster) and space-saving (5×–4600× less), while only sacrificing little or negligible prediction accuracy. An ablation study is also conducted to investigate the effectiveness of the number of super nodes and the graph update interval.

Keywords: Graph Neural Networks · Time-series Forecasting · Evolving Graph Modeling

1 Introduction

Graph Recurrent Neural Networks (GRNN) have proved successful for accurate multi-variate times-series forecasting in various applications such as traffic forecasting, crime trends prediction, and website view forecasting, among many others [3,10,17,21]. Existing GRNN models utilize pairwise relationships between time-series in graph structures by associating each node with a time-series. Nevertheless, most models experience quadratic growth in runtime and space demand as the number of node time-series increases [12,21], making them

D.-N. Yang et al. (Eds.): PAKDD 2024, LNAI 14650, pp. 201–212, 2024.
https://doi.org/10.1007/978-981-97-2266-2_16

impractical for large-scale data. Moreover, existing GRNN models only construct a static graph and use the same graphs across time, which fails to address the discrepancy of node connectivity between different time periods.

In this paper, we propose the Evolving Super Graph Neural Networks (ESGNN), which aim to address time inefficiency, scalability, and space limitations of existing graph-based time-series forecasting models for large-scale datasets. Our proposed Evolving Super Graph Neural Networks (ESGNN) address limitations of the current state-of-the-art methods, including: (1) Most GRNN multi-variate time-series forecasting methods rely on a graph structure built with external features such as geographic information. However, the availability of geo-information is not guaranteed, and it may not accurately reflect the actual connections between nodes. For instance, two closely located compute nodes may belong to different cloud clusters, resulting in weak or negligible correlation between them. Therefore, modeling their relation does not contribute to forecasting accuracy. On the other hand, connections between highly correlated nodes may be missing in the datasets and should be taken into consideration. In our model, we assume that compute nodes in the same computing cluster share correlated patterns, and we consequently derive the cluster information by a time-series similarity-based clustering algorithm (e.g., LSH). The geography-based graph construction of previous methods does not embed the node correlation relationships and loses useful information. By contrast, graphs in our model are built with a time-series metric where time-series observations are taken to derive the correlation values between nodes. (2) Current GRNN methods only rely on a static graph throughout the entire time span of the time-series, neglecting the dynamically evolving connections between nodes. In reality, nodes not only neighbor different nodes as time progresses but also have varying link weights between the same pairs of nodes. In our proposed ESGNN, we create a new evolving graph for every pre-selected time interval, which ensures the provisioned graphs reflect the most up-to-date connection status. (3) Previous GRNN models require high space complexity to store the graph since they construct graphs directly based on the target node time-series. In comparison, ESGNN first generates super nodes based on the target node time-series, then builds super graphs using these super nodes. With a much smaller number of super nodes compared to the number of target nodes, training on super nodes significantly reduces the demand of computation and memory. (4) Lastly, ESGNN mitigates the quadratic time complexity that many previous approaches suffer from. Previous related methods only handle small datasets with a few hundreds nodes, whereas ESGNN can provide predictions for tens of thousands of node time-series in a much shorter time. To the best of our knowledge, none of the existing GRNN models is driven by the large-scale nature of datasets for cloud workload resources. ESGNN tremendously reduces space usage and runtime, while at the same time achieving comparable forecasting accuracy.

2 Related Models

Classical time-series forecasting models such as ARIMA and exponential smoothing, are mainly statistical and often fall short of incorporating input features [2]. Recent deep learning models improve prediction accuracy by a large margin, especially when forecasting large-scale time-series [3,11]. Deep time-series forecasting models can be categorized into non-graph-based models [13,16,19,20] and graph-based models [10]. Non-graph-based models for cloud workload forecasting treat all time-series equivalently with no difference, which disadvantageously neglect the non-linear inter-relational dependency between compute nodes. In contrast to non-graph-based methods, Graph Recurrent Neural Networks (GRNN) predict time-series values with graph structures, which enables the non-linear inter-node relations modeling. Current state-of-the-art models capture inter-node relations by various techniques, such as attention, gated mechanisms, and multi-graph aggregation [4,7,8], but they fail to address significant issues such as graph scalability and the evolving node connection dynamics. For example, existing models [6,10,17] require $\mathcal{O}(|V|^2)$ time complexity to build graphs. When the number of nodes changes from a few hundred to tens of thousands, the time cost of graph construction alone worsens by 10,000 times, and even more when considering training runtime with the generated large graph. In our proposed Evolving Super Graph Neural Networks (ESGNN), we address the issue of graph scalability by utilizing clustering to build a super graph, which contains only a small number of super nodes. The number of super nodes is determined by the model users and used in the clustering stage, which enables the model to be trained with reduced space and time requirements. Clustering has been successful in allowing many nodes to share neural network parameters [22]. In combination with a small number of super nodes, our model demands much less time and space for training and inference. Lastly, the inter-node relationships are dynamic and subject to change over time, as nodes may collaborate with different groups of nodes at different moments. Depending on the data nature, recent deep graph models utilize the dynamics in various ways, such as aggregating the information of neighbor nodes at different timestamps [18], or continuously updating the current graph [5] when only a small number of nodes are updated, or taking graph snapshots as input [9]. Unlike the aforementioned methods, ESGNN constructs evolving graphs that provide more refined node dependency modeling.

3 Evolving Super Graph Neural Networks

In this section, we describe the workflow of our proposed Evolving Super Graph Neural Networks (ESGNN). We associate each target time-series with a node, as shown in Fig. 1 (a). In the cloud usage forecasting task, each time-series represents a resource type such as CPU or memory. We assume these nodes are connected based on their pattern correlations. Next, nodes with similar time-series patterns are grouped into clusters as shown in Fig. 1 (b). We build a super

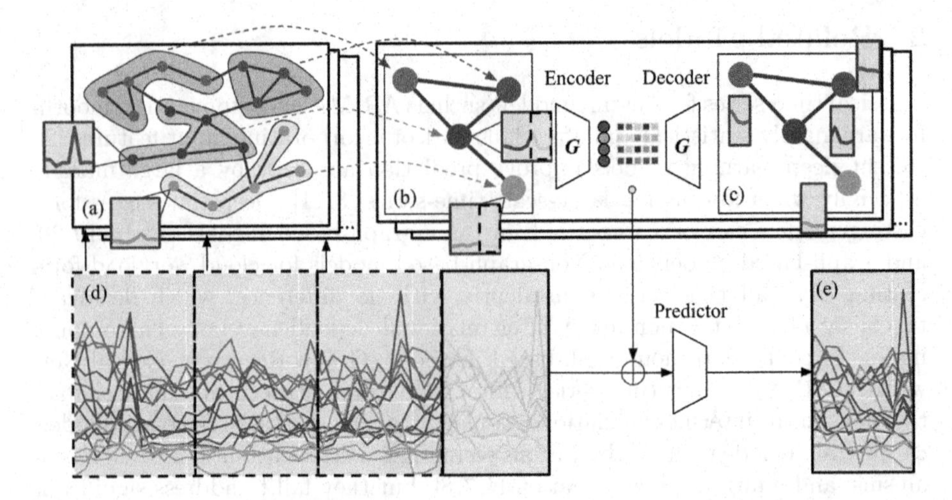

Fig. 1. Framework of our model. (a) Large-scale graphs before clustering: each graph comprises a period of node time-series. (b) Super graphs: each super graph is created by clustering nodes into super nodes. (c) An encoder-decoder structure is trained with super graphs to learn super node embeddings. (d) Time-series are sliced into periods to construct evolving super graphs per time interval. The testing data are masked off during training. (e) A predictor leverages both target time-series and encoded embeddings to deliver predictions. The masked-off data are used for evaluation.

node for each cluster and let the average time-series of time-series in each cluster represent them. Therefore, each super node is associated with an average time-series, or equivalently, a super time-series. ESGNN then constructs a super graph by connecting super nodes with super edges. Each super edge connects a pair of super nodes. We assign super edge weight by the similarity of the connected super time-series. As time-series evolve, nodes can belong to different super nodes at different time periods, and super graphs are dynamically updated with the latest period of time-series observations, as shown in Fig. 1 (d). Then, we leverage an evolving super graph encoder-decoder architecture that takes dynamic super graphs as input and trains in an end-to-end manner, as shown in Fig. 1 (c). The goal of this step is to learn the embeddings for super nodes. Finally, a predictor structure is trained upon the learned embeddings jointly with the target node time-series. The predictor uses the encoded embeddings from the super node time-series and the past target time-series to predict future values for the target time-series, as shown in Fig. 1 (e).

3.1 Preliminary Notations

We introduce symbols that are used in the paper. We form a time-series matrix $\mathbf{X} \in \mathbb{R}^{N \times T}$ containing N time-series of T time steps. We let the subscript denote the time-series index and superscript denote the timestamp, i.e., $\mathbf{x}_i \in \mathbb{R}^T$ indicates the i(th) time-series and $\mathbf{x}^t \in \mathbb{R}^N$ denotes all values at timestamp t. ESGNN creates evolving super graphs periodically. We let p denote the graph update interval. Hence, time-series values in the first p time steps $\mathbf{X}^{1:p}$ are used to construct the first super graph G_1, and time-series values in the second p time steps $\mathbf{X}^{p+1:2p}$ are used to construct the second super graph G_2, and so on. The last super graph may contain less than p time step values if T does not divide p evenly. Without loss of generality, we assume T divides p in this paper. Therefore, there are in total $S = \frac{T}{p}$ super graphs, denoted as $\mathcal{G} = \{G_1, G_2, \ldots, G_S\}$. Each super graph $G_i, i \in \{1, 2, \ldots, S\}$ is associated with a super node set and a super edge set, denoted as $G_i(V_i, E_i)$. The number of nodes in super graph G_i is denoted as $N_i = |V_i|$. With the aid of super graphs $\mathcal{G}$, our goal is to learn a function f that takes historical observations to simultaneously predict h-step ahead values for all time-series: $\left[X^{1:T}; \mathcal{G}\right] \xrightarrow{f(\cdot)} \left[\hat{X}^{T+1:T+h}\right]$.

3.2 Super Graph Construction

ESGNN constructs super graphs in two stages, namely, super node aggregation and super edge connection. Firstly, we build super nodes by clustering periods of time-series, such that a number of clusters are created where each cluster contains time-series with similar patterns. For each super node, a super node time-series is derived by averaging the time-series in the corresponding cluster. In the second stage, we connect super nodes by building super edges. Our model generates a dynamic super graph every p time steps, where p is the pre-selected time interval described in Sect. 3.1. Without loss of generality, we discuss the super graph construction for the i(th) period $\mathbf{X}^{(i-1)p+1:ip} \in \mathbb{R}^{N \times p}$, where $i \in \{1, 2, \ldots, S\}$ covers the whole observation time span.

Super Nodes Aggregation aims to cluster N nodes into N_i super nodes through a function denoted by ϕ. *K-Means* is one eligible algorithm for ϕ, in addition to alternative options such as X-Means [14]. K-Means is applied on N nodes with their individual node time-series of p time values. K-Means consists of three steps: initialization, assignment, and update. In the initialization step, $K = N_i$ nodes are randomly selected as centroids, denoted by $[M_1^0, M_2^0, \ldots, M_K^0]$, at the iteration 0. In the assignment step, each node is assigned to the closest centroid. We let A_u^t denote the assigned cluster for node u in the t(th) iteration, and let $[C_1^t, C_2^t, \ldots, C_K^t]$ denote the resulting K clusters, where $C_k^t = \{u|\ A_u^t = k\}$. The centroids are recalculated for each cluster during the update step: $M_k^t = \frac{1}{|C_k^t|} \sum_{u \in C_k^t} \mathbf{x}_u$. The node assignment step and the centroid update step are repeated until convergence. As the result of clustering, a number of K clusters $C_1, C_2, \ldots, C_K$ are generated. In ESGNN, a super node is created for each cluster. Let $\mathbf{x}'_k$ denote the k(th) super time-series: $\mathbf{x}'_k = \frac{1}{|C_k|} \sum_{u \in C_k} \mathbf{x}_u \in \mathbb{R}^p$. We let $\mathbf{X}' = [\mathbf{x}'_1, \mathbf{x}'_2, \ldots, \mathbf{x}'_{N_i}]^\intercal \in \mathbb{R}^{N_i \times p}$ denote all N_i derived super time-series.

Super Edge Connection aims to build edges among highly correlated super nodes. To quickly construct a super graph, we utilize Locality Sensitivity Hashing (LSH) to locate correlated super nodes. LSH assigns close super node time-series to various buckets in a linear time complexity. We connect super nodes within each bucket and use a time-series similarity metric ψ to set link weights. LSH generates a signature for each super time-series through a random matrix $\mathbf{B} \in [-1,1]^{p\times b}$, which maps super time-series onto b-dimensional vectors, with b being a small pre-selected number. Therefore, the matrix of b-dimensional vectors for all super time-series $\mathbf{X}^b \in \mathbb{R}^{N_i \times b}$ is calculated by $\mathbf{X}^b = \mathbf{X}'\mathbf{B}$. A binary *signature matrix* $\mathbf{X}^{sign} \in \mathbb{R}^{N_i \times b}$ of N_i time-series and b bits is derived by applying the sign function in an element-wise manner on $\mathbf{X}^b$:

$$\mathbf{X}_{i,j}^{sign} = \begin{cases} 1, & \mathbf{X}_{i,j}^b \geq 0 \\ 0, & \text{otherwise} \end{cases} \tag{1}$$

The resulting signatures are b-bit strings. Hence, there are at most $Q = 2^b$ unique strings ranging from $\underbrace{00\cdots0}_{b}$ to $\underbrace{11\cdots1}_{b}$. Super time-series with the same signatures are classified into the same bucket. We let $B_1, B_2, \ldots, B_S$ denote the Q different buckets, where two super nodes u and v are in the same bucket if and only if $\mathbf{X}_u^{sign} = \mathbf{X}_v^{sign}$. We link super nodes within the same bucket to form the edge set:

$$E = \{e_{u,v} |\ u, v \in B_i, u \neq v, i = \{1, 2, \ldots, Q\}\} \qquad e_{u,v} = \psi(\mathbf{x}'_u, \mathbf{x}'_v) \tag{2}$$

where ψ denotes a time-series similarity metric that takes two time-series as input and yields a similarity score. The selection of ψ is flexible and can naturally leverage any application-specific or state-of-the-art techniques. For instance, our approach can easily adopt Pearson Correlation Coefficient (PCC), Dynamic Time Warping (DTW) [12], among many other alternatives. Correspondingly, we derive an adjacency matrix $\mathbf{A} \in \mathbb{R}^{N_i \times N_i}$ for each super graph with $\mathbf{A}_{u,v} = e_{u,v}$. Leveraging LSH for build super edge construction brings two-fold benefits. Firstly, LSH has the advantage of efficient runtime, with a time complexity that is linear in the number of super nodes. Secondly, LSH helps to further reduce the number of super edges compared to constructing a complete graph construction with pairwise connections, which greatly improves the time complexity when training ESGNN. Other ways of graph construction (e.g., Radial Basis Function) are possible but take much more time.

4 Diffusion on Evolving Super Graphs

ESGNN learns from dynamically evolving super graphs through graph diffusion. Given super graphs $\mathcal{G}$, we adopt the graph diffusion layer from DCRNN to incorporate evolving graphs [10]. The diffusion layer, denoted by $g_{G,k}^{\Theta}$, maps P dimensions to Q dimensions with super graph G and super time-series $\mathbf{X}'$:

$$g_{G,k}^{\Theta}(\mathbf{X}') = \mathbf{a}\left(\sum_{j=1}^{P}\sum_{d=0}^{D-1} \mathbf{\Theta}(j,k,d,\mathbf{A})\mathbf{X}'\right) \tag{3}$$

$$\mathbf{\Theta}(j,k,d,\mathbf{A}) = \Theta_{j,k,d,1}\left(\mathbf{D}_O^{-1}\mathbf{A}\right)^d + \Theta_{j,k,d,2}\left(\mathbf{D}_I^{-1}\mathbf{A}^\intercal\right)^d \tag{4}$$

where Θ_* and $\mathbf{a}$ denote trainable parameters, D denotes the number of diffusion, $\mathbf{D}_I$ and $\mathbf{D}_O$ denote the in-degree and out-degree matrices of the adjacency matrix $\mathbf{A}$. Similar to DCRNN [10], we substitute the matrix multiplication in the Gated Recurrent Units (GRU) with the graph diffusion convolution:

$$\mathbf{r}^t = \sigma\left(g^r_{G_i}\left[\mathbf{X}'^t \oplus \mathbf{H}^{t-1}\right] + \mathbf{b}_r\right) \qquad \mathbf{u}^t = \sigma\left(g^u_{G_i}\left[\mathbf{X}'^t \oplus \mathbf{H}^{t-1}\right] + \mathbf{b}_u\right)$$
$$\mathbf{C}^t = \tanh\left(g^C_{G_i}\left[\mathbf{X}'^t \oplus \left(\mathbf{r}^t \odot \mathbf{H}^{t-1}\right)\right] + \mathbf{b}_u\right) \quad \mathbf{H}^t = \mathbf{u}^t \odot \mathbf{H}^{t-1} + (1 - \mathbf{u}^t) \odot \mathbf{C}^t$$

where G_i is the $i = \left\lceil \frac{t}{p} \right\rceil$(th) graph of timestamp t and period length p. Thus, different graphs are used in different periods for model training.

We design an encoder-decoder architecture with a modified GRU structure. The encoder consists of evolving super graph GRUs and derives encoded states, with the final state serving as the initialization for the decoder. The decoder, also composed of evolving super graph GRUs, generates predictions for the super time-series. The prediction on super time-series is therefore given by:

$$\hat{\mathbf{X}}'^t = \text{Decoder}\left(\text{Encoder}\left(\mathbf{X}'^{1:t-1}\right)\right) \tag{5}$$

We use Mean Absolute Error (MAE) as the loss function to train the neural network by minimizing the difference between the ground truth and the prediction.

$$Loss_{\text{super}} = \frac{1}{|\Omega_{train}|} \text{MAE}\left(\mathbf{X}'^t, \hat{\mathbf{X}}'^t\right) \qquad \Theta_*, \mathbf{a} = \underset{\Theta_*, \mathbf{a}}{\text{argmin}}\, Loss_{\text{super}} \tag{6}$$

where Ω_{train} denotes the set of training samples.

4.1 Predictor

ESGNN leverages a predictor to decode the learned super time-series embeddings. The predictor allows each original time-series to be modeled individually with their corresponding super time-series embeddings. To ensure a lightweight predictor, we utilize a single-layer network that takes the concatenation of target time-series and its super time-series as input. Let $\mathbf{X}'_{emb} = \text{Encoder}(\mathbf{X}')$ denote the encoded hidden states for all N_i super nodes. At any moment, a target time-series is associated with a super node, as described in Sect. 3.2. The super time-series embeddings from the latest super graph are used in the predictor since they are the closest to the prediction period. Let A^{last}_u denote the last assigned cluster of node u, a concatenation is derived by combining the two time-series $\mathbf{x}_{cat,u} = \left[\mathbf{x}_u \oplus \mathbf{x}'_{emb,A^{last}_u}\right]$. We further construct a matrix with all N concatenations $\mathbf{X}_{cat} = \left[\mathbf{x}_{cat,1}, \mathbf{x}_{cat,2}, \ldots, \mathbf{x}_{cat,N}\right]^{\mathsf{T}}$. Our predictor uses $\mathbf{X}_{cat}$ to yield forecasting for h steps ahead using parameters $\mathbf{W}$ and $\mathbf{b}$:

$$\hat{\mathbf{X}} = \mathbf{W}\mathbf{X}_{cat} + \mathbf{b} \tag{7}$$

Similarly to Eq. 6, the predictor model is trained with the MAE loss function.

$$Loss = \frac{1}{|\Omega_{train}|} \text{MAE}\left(\mathbf{X}^t, \hat{\mathbf{X}}^t\right) \qquad \mathbf{W}, \mathbf{b} = \underset{\mathbf{W}, \mathbf{b}}{\text{argmin}}\, Loss \tag{8}$$

5 Experiments on Large-Scale Datasets

We design experiments for following research questions: (1) How accurate is the prediction result given by our ESGNN model compared to previous methods? (2) How much runtime does ESGNN model take compared to other models? (3) How much space is saved with ESGNN?

ESGNN is evaluated on two cloud trace datasets from Google and Adobe [15]. The Google trace dataset records the activities of a cluster of 12, 224 machines over a period of 2 days in 2011. The measurements of CPU usage, maximum CPU usage, and memory usage are collected at 5-minute intervals, resulting in a total of 36, 672 time-series. The Adobe trace dataset records the activities of a cluster of 795 machines. The CPU usage for each task is recorded every 30 minutes. For both datasets, we obtain time-series of length 576 and split them chronologically into 50% : 25% : 25% for training, validation, and testing, respectively. In K-Means, we set the number of super nodes as $K = 100$ for both the Google and Adobe datasets across all time periods. The length of the binary strings in the LSH algorithm is selected as $b = 12$. We set the graph update time interval as $p = 48$ and generate 6 super graphs in total. The number of units in each graph diffusion layer is set as 16. We use a rolling window of length 24 to create data samples. For each data sample in the window, the first 12 values are conditioned to predict the next 12 values. We compare our model against state-of-the-art large-scale models, including NBEATS [13], DeepAR [16], MQRNN [20] and DF [19]. All of these models are implemented using Gluonts [1] with the suggested hyperparameter selection from the DF paper [19]. All experiments are run with 16 CPUs of 2.40 GHz, 37.2 GB memory, and a 12 GB NVIDIA GPU.

5.1 Forecasting Result and Analysis

We report the Mean Absolute Error (MAE) and Root Mean Square Error (RMSE) for 3, 6, and 12 steps ahead prediction of each model in Table 1. The best performance is highlighted in **bold** and the second best performance is shown in *italic*. For the Google dataset, ESGNN achieves the next-to-best result for 3 and 6 steps ahead predictions and the best result for 12 steps ahead prediction. We observe that DF performs the worst on the Google dataset which may be attributed to the small number of global factors in the DF model, since it implies a strong assumption that all time-series are related to each other and share global factors. For experiments with the Adobe dataset, our method ESGNN achieves the next-to-best result for the 3 steps ahead prediction and the best forecasting for the 6 and 12 steps ahead horizon. Overall, our proposed ESGNN demonstrates competitive accuracy compared to state-of-the-art models. It is worth

Table 1. Prediction performance on Google and Adobe datasets. MAE and RMSE are reported on 3, 6 and 12 steps ahead predictions.

Data	Method	3 steps ahead		6 steps ahead		12 steps ahead	
		MAE	RMSE	MAE	RMSE	MAE	RMSE
Google	ESGNN (Ours)	*0.0687*	*0.1766*	*0.0781*	*0.1955*	**0.0867**	**0.2019**
	NBEATS	**0.0590**	**0.1294**	**0.0711**	**0.1498**	*0.0889*	*0.2893*
	DeepAR	0.1172	0.3183	0.1413	0.3326	0.1576	0.3763
	MQRNN	0.1684	0.2576	0.1787	0.2602	0.1837	0.3399
	DF	0.3691	0.5059	0.3864	0.5283	0.3947	0.5802
Adobe	ESGNN (Ours)	*0.0140*	*0.0302*	**0.0175**	**0.0364**	**0.0232**	**0.0455**
	NBEATS	**0.0107**	**0.0187**	*0.0221*	*0.0429*	*0.0274*	*0.0530*
	DeepAR	0.0199	0.0628	0.0340	0.0784	0.0424	0.0852
	MQRNN	0.1066	0.1739	0.1099	0.1754	0.1195	0.1814
	DF	0.0972	0.1329	0.1016	0.1474	0.1141	0.1590

noting that although NBEATS outperforms ESGNN at certain horizons, our model requires less space and has a shorter training time, as discussed in the next section.

5.2 Runtime and Space Usage Analysis

We report the training runtime of our ESGNN against baselines in Fig. 2 for both the Google dataset and the Adobe dataset. The training runtime measures the time from the start to the end of training for all time-series and all training samples. To ensure a fair comparison, we include the time for super graph construction in ESGNN. The result shows that ESGNN is significantly faster than baselines regarding training runtime, namely, more than ten times faster than NBEATS (15.9×), DeepAR (14.6×), and DF (14.3×) on Google dataset, and more than three times faster than MQRNN (3.2×). A similar speedup result can be observed with the Adobe dataset, as NBEATS (40.5×), DeepAR (37.2×), MQRNN (8.1×), and DF (36.3×).

In addition, we report the model space usage in terms of space ratio compared to the baselines, as shown in Table 2. We notice that NBEATS takes much more parameters than other methods, due to its multi-layer residual block stacking mechanism, where a great number of parameters are needed for each block. In contrast, our proposed ESGNN requires fewer parameters and significantly saves space usage compared to other models. Specifically, ESGNN uses 4600× less space than NBEATS, 5× less space than DeepAR, 14× less space than MQRNN, and 1200× less space than DF.

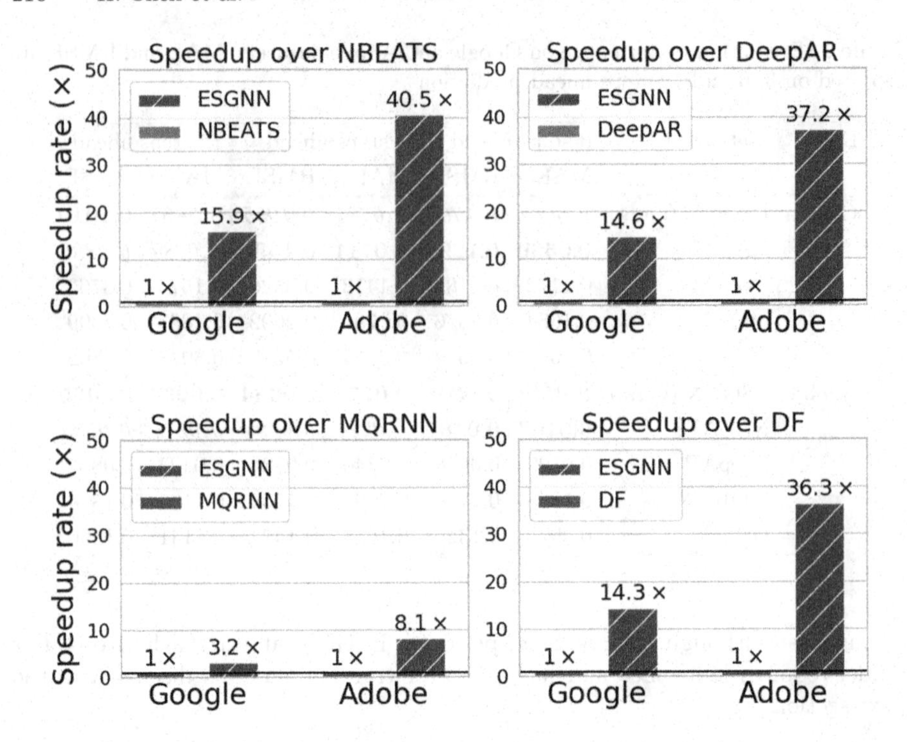

Fig. 2. ESGNN training speedup over baselines on the Google and Adobe datasets. (e.g., ESGNN is 15.9 times faster over NBEATS on Google dataset)

Table 2. Model space complexity in number of parameters.

Method	ESGNN (ours)	NBEATS	DeepAR	MQRNN	DF
#parameters	5,357	24,830,628	26,844	76,602	6,483,211

5.3 Ablation Study

We conduct extensive experiments to investigate the effectiveness of number of super nodes (i.e. K) and period length (i.e., p) in ESGNN. To investigate the effectiveness of number of super nodes, we keep all other hyperparameters the same while varying the number of super nodes K in $[10, 100, 1000]$. We report the average MAE across 12 steps ahead prediction and the runtime, as shown in Fig. 3 (Left). As indicated in the figure, the prediction loss decreases when the number of super nodes K increases from 10 to 100 and 1000. On the one hand, having more super nodes allows the model to extract time series patterns in a more refined granularity, which contributes to more accurate prediction. On the other hand, it takes more time to generate larger super graphs and more time to train upon them. To study the impact of different period length p, we ran experiments with p in $[12, 48, 144]$ while keeping other hyperparameters unchanged. The selected graph update intervals correspond to 1, 4, and 12 h.

Similar to the previous ablation study, we report the average MAE across 12 steps ahead prediction and the runtime for different time intervals, as shown in Fig. 3 (Right). As observed in the figure, the performance of ESGNN is better with a smaller graph update interval. This may be ascribed to that super graphs with smaller time intervals is better at helping ESGNN leverage the mutual relationships among time-series.

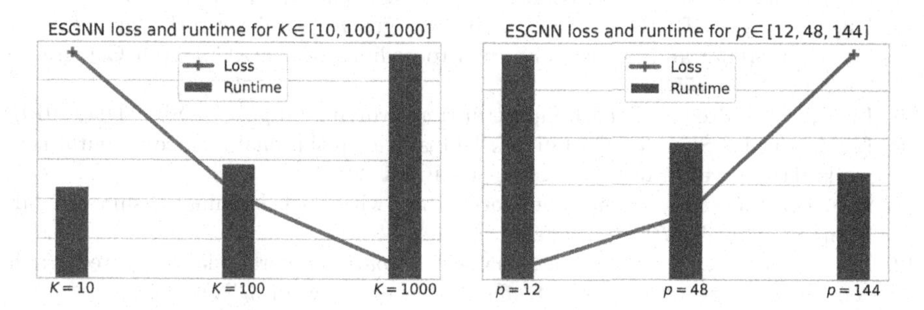

Fig. 3. (Left): Prediction loss (line) and runtime (bars) of $K \in \{10, 100, 1000\}$ for ESGNN. As the number of super nodes K increases, the prediction loss decreases while the required runtime increases. (Right): Prediction performance (line) and runtime (bar) of $p \in \{12, 48, 144\}$ for ESGNN.

6 Conclusion

In this paper, we propose Evolving Super Graph Neural Networks (ESGNN) to address the challenges of time and space complexity in time-series forecasting. ESGNN significantly reduces the number of nodes in the super graph, which consequentially reduces the runtime and space requirements. As a result, ESGNN exhibits substantial advantages in runtime and space usage compared to previous state-of-the-art models, without compromising prediction accuracy. We also conducted an ablation study to examine the effectiveness of the number of super nodes and the graph update interval. Our findings show that increasing the number of super nodes and having more frequent graph updates improves prediction performance, but comes at the cost of longer training time.

References

1. Alexandrov, A., et al.: GluonTS: probabilistic and neural time series modeling in Python. J. Mach. Learn. Res. **21**(116), 1–6 (2020)
2. Chatfield, C.: Time-Series Forecasting. Chapman and Hall/CRC, New York (2000)
3. Chen, H., Eldardiry, H.: Graph time-series modeling in deep learning: a survey. ACM Trans. Knowl. Discov. Data **18**, 1–35 (2023)
4. Chen, H., Rossi, R.A., Mahadik, K., Kim, S., Eldardiry, H.: Graph deep factors for forecasting with applications to cloud resource allocation. In: Proceedings of the 27th ACM SIGKDD Conference on Knowledge Discovery & Data Mining (2021)

5. Chen, X., Wang, J., Xie, K.: TrafficStream: a streaming traffic flow forecasting framework based on graph neural networks and continual learning. In: Proceedings of the Thirtieth International Joint Conference on Artificial Intelligence (2021)
6. Chen, Y., Segovia, I., Gel, Y.R.: Z-GCNets: time zigzags at graph convolutional networks for time series forecasting. In: ICML (2021)
7. Du, B., Yuan, C., Wang, F., Tong, H.: Geometric matrix completion via Sylvester multi-graph neural network. In: Proceedings of the 32nd ACM International Conference on Information and Knowledge Management, pp. 3860–3864 (2023)
8. Du, L., et al.: GBK-GNN: gated Bi-Kernel graph neural networks for modeling both homophily and heterophily. In: Proceedings of the ACM Web Conference 2022, pp. 1550–1558 (2022)
9. Li, J., et al.: Predicting path failure in time-evolving graphs. In: SIGKDD (2019)
10. Li, Y., Yu, R., Shahabi, C., Liu, Y.: Diffusion convolutional recurrent neural network: data-driven traffic forecasting. In: ICLR (2018)
11. Lim, B., Zohren, S.: Time-series forecasting with deep learning: a survey. Phil. Trans. R. Soc. A **379**(2194), 20200209 (2021)
12. Lu, B., Gan, X., Jin, H., Fu, L., Zhang, H.: Spatiotemporal adaptive gated graph convolution network for urban traffic flow forecasting. In: CIKM (2020)
13. Oreshkin, B.N., Carpov, D., Chapados, N., Bengio, Y.: N-BEATS: neural basis expansion analysis for interpretable time series forecasting. In: ICLR (2020)
14. Pelleg, D., Moore, A.W., et al.: X-Means: extending K-Means with efficient estimation of the number of clusters. In: ICML, vol. 1, pp. 727–734 (2000)
15. Reiss, C., Wilkes, J., Hellerstein, J.L.: Google cluster-usage traces: format+schema, pp. 1–14. Google Inc., White Paper (2011)
16. Salinas, D., Flunkert, V., Gasthaus, J., Januschowski, T.: DeepAR: probabilistic forecasting with autoregressive recurrent networks. Int. J. Forecast. **36**, 1181–1191 (2020)
17. Shang, C., Chen, J., Bi, J.: Discrete graph structure learning for forecasting multiple time series. In: International Conference on Learning Representations (2020)
18. Tran, A., Mathews, A., Ong, C.S., Xie, L.: Radflow: a recurrent, aggregated, and decomposable model for networks of time series. In: The Web Conference (2021)
19. Wang, Y., Smola, A., Maddix, D., Gasthaus, J., Foster, D., Januschowski, T.: Deep factors for forecasting. In: ICML (2019)
20. Wen, R., Torkkola, K., Narayanaswamy, B., Madeka, D.: A multi-horizon quantile recurrent forecaster. arXiv preprint arXiv:1711.11053 (2017)
21. Wu, Z., Pan, S., Long, G., Jiang, J., Chang, X., Zhang, C.: Connecting the dots: multivariate time series forecasting with graph neural networks. In: KDD (2020)
22. Xing, Y., et al.: Learning hierarchical graph neural networks for image clustering. In: Proceedings of the IEEE International Conference on Computer Vision (2021)

Unlearnable Examples for Time Series

Yujing Jiang[1(✉)], Xingjun Ma[2], Sarah Monazam Erfani[1], and James Bailey[1]

[1] Faculty of Engineering and Information Technology, The University of Melbourne, Parkville, Australia
yujingj@student.unimelb.edu.au, {sarah.erfani,baileyj}@unimelb.edu.au
[2] School of Computer Science, Fudan University, Shanghai, China
xingjunma@fudan.edu.cn

Abstract. Unlearnable examples (UEs) refer to training samples modified to be unlearnable to Deep Neural Networks (DNNs). These examples are usually generated by adding error-minimizing noises that can fool a DNN model into believing that there is nothing (no error) to learn from the data. The concept of UE has been proposed as a countermeasure against unauthorized data exploitation on personal data. While UE has been extensively studied on images, it is unclear how to craft effective UEs for time series data. In this work, we introduce the first UE generation method to protect time series data from unauthorized training by deep learning models. To this end, we propose a new form of error-minimizing noise that can be *selectively* applied to specific segments of time series, rendering them unlearnable to DNN models while remaining imperceptible to human observers. Through extensive experiments on a wide range of time series datasets, we demonstrate that the proposed UE generation method is effective in both classification and generation tasks. It can protect time series data against unauthorized exploitation, while preserving their utility for legitimate usage, thereby contributing to the development of secure and trustworthy machine learning systems.

Keywords: Time Series Analysis · Unlearnable Example

1 Introduction

The rapid advancement of deep learning and large models is largely driven by the vast amounts of data "freely" available on the Internet. While there has been significant research aimed at training deep learning models with privacy preservation [1,22,23,30,31], these approaches still neglect the necessity to obtain users' consent to use their data. Recent works have proposed useful tools such as Fawkes [29] to address this gap by promoting consent-based data utilization and protection. Yet, the issue remains unresolved. Rising public concerns stem from several instances where personal data, harvested from the Internet without consent, has been utilized to train commercial machine learning models [10]. Concerns now encompass not only images but also time series and multi-modal data. This

D.-N. Yang et al. (Eds.): PAKDD 2024, LNAI 14650, pp. 213–225, 2024.
https://doi.org/10.1007/978-981-97-2266-2_17

broadening scope underscores the need for thorough data protection strategies, particularly in the relatively underexplored field of time series data.

In response to privacy concerns, a number of data protection techniques have been developed including secure release and protective data poisoning. Among those works, protective data poisoning techniques have become increasingly attractive as they allow users to actively add poisoning or adversarial noise into their data (like selfies) before posting them on online social media platforms to protect data exploits. Recently, more advanced data protection techniques such as Unlearnable Examples (UEs) [8,11,25,39] have been proposed which can make (image) data unlearnable to machine learning models. Contrasting this with conventional data protection techniques that simply obscure identifiable data, UEs ensure that a DNN trained on such examples performs no better than random guessing on standard test examples.

Existing research on either data poisoning-based data protection or UEs has primarily focused on image-based applications, overlooking the significance of time series data which is vital in applications such as financial forecasting [2], health monitoring [20], energy prediction [7], and transportation [18]. Given its distinct characteristics and broad applications, there is an urgent need for time series data protection methods. Image-oriented data protection methods might not translate well to time series data due to their dynamic and sequential nature [13]. Although the concept of UEs has predominantly been confined to computer vision, our research will demonstrate that this concept can also be effectively extended to time series applications.

Existing methods developed for image-based UEs often apply unlearnable noise across the whole image. However, this approach is less suitable for time series data, which is inherently sequential and often requires interventions in certain segments instead of the entire dataset. Given that a short segment in time series data can hold critical information about a particular process or entity, the direct application of image-based UE techniques to time series data encounters significant challenges. Recognizing these limitations, we propose to make only a fraction, i.e., the most sensitive or crucial part, of the time series data unlearnable. This allows for the protection of specific data segments while maintaining the usability of the remainder, balancing security with data integrity.

In this work, we extend the concept of UEs to time series data and propose a novel and effective UE generation method. Our contributions are as follows:

- We introduce a new form of error-minimizing noise that can be applied *selectively* to segments of time series. This noise is imperceptible to humans, preserving the overall utility of the data while ensuring its primary purpose of rendering the data unlearnable to DNNs.
- We propose a novel unlearnable noise generator that can mitigate the potential risk of the underlying time series data being recognized or trained by either classification or generative models. By applying this noise, we effectively create a layer of protection around the data, making it ineffective for exploitation by AI technologies, while preserving its value for legitimate use.

- We conduct empirical studies to demonstrate the effectiveness of our method in generating unlearnable examples. Our evaluation covers a broad range of time series datasets, showcasing its versatility and robustness.

2 Related Work

In this section, we briefly review the most relevant data protection methods including data poisoning, adversarial attacks, and unlearnable examples.

2.1 Data Poisoning

Data poisoning attacks aim to weaken a model's performance by altering training data. Such attacks on Support Vector Machines (SVM) were shown by [3]. Koh et al. [14] expanded this, targeting influential training samples in DNNs with adversarial noise. This was later adopted into an end-to-end framework [21]. The work "Poison Frogs!" presents a clean-label poisoning technique that retains correct labels, making the attack more insidious [26]. Backdoor attacks, another variant of data poisoning techniques, involve embedding a hidden trigger pattern into the training dataset. Despite this manipulation, these attacks will not have a detrimental impact on the model's performance when evaluated on benign (clean) data [13,17,40]. Our work diverges from these approaches by generating unlearnable examples using imperceptible noise to effectively "bypass" the training process of DNNs, rendering them incapable of learning from the altered data, thereby offering a more robust strategy for data protection.

2.2 Adversarial Attack

Adversarial attacks are techniques designed to deceive machine learning models, especially DNNs, by injecting minor, often imperceptible noise that can lead models to make different predictions. The aim is to identify the minimal input modification causing misclassification or heightened prediction error. Extensive research has established adversarial examples that can deceive DNNs during the testing phase [4,5,9,15,19,27,32]. In these attacks, the adversary identifies a form of error-maximizing noise that significantly increases the model's prediction error. In response to the vulnerabilities exposed by adversarial attacks, adversarial training has emerged as the most robust countermeasure [19,28,33,34,36,38]. This training strategy is formulated as a *min-max* optimization problem, where the objective is to minimize the model's vulnerability to error-maximizing noise while maximizing its performance on clean data.

2.3 Unlearnable Examples

In contrast to adversarial examples, which focus on error-maximizing noise, unlearnable examples (UEs) pursue the opposite direction by identifying minimal noise that reduces the model's error through a *min-min* optimization process.

In this regard, Huang et al. [11] proposed the concept of UE, aimed at making training data ineffective for DNNs. Similarly, Yuan et al. [37] introduced Neural Tangent Generalization Attacks (NTGAs), a method that proficiently conducts generalization attacks on DNNs without requiring explicit knowledge about the learning model. Fu et al. [8] identified privacy limitations using error-minimizing noise and introduced robust error-minimizing noise via a *min-min-max* optimization. This limits adversarial learners from gleaning dataset information. Ren et al. [25] introduced transferable UEs that can improve their data-wise transferability. Based on this, Zhang et al. [39] proposed Unlearnable Clusters (UCs), offering a versatile approach to create UEs adaptable to various label exploitations. On the other hand, several countermeasures have been proposed against unlearnable examples, such as UEraser [24] that uses error-maximizing data augmentation, and Jiang et al. [12] propose a method to revert unlearnable samples to learnable ones. Our work expands the UE from the image domain to the time series domain across classification and generation tasks. Our approach can target the specified segments of data and make them unlearnable, thereby safeguarding the sensitive time series data against misuse and exploitation.

3 Error-Minimizing Noise for Time Series

In this section, we introduce our proposed method for generating error-minimizing noise *selectively* on segments of time series data.

3.1 Objective

In this paper, we primarily focus on applications related to time series classification and generation tasks. The models of interest in this domain are Recurrent Neural Networks (RNNs). The goal of our research is to protect time series samples that contain sensitive information in the public domain from being exploited by RNNs to ensure sensitive details are not inadvertently learned by machine learning models. Consequently, the defender's objectives are twofold. First, given the open accessibility of the data, it is imperative to inhibit deep learning models (RNNs) from processing or learning from this sensitive information. Second, these protective measures should not adversely affect the model's ability to generalize or perform its intended functions using non-sensitive information.

3.2 Threat Model

The defenders (data subjects) are aware of the general characteristics of the dataset into which their data will be collected and incorporated. This knowledge may include aspects such as the type of data, its source, and its intended application. While the defenders lack the authority to directly access or modify the dataset, they have the ability to access or alter their own individual data within it. Additionally, defenders are aware of the architecture of the DNNs being employed, but they lack information on more granular details such as the exact training procedure, optimization methods, or hyperparameters. This setting simulates the real-world scenario where defenders are often equipped with only partial information and lack full access or a complete understanding of the system. The defenders seek to safeguard their sensitive information from unauthorized exploitation by introducing error-minimizing noise into the time series data. This addition of noise is designed to render only the sensitive portions of the data unlearnable for machine learning models. Given the defenders' limited knowledge of the exact training model, the noise introduced should be adaptable across various machine learning models. Defenders, therefore, create noise based on a model they estimate to be close to the real one. This estimated model aids in crafting noise that remains effective across different architectures.

3.3 Challenges

Building upon the established concept of unlearnable examples in image data [8,11,25,39], our research extends this technique to time series data. We aim to create specific unlearnable (error-minimizing) noise that can be added to time series samples, hindering DNNs from effectively learning from these modified samples. While most existing methods focus on images and CNNs, our approach focuses on time series and RNNs. RNNs operate by processing sequential elements and retaining information from prior elements, and this iterative, memory-like nature of RNNs poses unique challenges in generating unlearnable noise. Unlike image data, where noises added to different locations are more independent, noises in time series data can be highly interconnected across the sequence, interfering with each other. Thus, a change at one single time point can cascade effects throughout the sequence. Given the memory mechanism of RNNs, even slight perturbations can amplify in later stages, greatly affecting the final output. Hence, creating error-minimizing noise for RNNs requires a novel approach that ensures the targeted segments of data are unlearnable while preserving the integrity and semantic value of the remaining parts.

3.4 Problem Formulation

Consider a time series sequence x_i, indexed by time t, which can be formally represented as $x = \{\mathbf{x}_0, \mathbf{x}_1, \ldots, \mathbf{x}_{t-1}, \mathbf{x}_t\}$. This sequence is processed through an RNN model for classification task that yields $y_i = f_\theta(x_i)$, where y_i serves as a class probability vector in the context of time series classification, or as a generated sequence for sequence generation. θ represents the model's learnable parameters, which govern the transformation f. Training the RNN model is to minimize its empirical error on the training samples, which can be achieved via empirical risk minimization (ERM). The optimization problem can be formulated as follows:

$$\min_\theta \mathbb{E}_{(x_i,y_i)\in D}\ell(f_\theta(x_i), y_i). \tag{1}$$

where D represents the training data and ℓ is the loss function that quantifies the dissimilarity between the model's output and the true target. To ensure minimal or negligible updates to the model parameters for a given time series sample, we introduce an error-minimizing noise denoted as δ. The primary objective of incorporating this noise is to significantly reduce the training loss of a sample when noise has been added to it. This noise term is designed to have the same dimensional structure as the input sample x, resulting in a sequence of noise values $\delta = \{\delta_0, \delta_1, ..., \delta_{t-1}, \delta_t\}$. Consequently, when considering the time series sample x_i in conjunction with its sample-specific error-minimizing noise δ_i, the combined effect can be mathematically expressed as follows:

$$\ell(f_\theta(x_i + \delta_i), y_i) \to 0. \tag{2}$$

The objective of this perturbation is to drive the loss $\ell(\cdot)$ towards zero. By doing so, the noise serves to minimize the discrepancy between the RNN's output and the actual target y_i. Consequently, the model is tricked into learning nothing from these perturbed samples.

3.5 A Straightforward Baseline Approach

A baseline method can be established for the generation of unlearnable examples in time series data by leveraging the concept of unlearnable examples as described in [11]. During the training phase of a basic noise generator, denoted as f'_θ, the system aims to solve the optimization problem as stated in Eq. 3.

$$\min_\theta \frac{1}{n} \sum_{i=1}^{n} \min_{\|\delta_i\| \le \rho_u} \ell(f'_\theta(x_i + \delta_i), y_i). \tag{3}$$

The generation of an unlearnable example, represented as (x', y), is accomplished using the trained noise generator f'_θ. This transformed data point is formally defined in Eq. (4).

$$x' = x + \arg \min_{\|\delta\| \le \rho_u} \ell(f'_\theta(x + \delta), y). \tag{4}$$

Given the sequential nature of RNNs, Backpropagation Through Time (BPTT) will be used where the network will be unrolled to match the length of the time series data. The calculation of the loss with respect to this unrolled RNN model takes into account these hidden states, allowing for a more detailed understanding of how each temporal data point in the sequence influences the overall loss.

3.6 Controllable Noise on Partial Time Series Samples

A significant limitation of directly translating image-based methods to time series data is the inability to localize and control the region of noise application.

In this case, noise tends to be distributed uniformly across the entire sequence. In the context of fixed-sized inputs, such as images, this uniform distribution is generally acceptable because the noise can be easily processed and interpreted within a consistent framework. However, given that RNN models process time series data in sequential order across the time regions, the effectiveness of this noise is not uniform across different temporal segments. Consequently, some portions of the time series will be more affected than others, leading to inconsistent training and prediction outcomes.

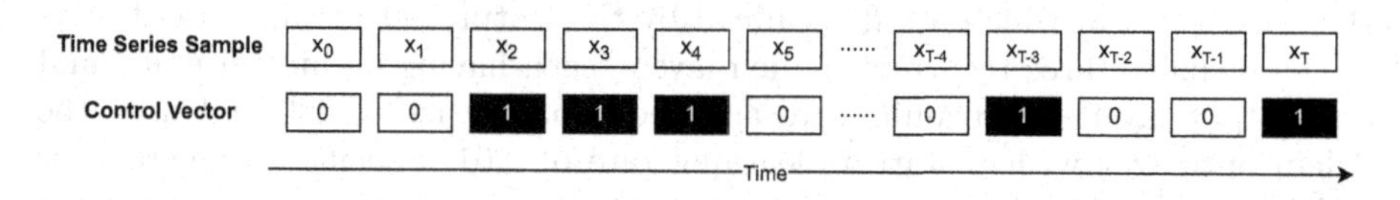

Fig. 1. Illustration of the control vector applied on a time series sample of length T. Data protection is indicated when the control vector highlights particular time stamps with a value of 1 (marked in black).

We propose a novel control vector, denoted as v, that highlights regions within the samples that should be "protected" from data exploitation. This concept is depicted in Fig. 1. As an example, consider a dialogue that comprises speech data from multiple individuals. If there is a need to protect the speech of a specific individual, their corresponding temporal segments can be distinctly marked using the control vector. To achieve this, we selectively add error-minimizing noise to the targeted segments of the time series data. Our primary objective is to reduce the training loss associated with these specified regions of the time series samples by solving the following optimization problem:

$$\min_{\theta} \frac{1}{n} \sum_{i=1}^{n} \min_{\|\delta_i\| \leq \rho_u} |\ell(f'_\theta(x_i + \delta_i \odot v_i), y_i) - \alpha \cdot \ell(f'_\theta(x_i \odot (1 - v_i)), y_i)|, \tag{5}$$

where $\odot$ represents element-wise multiplication on two vectors, and $|x|$ represents the absolute value of x. Specifically, our objective is to ensure that the training loss incurred by the sample with noise added to the targeted region $(x_i + \delta_i \odot v_i)$ is equivalent to the training loss when the target region is completely omitted from the time series sample $(x_i \odot (1 - v_i))$. This is achieved by minimizing the absolute difference between these two loss terms. By aligning the loss from noise addition to that of complete removal on the partial time series sample, we ensure that the model does not derive any insights from the target regions of the sample, while preserving the consistent learning patterns from the other segments. In summary, our method endeavors to provide a new solution that bridges the gap between conventional error-minimizing noise generation methods and the unique requirements of time series data.

4 Experiments

In this section, we evaluate our proposed controllable error-minimizing noise in both time series classification and sequence generation tasks.

4.1 Experiment Setup

For our experiments on time series classification tasks, we use a simple RNN architecture as the backbone model. This architecture consists of an input layer, the dimensions of which are determined by the feature set of the dataset. The model includes three recurrent hidden layers, each having 64 hidden units, and one output layer. For training, we adopted a batch size of 256 and used the Adam optimizer with a starting learning rate of 0.01. Specific parameters for RNN training included a 0.01 learning rate for noise generation (γ), a maximum noise magnitude set to $0.05 \times max_{\text{magnitude}}$ per sample (ρ_u), a trade-off parameter of 1 (α), a warm-start duration of 5 epochs ($T_{\text{warm\_start}}$), and a total training epoch of 50 (T_{training}). We use the model checkpoint at the 55th epoch as the final error-minimizing noise generator. Subsequently, we applied three different noise configurations with the control vector v, covering 20%, 50%, and 100% of the sample with 10% non-overlapping consecutive segments. The positioning of these segments is selected randomly for every sample.

We use ten unique time series datasets, including six univariate datasets from the UCR Archive and four multivariate datasets from the MTS Archive. We also employ two baseline methods including masking and universal adversarial perturbation (UAP) [16]. In our approach, we use masking to hide specific segments within time series samples. We randomly choose segments covering 50% of each sample, dividing them into five non-overlapping regions, each spanning 10% of the sample. This masking serves as a baseline to gauge the model's performance with significant data absence. To ensure equitable comparison, the adversarial

Table 1. Performance degradation results of various noise types introduced into the training data. Datasets D_1 through D_6 are univariate, sourced from the UCR Archive; whereas D_7 to D_{10} are multivariate from the MTS Archive. The 2nd column, labeled as **Clean**, depicts the accuracy of models trained on benign data.

Dataset	Clean	Masking	Universal	$\text{Ours}_{(20\%)}$	$\text{Ours}_{(50\%)}$	$\text{Ours}_{(100\%)}$
(D_1) BirdChicken	96.0%	80.9% (−15.1%)	39.1% (−56.9%)	19.3% (−76.7%)	12.1% (−83.9%)	8.8% (−87.2%)
(D_2) ECG5000	94.6%	78.4% (−16.2%)	41.6% (−53.0%)	16.9% (−77.7%)	11.5% (−83.1%)	7.4% (−87.2%)
(D_3) Earthquakes	72.5%	65.1% (−7.4%)	27.7% (−44.8%)	10.7% (−61.8%)	6.4% (−66.1%)	3.1% (−69.4%)
(D_4) ElectricDevices	72.3%	63.3% (−9.0%)	22.7% (−49.6%)	10.5% (−61.8%)	7.6% (−64.7%)	5.0% (−67.3%)
(D_5) Haptics	50.2%	29.0% (−21.2%)	17.5% (−32.7%)	13.4% (−36.8%)	8.3% (−41.9%)	6.3% (−43.9%)
(D_6) PowerCons	88.2%	68.4% (−19.8%)	37.0% (−51.2%)	15.6% (−72.6%)	9.3% (−78.9%)	4.9% (−83.3%)
(D_7) ArabicDigits	99.4%	83.0% (−16.4%)	26.1% (−73.3%)	9.4% (−90.0%)	4.7% (−94.7%)	2.1% (−97.3%)
(D_8) ECG	87.4%	74.2% (−13.2%)	20.2% (−67.2%)	8.9% (−78.5%)	5.6% (−81.8%)	2.6% (−84.8%)
(D_9) NetFlow	89.4%	78.5% (−10.9%)	16.7% (−72.7%)	6.7% (−82.7%)	3.2% (−86.2%)	1.9% (−87.5%)
(D_{10}) UWave	93.4%	80.8% (−12.6%)	26.4% (−67.0%)	10.2% (−83.2%)	7.9% (−85.5%)	4.0% (−89.4%)

perturbation was capped at $0.05 \times max_{\text{magnitude}}$ and integrated into 50% of every sample, specifically at the same regions chosen for masking.

4.2 Against Classification Models

Our experimental results for the controllable unlearnable noise generator, featuring three configurations and two baseline methods, are presented in Table 1. Using the masking baseline, we noticed a 14.18% average drop in accuracy compared to the clean model. The unmasked segments, retaining key features, possibly account for the limited decline. With the time series UAP, the accuracy decrease averaged 56.84%. Remarkably, our proposed noise method, targeting only 20% of samples, achieved a more significant average accuracy drop of 72.18%, emphasizing its efficacy over the UAP. With 50% targeting, as in the baselines, accuracy fell to 7.66%, marking a 76.68% reduction from the clean model. Additionally, our error-minimizing noise demonstrates more significant impacts on multivariate datasets, which capture the interactions and relationships across multiple variables. This multi-dimensionality allows the unlearnable noise to envelop both primary and subtle features. As a result, it can reduce the training loss more effectively, obscuring the genuine data patterns.

Taking a closer look at the accuracy drops, we found that increasing the amount of unlearnable noise does not linearly decrease classification accuracy. This suggests a diminishing return on increasing noise levels, indicating that beyond a certain threshold, the addition of more unlearnable noise might not yield significantly enhanced privacy protections. This observation implies that introducing noise to only a segment of the time series might be the most beneficial strategy. By targeting only a small part of the samples, one can achieve the desired reduction in training effectiveness without compromising the entire dataset. The results highlight the efficacy of our method, showcasing its adaptability in safeguarding time series data privacy, especially potent against data misuse in classification tasks.

4.3 Against Generative Models

We extend the evaluation to assessing the application of our proposed unlearnable noise in the context of time series generation tasks. We employ 8 multivariate time series datasets for this study, encompassing a range of classes and sample sizes. For the task of data generation, we apply two time series generative models: the Recurrent GAN (RGAN) [6] and Quant GAN (QGAN) [35]. These models are then used to generate synthetic data for the first class (class 0) of each dataset. We follow the training procedure stated in the original papers. The noise is configured to perturb 50% of the samples in the target class, and every selected sample is entirely perturbed by the noise.

We apply the *Train on Synthetic, Test on Real* (TSTR) [6] approach to test the effectiveness of our proposed unlearnable noise. Specifically, we first train a GAN model with data perturbed by unlearnable noise, then train a classifier model using data generated by the GAN and subsequently test it on a separate set of genuine samples. In this experiment, we subset all samples from the first class (class 0) of each dataset and then feed them for GAN training. The objective is to minimize the generator's reconstruction loss on the entire sample. Then, we train the time series classifiers using the generated synthetic samples, using Long Short-Term Memory (LSTM) and Fully Convolutional Network (FCN).

Table 2. Classification accuracy of real or synthetic time series samples using the "Train on Synthetic, Test on Real" (TSTR) approach. The 2nd column, labeled as **Real**, depicts the accuracy of classification models trained and tested on benign data. The columns presented as **Model**$_c$ use clean data to train the generative model. The columns presented as **Model**$_n$ use unlearnable data to train the generative model.

Dataset	Network	Real	RGAN$_c$	RGAN$_n$	QGAN$_c$	QGAN$_n$
(D_7)	FCN	99.6%	75.2%	6.2%	78.6%	8.4%
	LSTM	98.4%	83.4%	4.2%	81.4%	2.6%
(D_8)	FCN	91.2%	77.6%	7.4%	76.0%	7.8%
	LSTM	89.4%	72.0%	3.0%	73.6%	5.8%
(D_9)	FCN	94.6%	75.4%	11.6%	77.0%	10.6%
	LSTM	90.1%	74.0%	6.8%	75.6%	3.4%
(D_{10})	FCN	95.0%	82.0%	10.5%	84.0%	13.6%
	LSTM	93.0%	86.0%	5.2%	88.0%	6.2%
(D_{11})	FCN	94.2%	80.4%	8.4%	83.2%	9.6%
	LSTM	95.0%	76.4%	3.8%	82.4%	5.8%
(D_{12})	FCN	78.9%	54.2%	6.2%	58.0%	8.4%
	LSTM	76.0%	57.8%	2.8%	60.8%	4.0%
(D_{13})	FCN	86.4%	68.6%	11.6%	73.6%	10.8%
	LSTM	75.0%	64.2%	6.0%	72.4%	5.2%
(D_{14})	FCN	71.0%	54.6%	9.5%	61.2%	10.2%
	LSTM	64.0%	49.0%	5.4%	56.0%	4.8%

The experimental results shown in Table 2 demonstrate a significant drop in performance when adding unlearnable noise to 50% of the training samples. The average classification accuracy drops below 10%, marking an average reduction of over 60% when compared to the results obtained for clean data training. Note that, while the noise is introduced into only 50% of the samples within a specific class, it has the capability to render the entire class unlearnable (non-generative) against the sequence generation model. This implies that our proposed noise has great potential to be applied to protect sensitive samples from being learned during the training of a generative model, preventing the model from recreating or understanding the sensitive or private aspects of the original data.

5 Conclusion

In this work, we have studied the problem of protecting time series data against unauthorized exploitations. We extended the concept of Unlearnable Examples (UEs) from the image domain to the time series domain and proposed a novel method specifically designed for generating unlearnable noise for time series. The proposed method leverages a novel min-min bilevel optimization framework alongside a control vector, enabling the creation of unlearnable noise targeted at the most sensitive parts of a time series. This approach can be selectively used on specific segments of the time series data. Through extensive experiments on both time series classification and generation tasks, we demonstrated the effectiveness of our method across different datasets. Our work could help individuals and organizations protect their time series data from being exploited (without permission) in the development of commercial models.

References

1. Abadi, M., et al.: Deep learning with differential privacy. In: SIGSAC (2016)
2. Barra, S., Carta, S.M., Corriga, A., Podda, A.S., Recupero, D.R.: Deep learning and time series-to-image encoding for financial forecasting. IEEE/CAA J. Automat. Sinica **7**(3), 683–692 (2020)
3. Biggio, B., Nelson, B., Laskov, P.: Poisoning attacks against support vector machines. arXiv preprint arXiv:1206.6389 (2012)
4. Carlini, N., Wagner, D.: Towards evaluating the robustness of neural networks. In: SP (2017)
5. Croce, F., Hein, M.: Reliable evaluation of adversarial robustness with an ensemble of diverse parameter-free attacks. arXiv preprint arXiv:2003.01690 (2020)
6. Esteban, C., Hyland, S.L., Rätsch, G.: Real-valued (medical) time series generation with recurrent conditional gans. arXiv e-prints pp. arXiv–1706 (2017)
7. Feng, Y., Duan, Q., Chen, X., Yakkali, S.S., Wang, J.: Space cooling energy usage prediction based on utility data for residential buildings using machine learning methods. Appl. Energy **291**, 116814 (2021)
8. Fu, S., He, F., Liu, Y., Shen, L., Tao, D.: Robust unlearnable examples: protecting data against adversarial learning. arXiv preprint arXiv:2203.14533 (2022)
9. Goodfellow, I.J., Shlens, J., Szegedy, C.: Explaining and harnessing adversarial examples. In: ICLR (2015)
10. Hill, K.: The secretive company that might end privacy as we know it (2020)
11. Huang, H., Ma, X., Erfani, S.M., Bailey, J., Wang, Y.: Unlearnable examples: making personal data unexploitable. In: ICLR (2020)
12. Jiang, W., Diao, Y., Wang, H., Sun, J., Wang, M., Hong, R.: Unlearnable examples give a false sense of security: piercing through unexploitable data with learnable examples. arXiv preprint arXiv:2305.09241 (2023)
13. Jiang, Y., Ma, X., Erfani, S.M., Bailey, J.: Backdoor attacks on time series: a generative approach. In: SaTML (2023)
14. Koh, P.W., Liang, P.: Understanding black-box predictions via influence functions. In: ICML (2017)
15. Kurakin, A., Goodfellow, I., Bengio, S.: Adversarial machine learning at scale. arXiv preprint arXiv:1611.01236 (2016)

16. Li, J., et al.: Universal adversarial perturbations generative network for speaker recognition. In: ICME (2020)
17. Liu, Y., Ma, X., Bailey, J., Lu, F.: Reflection backdoor: a natural backdoor attack on deep neural networks. In: Vedaldi, A., Bischof, H., Brox, T., Frahm, J.-M. (eds.) Computer Vision – ECCV 2020, pp. 182–199. Springer, Cham (2020). https://doi.org/10.1007/978-3-030-58607-2_11
18. Ma, T., Antoniou, C., Toledo, T.: Hybrid machine learning algorithm and statistical time series model for network-wide traffic forecast. Transport. Res. Part C: Emerg. Technol. **111**, 352–372 (2020)
19. Madry, A., Makelov, A., Schmidt, L., Tsipras, D., Vladu, A.: Towards deep learning models resistant to adversarial attacks. In: ICLR (2018)
20. Maweu, B.M., Shamsuddin, R., Dakshit, S., Prabhakaran, B.: Generating healthcare time series data for improving diagnostic accuracy of deep neural networks. IEEE Trans. Instrum. Meas. **70**, 1–15 (2021)
21. Muñoz-González, L., et al.: Towards poisoning of deep learning algorithms with back-gradient optimization. In: AISec (2017)
22. Phan, N., Wang, Y., Wu, X., Dou, D.: Differential privacy preservation for deep auto-encoders: an application of human behavior prediction. In: AAAI (2016)
23. Phan, N., Wu, X., Hu, H., Dou, D.: Adaptive laplace mechanism: differential privacy preservation in deep learning. In: ICDM (2017)
24. Qin, T., Gao, X., Zhao, J., Ye, K., Xu, C.Z.: Learning the unlearnable: adversarial augmentations suppress unlearnable example attacks. arXiv preprint arXiv:2303.15127 (2023)
25. Ren, J., Xu, H., Wan, Y., Ma, X., Sun, L., Tang, J.: Transferable unlearnable examples. In: ICLR (2022)
26. Shafahi, A., et al.: Poison frogs! targeted clean-label poisoning attacks on neural networks. In: NeurIPS (2018)
27. Shafahi, A., Najibi, M., Xu, Z., Dickerson, J., Davis, L.S., Goldstein, T.: Universal adversarial training. In: AAAI (2020)
28. Shan, S., Ding, W., Wenger, E., Zheng, H., Zhao, B.Y.: Post-breach recovery: protection against white-box adversarial examples for leaked DNN models. In: ACM SIGSAC Conference on Computer and Communications Security (2022)
29. Shan, S., Wenger, E., Zhang, J., Li, H., Zheng, H., Zhao, B.Y.: Fawkes: protecting personal privacy against unauthorized deep learning models. In: USENIX-Security (2020)
30. Shokri, R., Shmatikov, V.: Privacy-preserving deep learning. In: SIGSAC (2015)
31. Shokri, R., Stronati, M., Song, C., Shmatikov, V.: Membership inference attacks against machine learning models. In: SP (2017)
32. Szegedy, C., et al.: Intriguing properties of neural networks. In: ICLR (2014)
33. Wang, Y., Ma, X., Bailey, J., Yi, J., Zhou, B., Gu, Q.: On the convergence and robustness of adversarial training. In: ICML, pp. 6586–6595 (2019)
34. Wang, Y., Zou, D., Yi, J., Bailey, J., Ma, X., Gu, Q.: Improving adversarial robustness requires revisiting misclassified examples. In: ICLR (2020)
35. Wiese, M., Knobloch, R., Korn, R., Kretschmer, P.: Quant gans: deep generation of financial time series. Quantitative Finance **20**(9), 1419–1440 (2020)
36. Wu, D., Xia, S.T., Wang, Y.: Adversarial weight perturbation helps robust generalization. Adv. Neural Inf. Process. Syst. **33** (2020)
37. Yuan, C.H., Wu, S.H.: Neural tangent generalization attacks. In: International Conference on Machine Learning, pp. 12230–12240. PMLR (2021)
38. Zhang, H., Yu, Y., Jiao, J., Xing, E.P., Ghaoui, L.E., Jordan, M.I.: Theoretically principled trade-off between robustness and accuracy. In: ICML (2019)

39. Zhang, J., et al.: Unlearnable clusters: towards label-agnostic unlearnable examples. In: CVPR (2023)
40. Zhao, S., Ma, X., Zheng, X., Bailey, J., Chen, J., Jiang, Y.G.: Clean-label backdoor attacks on video recognition models. In: CVPR (2020)

Learning Disentangled Task-Related Representation for Time Series

Liping Hou, Lemeng Pan(✉), Yicheng Guo, Cheng Li, and Lihao Zhang

IT Innovation and Research Center, Huawei Technologies, Shenzhen, China
{houliping1,panlemeng,guoyicheng3,licheng81,zhanglihao2}@huawei.com

Abstract. Multivariate time series representation learning employs unsupervised tasks to extract meaningful representations from time series data, enabling their application in diverse downstream tasks. However, despite the promising advancements in contrastive learning-based representation learning, the study of task-related feature learning is still in its early stages. This gap exists because current unified representation learning frameworks lack the ability to effectively disentangle task-related features. To address this limitation, we propose DisT, a novel contrastive learning-based method for efficient task-related feature learning in time series representation. DisT disentangles task-related features by incorporating feature network structure learning and contrastive sample pair selection. Specifically, DisT incorporates a feature decoupling module, which prioritizes global features for time series classification tasks, while emphasizing periodic and seasonal features for forecasting tasks. Additionally, DisT leverages contrastive loss and task-related feature loss to adaptively select data augmentation methods, preserving task-relevant shared information between positive samples across different data and tasks. Experimental results on various multivariate time-series datasets including classification and forecasting tasks show that DisT achieves state-of-the-art performance.

Keywords: Time Series · Representation Learning · Data Mining

1 Introduction

Time series data plays a pivotal role in various domains, including IT system operation, financial forecasting, anomaly detection, and climate modeling [8]. Consequently, learning effective representations for multivariate time series is a critical task. However, it is challenging due to the diversity and complexity of time series data, the absence of uniform semantic definition patterns, and intricate multivariate dependency relationships. Inspired by the success of contrastive learning in Computer Vision (CV) and Natural Language Processing (NLP), researchers have turned to contrastive learning for time series representation learning. Multi-task representation learning frameworks have gained significant attention due to their generalizability.

D.-N. Yang et al. (Eds.): PAKDD 2024, LNAI 14650, pp. 226–238, 2024.
https://doi.org/10.1007/978-981-97-2266-2_18

For instance, T-Loss [7] introduces an unsupervised method for learning universal representations of multivariate time series using dilated causal convolution and subsequence consistency. TNC [18] effectively learns representations for non-stationary time series by leveraging the time consistency principle and adjacent windows as positive samples. TS2Vec [25] presents an unsupervised approach for learning contextual representations at different semantic levels, selecting overlapping subsequences as positive sample pairs. However, these multi-task learning frameworks face two unresolved issues that require further attention. **(i) Mixed representation learning**. As discussed by Bengio et al. [3], a good representation should be able to disentangle various explanatory factors, enabling robustness to complex and structurally rich variations. Different time series tasks require dissimilar key features, such as seasonality, trend, amplitude, or a blend of these factors. Specifically, classification tasks prioritize capturing high-level global features, whereas forecasting tasks specifically target trend and seasonal features. However, existing unified representation extraction frameworks [14,25] do not explicitly consider feature decoupling, therefore tend to produce mixed task-related representations together at the expense of optimal performance on downstream tasks. Task-specific frameworks, such as CoST [23] and LaST [22] suggested disentangled trend and seasonality are significant task-related features for forecasting tasks. However, these frameworks do not demonstrate cross-task generalizability. **(ii) Fixed positive pairs selection**. The shared information between contrast pairs affects representation learning. Therefore, the design of positive pairs is another pivotal strategy to disentangle task-related features. An unified and fixed contrast pair selection strategy for all data and tasks is inadequate and lacks robustness.

To address these challenges, we propose a method to learn Disentangled Task-Related (DisT) representation. DisT combines decoupled structural feature extraction with flexible augmentation method selection to extract key features based on the task and data scenario. By combining contrastive loss and task-related feature loss, DisT dynamically selects data augmentation methods to construct augmented data, ensuring that the augmented data retains task-relevant characteristics. This facilitates the learning of task-related features. This paper makes the following contributions:

1. An innovative time series representation learning framework DisT is proposed, which focuses on disentangling features relevant to different tasks within time series data.
2. To achieve feature decoupling, we introduce a module designed to handle task-related features, incorporating a global feature component for classification tasks and a decomposition component for prediction tasks. We propose classification loss and forecasting loss to supervise the learning of task-related features based on the decoupled representations.
3. A task-adaptive augmentation selection learner based on task-related feature loss is constructed, aiming to retain more task-specific features between positive pairs. Therefore, enhancing the disentanglement and learning of task-relevant features.

2 Related Work

Contrastive learning, as a significant branch of unsupervised learning, has received considerable attention. Recent research in contrastive learning has focused on optimizing the quality of representation learning from various perspectives, such as capturing semantic information at different granularities [25], decomposing data into several components to attain interpretability [22,23], and designing various positive pairs [6,7,25]. **Time series decomposition** is a classic approach in time series analysis, widely utilized in time forecasting [4]. Recent studies [22,23,26] have applied time series decomposition to representation learning, extracting seasonality and trend features that are beneficial for forecasting tasks. These findings demonstrate the significance of flexible representation learning. Additionally, in the field of CV representation learning, several studies [17,21] have emphasized the importance of task-related features. However, how to decompose time series and decouple task-specific features within a unified representation learning framework to facilitate the learning of task-relevant features is still a research area that has not been intensively explored.

Data augmentation is a typical way for generate positive pairs, which also regarded as various views of inputs. Given the limitations of setting up views manually, previous work [13] have focused on designing generative views with greater flexibility, and significant progress has been made. Inspired by the success in CV, contrastive learning with data augmentation has also attracted attention in time series. TS-TCC [6] generates two different but related views through handcrafted augmentation ways, which is the most natural way. Previous study [11] introduces data augmentation methods in time series and analyzed their impact on prediction, classification, and anomaly detection. However, research on the influence of augmentation methods on representation learning is not covered. LEAVES [24] has devoted to explore the automatic way of data augmentation. However, the semantics of time series are hidden in human indistinguishable temporal relationships and the variability of data distribution, it is very challenging for generative augmentations to accurately capture key features of the original time series data.

3 The Proposed Method

3.1 Overview

Contrastive learning aims to maximize the mutual information between the representations of positive pairs, which are served as views of the input data and denoted as v and $\hat{v}$. Positive pairs v and $\hat{v}$ are typically obtained by applying various data augmentation methods. The overall architecture of DisT is shown in Fig. 1, given a series of inputs $\mathcal{X} = \{x_1, x_2, \cdots, x_N\}$, and its corresponding augmented counterpart $\mathcal{X}' = \{x'_1, x'_2, \cdots, x'_N\}$, a randomly cropped sequence from both x_i and x'_i at the same position is selected as the views v and $\hat{v}$ for the encoder as the input, where $\hat{v} = \mathcal{Q}(x)$ and $\mathcal{Q}$ denotes the augmentation function. After extracting a comprehensive set of generic features using the encoder,

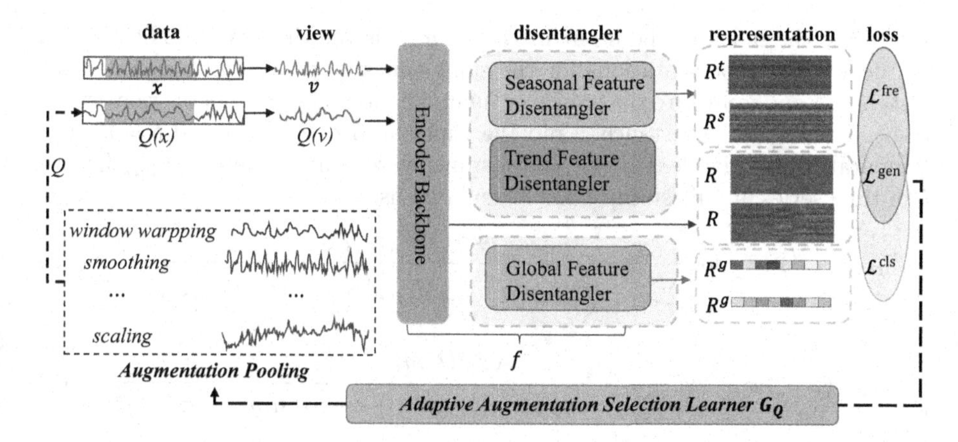

Fig. 1. The overall architecture of DisT in this paper. Decoupling and learning task-related features are achieved at both the structural and view levels. The augmented data is generated by an augmentation method learner, which is optimized based on the loss of task-related features.

the feature decoupling module effectively disentangles the task-related features. Subsequently, the training process involves computing generic and task-specific contrastive loss to supervise the network training. The training procedure consists of two stages. In the first stage, the network is cold started by using only the smoothing data augmentation method. In the second stage, the data augmentation selector is supervised by both generic and task-specific losses to facilitate learning. The selector then adaptively chooses the optimal data augmentation method for different tasks. This adaptive selection ensures the sharing of sufficient task representation between v and $\hat{v}$.

3.2 Task-Relevant Feature Disentangled

Task-Related Features of the Classification. The features related to the classification task are detailed in sugfig (b) Fig. 2. When v and $\hat{v}$ are input to the encoder backbone, the representations $\mathcal{R}$ and $\hat{\mathcal{R}}$ are obtained. For the classification task, a global feature module is embedded to decouple critical features related to the task. The architecture of this module is shown in Fig. 1 and is represented as $\mathcal{R}^g = \text{MaxPool}(\mathcal{R})$ and $\hat{\mathcal{R}^g} = \text{MaxPool}\left(\hat{\mathcal{R}}\right)$, and $\mathcal{R}^g, \hat{\mathcal{R}^g} \in \mathbb{C}^{B\times 1\times D}$, where B and D are instance number of one batch and feature dimension. Following previous works [25], hierarchical loss is used as contrastive loss, which consists of temporal contrastive loss and instance-wise contrastive loss based on InfoNCE. Let z and z' be representations, $z, z' \in \mathbb{C}^{N\times T\times D}$, where N, T, and D are instance number, time sequence length, and feature dimension, respectively. The hierarchical loss [25] is calculated as:

$$\mathcal{L}(z, z') = \frac{1}{NT}\sum_{i=1}^{N}\sum_{t=1}^{T}\left(\mathcal{L}_{\text{temp}}(z, z') + \mathcal{L}_{\text{inst}}(z, z')\right), \quad (1)$$

where i is the index of the input instance and t is the timestamp. The $z_{i,t}$ and $z'_{i,t}$ denote the representations for the same timestamp t but two views of one instance. $\mathbb{1}$ and $sim()$ are the indicator function and inner product of two vectors, respectively. Set the instance j and the timestamp t' to $i \neq j$ and $t \neq t'$. The temporal contrastive loss $\mathcal{L}_{\text{temp}}$ and instance-wise contrastive loss $\mathcal{L}_{\text{inst}}$ for the i-th time series at timestamp t are formulated as:

$$\mathcal{L}_{\text{temp}}(z, z') = -\log \frac{\exp\left(sim(z_{i,t}, z'_{i,t})\right)}{\sum_{t' \in T}\left(\exp\left(sim(z_{i,t}, z'_{i,t'})\right) + \exp\left(sim(z_{i,t}, z_{i,t'})\right)\right)}, \quad (2)$$

$$\mathcal{L}_{\text{inst}}(z, z') = -\log \frac{\exp\left(sim(z_{i,t}, z'_{i,t})\right)}{\sum_{j=1}^{B}\left(\exp\left(sim(z_{i,t}, z'_{j,t})\right) + \exp\left(sim(z_{i,t}, z_{j,t})\right)\right)}. \quad (3)$$

In DisT, $\mathcal{R}$ and $\hat{\mathcal{R}}$ are universal feature indirectly obtained by backbone encoder, and $\mathcal{R}, \hat{\mathcal{R}} \in \mathbb{C}^{B \times T \times D}$. Output dimension is set to 320 in this paper, and α is a parameter of classification specific task. The classification task specific loss is denoted as:

$$\mathcal{L}^{\text{cls}} = \mathcal{L}(\mathcal{R}, \hat{\mathcal{R}}) + \alpha \mathcal{L}(\mathcal{R}^g, \hat{\mathcal{R}}^g). \quad (4)$$

In the classification tasks, enhancing a global feature loss between views effectively disentangles the task-relevant global features from the latent space. $\mathcal{L}$ defined in Eq. (1), and $\mathcal{L}(\mathcal{R}, \hat{\mathcal{R}})$ is the calculated as the general feature loss $\mathcal{L}^{\text{gen}}$.

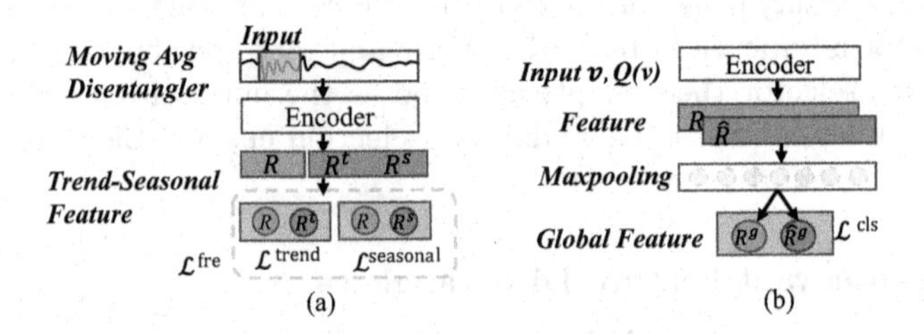

Fig. 2. The module details for decoupling the features.

Task-Related Features of the Forecasting. Previous studies have demonstrated that seasonal and trend terms are crucial task-related features for forecasting task [23]. The disentangler for forecasting task utilizes a moving average kernel on the input sequence to accurately extract the trend-cyclical component of the time series, follow DLinear [26]. As details illustrated in Fig. 2 (a), in the context of the forecasting task's features, the input data v is decoupled into trend component v^t and seasonal component v^s. Subsequently, seasonal feature R^s and trend feature R^t are obtained after encoding v^t and v^s. The loss computation associated with the forecasting task can be described as follows:

$$\mathcal{L}^{\text{fre}} = \mathcal{L}(\mathcal{R}, \hat{\mathcal{R}}) + \left(\beta_s \mathcal{L}(\mathcal{R}, \mathcal{R}^s) + \beta_t \mathcal{L}(\mathcal{R}, \mathcal{R}^t)\right), \quad (5)$$

where β_s and β_t are adjusted parameters of trend and seasonal features.

3.3 Task-Adaptive Augmentation Selection

As discussed by Tian et al. [17], the positive pairs should contain minimal irrelevant features and maximal task-related features for representation learning. Task-related features vary across different tasks and data, and the choice of data augmentation methods leads to varying effects. Hence, it is critical to choose the augmentation method. One of the pivotal principles of ensuring the effectiveness of representation is that the views share sufficient task-related information. Based on the aforementioned principle, this study proposes a task-adaptive augmentation selection learner. The underlying concept is to maximize the preservation of task-related features when transforming the data to maintain the fidelity between views, which is denoted as $\max_f I\left(\mathcal{R}, \hat{\mathcal{R}}\right)$, where $\hat{\mathcal{R}} = f\left(\mathcal{Q}(v)\right)$, I represents mutual information and $v' = \mathcal{Q}(v)$, f and $\mathcal{Q}$ are encoding and augmentation functions, respectively.

Subsequently, we will present how to obtain the optimal $\mathcal{Q}$. In this paper, a learner G_Q (three linear layers + ReLU + Softmax) is constructed to learn the weight parameters of all augmentation methods in the pool, and $G_Q \to Q \to v$. Tian et al. [17] have demonstrated that the mutual information between v and $\mathcal{Q}(v)$ should strike a balance between being neither excessive nor insufficient, aiming to maintain a certain level of fidelity while preserving diversity. Task-related features play a significant role in representation learning, influencing downstream tasks. Therefore, maximizing the similarity of task-related features between v and $\mathcal{Q}(v)$ is desirable. Meanwhile, minimizing the mutual information between v and $\mathcal{Q}(v)$ helps to avoid excessive similarity that may cause noise, over-fitting and inclusion of irrelevant features. This process allows for noise suppression while preserving critical information and can be represented as:

$$\arg\min_{\mathcal{Q}} I\left(v, \mathcal{Q}\left(v\right)\right) + \arg\max_{\mathcal{Q}} I_t\left(v, \mathcal{Q}\left(v\right)\right) \tag{6}$$

We use InfoNCE [15] to replace mutual information bound calculation, the optimization objective of learner can be formulated as follows:

$$\mathcal{L}^{\text{aug}} = \mathcal{L}_{InfoNCE}(v^{tr}, (\mathcal{Q}(v))^{tr}) - \lambda \mathcal{L}_{InfoNCE}(v, \mathcal{Q}(v)). \tag{7}$$

v^{tr} and $(\mathcal{Q}(v))^{tr})$ denotes the global, seasonal and trend features of data described in Sect. 3.2. The outputs of G_Q are the weights of various augmentation methods. Through their weighted combination, $\mathcal{Q}$ is obtained, ensuring the preservation of a maximum amount of task-related features.

It is worth noting that there are two training stages, in the initial training stage (first $\frac{1}{3}$ of total epochs), we employ a low-pass smoothing augmentation method to cold-start the network. In the second stage, the augmented data generated by task-adaptive augmentation learner is utilized. The two-stage training approach ensures that the shared information between views is task-relevant and diverse. This approach helps the model to balance between task-related and general features, which improves the representation learning capability of the model. Figure 3 shows the features obtained through DisT and the global features, demonstrating the superior feature extraction capability of DisT.

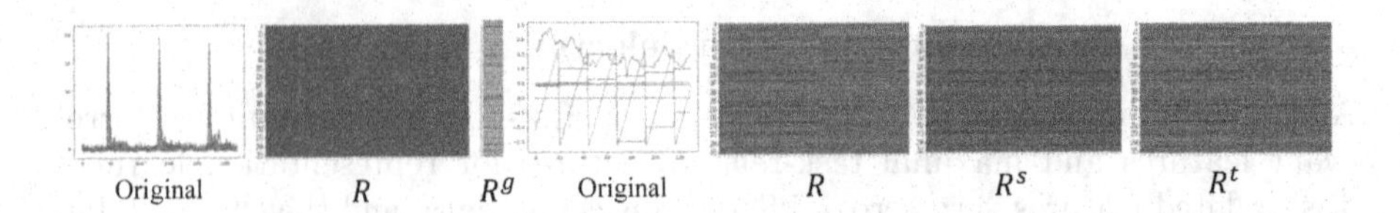

Fig. 3. Visualization of two multivariate time series samples, features and task-related features. Global feature R^g, seasonal feature R^s and trend feature R^t are decoupled from general feature R.

4 Experiments and Discussions

4.1 Datasets and Implementation Details

UEA (University of East Anglia) [1] contains a collection of 30 datasets with a wide range of cases, dimensions, and time-series lengths for multivariate time-series classification, which is widely recognized as one of the most comprehensive classification benchmark. **ETT** (Electricity Consumption and Temperature Time-series) [28] is a publicly dataset for time series forecasting. For our experiments, we employ three sub-datasets, namely ETTh1, ETTh2, and ETTm1. The dataset spans a diverse range of time intervals with ETTh1 and ETTh2 recorded at hourly intervals, and ETTm1 recorded at 15-minute intervals.

We adopted an unsupervised representation learning framework TS2Vec [25] as our baseline. The dilated convolutions and input projection layer in the encoder model were retained, while the timestamp masking layer was removed. All experiments were performed using PyTorch-1.8 on one Tesla V 100 GPU with 32G memory, while the operating system was CentOS 7.5. We follow previous work [25] and use supervised methods to train SVM classifiers and linear regressors respectively for classification and forecasting tasks. Same as previous work, different forecasting steps H were set on various datasets, $H \in \{24, 48, 168, 336, 720\}$ in ETTh1, ETTh2 and $H \in \{24, 48, 96, 288, 672\}$ in ETTm1. Mean squared error (MSE) and mean absolute error (MAE) were used as evaluation criteria. In the experiments for classification task, we selected appropriate values for the parameter α in Eq. (4) for each scenario in $[0.1, 0.5, 0.05]$. As for the forecasting tasks, due to the similarity and uniformity of the data sources and the stability of features, the parameter β_s and β_t in Eq. (5) for the forecasting task was set to 0.005 and 0.025, and λ in Eq. (7) is 0.5. All parameters are set based on empirical knowledge and a series of pilot experiments. The augmentation methods including smoothing, scaling, and window slicing with wrapping [11], STL decomposition [4], cutout, subsequence [25], and augmentation in frequency domain aaft[16]. The smoothing process involves two steps: applying a sliding window with a size of 3 for mean downsampling, and then upsampling to restore the original time series. The window slicing and wrapping augmentation involve initially slicing the time series using windows with a slicing ratio of 0.95 and then wrapping it back to its original length. The STL decomposition method includes two augmentations. One focuses on preserving the trend component of the original data

$\mathcal{P}(\lambda_1, \lambda_2, \lambda_3) \sim (0, 1, 0.5)$, while the other emphasizes the preservation of the seasonal component $\mathcal{P}(\lambda_1, \lambda_2, \lambda_3) \sim (1, 0, 0.5)$. Scaling scales samples by a factor randomly sampled from a Gaussian distribution $\mathcal{N}(0, 0.1)$. The cropping operation replaces 10% of randomly sampled input values with zeros. Subsequence refers to extracting a sub sequence from input.

Table 1. The ablation study on 30 UEA classification datasets, employing various task-related feature losses and augmentation selection strategies.

Setting	TR-F			TA-A		
	only $\mathcal{L}^{gen}$	use $\mathcal{L}^{fre}$	use $\mathcal{L}^{cls}$	Random Choice	Random Sum	**DisT**
30 Avg. (Acc.)	70.4%	69.2%	**70.9%**	70.2%	70.1%	**70.8%**

Table 2. The ablation study of all components of the DisT.

Setting		Forecasting				Classification			
		ETTm1		ETT Avg.		SCP1.	ERing	Handw.	30 Avg.
TR-F	TA-A	MSE	MAE	MSE	MAE	Acc. (%)	Acc. (%)	Acc. (%)	Acc. (%)
–	–	0.628	0.553	0.994	0.712	81.2	87.4	51.5	70.4
✓	–	0.623	0.556	0.979	0.708	83.0	89.2	52.8	70.9
✓	✓	**0.578**	**0.531**	**0.972**	**0.703**	**90.4**	**92.6**	**55.6**	**71.8**

4.2 Ablation Analysis

The ablation experiments on the feature disentanglement method proposed in this paper are presented in Table 1 and Table 2. In the tables, 'TR-F' refers to the task-related feature disentanglement module, and 'TA-A' represents task-adaptive augmentation selection strategy. As shown in Table 1, to demonstrate the effectiveness of the feature disentangled module, we compare the model performance on classification task when different losses (general contrastive loss $\mathcal{L}^{gen}$, forecasting task-related loss $\mathcal{L}^{fre}$, and classification task-related loss $\mathcal{L}^{cls}$ are adopted. The results confirm that the classification of task-related loss is more effective. In order to analyze the effectiveness of the proposed task-adaptive augmentation selection method, we design two random selection strategies for comparison. "Random Choice" represents randomly selecting a method from the augmentation method pool, while 'Random Sum' represents randomly generating weights for each method in the augmentation method pool. Compared to the random strategy, the average classification results improved by 0.8%, indicating the effectiveness of the task-adaptive augmentation selection strategy. Table 2 demonstrates the effectiveness of the two proposed task-related feature learning methods compared to the baseline. "SCP1." and "Handw." represent SelfRegulationSCP1 and Handwriting dataset in the UEA benchmark, while "30

Table 3. Detailed results with other state-of-the-arts on multivariate classification UEA dataset.

Dataset	DisT	InfoTS [14]	TS2Vec	T-Loss	TCN [19]	TS-TCC	TST [27]
ArticularyWordRecognition	**0.993**	0.987	0.987	0.943	0.973	0.953	0.977
AtrialFibrillation	**0.333**	0.200	0.200	0.133	0.133	0.267	0.067
BasicMotions	**1.000**	0.975	0.975	**1.000**	0.975	**1.000**	0.975
CharacterTrajectories	0.994	0.974	**0.995**	0.993	0.967	0.985	0.975
Cricket	**1.000**	0.992	0.972	0.972	0.958	0.917	**1.000**
DuckDuckGeese	0.500	0.540	**0.680**	0.650	0.460	0.380	0.620
EigenWorms	0.798	0.733	**0.847**	0.840	0.840	0.779	0.748
Epilepsy	0.949	**0.971**	0.964	**0.971**	0.957	0.957	0.949
ERing	0.926	**0.949**	0.874	0.133	0.852	0.904	0.874
EthanolConcentration	0.293	0.281	**0.308**	0.205	0.297	0.285	0.262
FaceDetection	0.524	0.534	0.501	0.513	0.536	**0.544**	0.534
FingerMovements	0.540	**0.630**	0.480	0.580	0.470	0.460	0.560
HandMovementDirection	**0.392**	**0.392**	0.338	0.351	0.324	0.243	0.231
Handwriting	**0.556**	0.452	0.515	0.451	0.249	0.498	0.225
Heartbeat	0.717	0.722	0.683	0.741	0.746	**0.751**	0.746
JapaneseVowels	0.984	0.984	0.984	**0.989**	0.978	0.93	0.978
Libras	**0.883**	**0.883**	0.867	**0.883**	0.817	0.822	0.656
LSST	0.542	0.591	0.537	0.509	**0.595**	0.474	0.408
MotorImagery	0.510	0.500	0.510	0.580	0.500	**0.610**	0.500
NATOPS	0.912	0.922	**0.928**	0.917	0.911	0.822	0.850
PEMS-SF	0.728	0.705	0.682	0.676	0.699	0.734	**0.74**
PenDigits	**0.989**	0.980	**0.989**	0.981	0.979	0.974	0.560
PhonemeSpectra	0.223	0.200	0.233	0.222	0.207	**0.252**	0.085
RacketSports	0.855	**0.864**	0.855	0.855	0.776	0.816	0.809
SelfRegulationSCP1	**0.904**	0.843	0.812	0.843	0.799	0.823	0.754
SelfRegulationSCP2	0.561	0.572	**0.578**	0.539	0.550	0.533	0.550
SpokenArabicDigits	0.982	**0.991**	0.988	0.905	0.934	0.970	0.923
StandWalkJump	**0.533**	**0.533**	0.467	0.333	0.400	0.333	0.267
UWaveGestureLibrary	**0.928**	0.900	0.906	0.875	0.759	0.753	0.575
InsectWingbeat	**0.491**	0.443	0.466	0.156	0.469	0.264	0.105
Avg. Acc. on 30 datasets	**0.718**	0.714	0.704	0.658	0.670	0.668	0.617

Avg." denotes the average classification accuracy across all 30 UEA classification datasets. The results highlight the effectiveness of individual components, but when all components are combined, DisT proves its superior performance across different datasets and tasks.

4.3 Results on Classification Tasks

The classification results for the 30 datasets in UEA are listed in Table 3. It is noticeable that certain datasets, like AtrialFibrillation, have features that are hard to mine. For such datasets, the optimal augmentation selection learner enables the model to focus on learning complex task-relevant features and greatly improves performance. Although all 30 datasets in the UEA dataset are classification tasks with known task-specific features disentangled by the Classification Feature Disentangler, it should be noted that the UEA dataset comprises various types of time series data, such as sensor data, audio, human activities, etc. In different scenarios, the crucial details on which the classification relies may vary, including amplitude, frequency, and local fluctuations. Therefore, fixed view generation methods are not robust for this diversity. The task-adaptive augmentation selection learner provides a solution to the aforementioned problem by allowing for the generation of appropriate positive samples for data from different tasks and scenarios. Additionally, Fig. 4 illustrates the critical difference diagram [5] of the Nemenyi test for all datasets.

4.4 Results on Forecasting Tasks

The experimental results of multivariate forecasting is listed in Table 4, comparative forecasting methods including TS2Vec [25], TNC [2], MoCo [9], CPC [15], Informer [28], LogTrans [12], and LSTnet [10]. Forecasting tasks mainly rely on

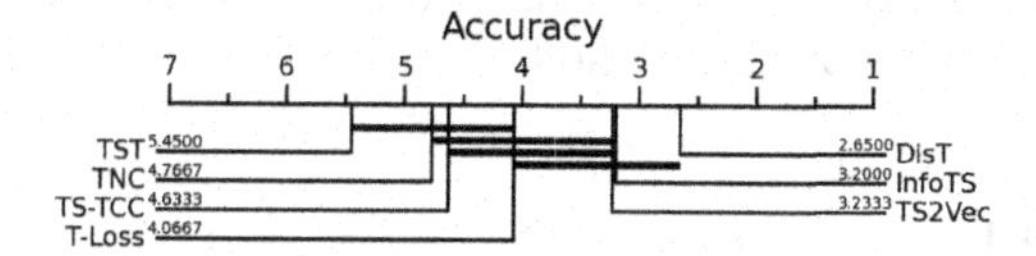

Fig. 4. Critical Difference (CD) diagram of various methods on time series classification tasks.

Table 4. Comparison of detailed results with other state-of-the-arts on multivariate forecasting task.

Method		Representation Learning										End-to-end Forecasting					
		DisT		TS2Vec		TNC		MoCo		CPC		Informer		LogTrans		LSTnet	
Data	H	MSE	MAE	MSE	MAE	MSE	MAE	MSE	MAE	MSE	MAE	MSE	MAE	MSE	MAE	MSE	MAE
ETTh1	24	**0.565**	0.535	0.599	**0.534**	0.708	0.592	0.623	0.555	0.728	0.600	0.577	0.549	0.686	0.604	1.293	0.901
	48	**0.605**	0.560	0.629	**0.555**	0.749	0.619	0.669	0.586	0.774	0.629	0.685	0.625	0.766	0.757	1.456	0.960
	168	**0.744**	0.642	0.755	**0.636**	0.884	0.699	0.820	0.674	0.920	0.714	0.931	0.752	1.002	0.846	1.997	1.214
	336	0.904	0.723	**0.907**	**0.717**	1.020	0.768	0.981	0.755	1.050	0.779	1.128	0.873	1.362	0.952	2.655	1.369
	720	**1.044**	0.795	**1.048**	0.790	1.157	0.830	1.138	0.831	1.160	0.835	1.215	0.896	1.397	1.291	2.143	1.380
ETTh2	24	0.506	0.513	**0.398**	**0.461**	0.612	0.595	0.444	0.495	0.551	0.572	0.720	0.665	0.828	0.750	2.742	1.457
	48	0.687	0.622	**0.580**	**0.573**	0.840	0.716	0.613	0.595	0.752	0.684	1.457	1.001	1.958	1.071	3.567	1.687
	168	**1.855**	**1.046**	1.901	1.065	2.359	1.213	1.791	1.034	2.452	1.213	3.489	1.515	4.070	1.681	3.242	2.513
	336	**2.269**	**1.189**	2.304	1.215	2.782	1.349	2.241	1.186	2.664	1.304	2.723	1.340	3.875	1.763	2.544	2.591
	720	**2.508**	**1.358**	2.650	1.373	2.753	1.394	2.425	1.292	2.863	1.399	3.467	1.473	3.913	1.552	4.625	3.709
ETTm1	24	**0.404**	**0.419**	0.443	0.436	0.522	0.472	0.458	0.444	0.478	0.459	0.323	0.369	0.419	0.412	1.968	1.170
	48	**0.513**	**0.485**	0.582	0.515	0.695	0.567	0.594	0.528	0.641	0.550	0.494	0.503	0.507	0.583	1.999	1.215
	96	**0.578**	**0.533**	0.622	0.549	0.731	0.595	0.621	0.553	0.707	0.593	0.678	0.614	0.768	0.792	2.762	1.542
	288	**0.650**	**0.579**	0.709	0.609	0.818	0.649	0.700	0.606	0.781	0.644	1.056	0.786	1.462	1.320	1.257	2.076
	672	**0.748**	**0.639**	0.786	0.655	0.932	0.712	0.821	0.674	0.880	0.700	1.192	0.926	1.669	1.461	1.917	2.941
Avg. ACC		**0.972**	**0.703**	0.994	0.712	1.171	0.785	0.996	0.721	1.160	0.778	1.342	0.859	1.635	1.053	2.411	1.782

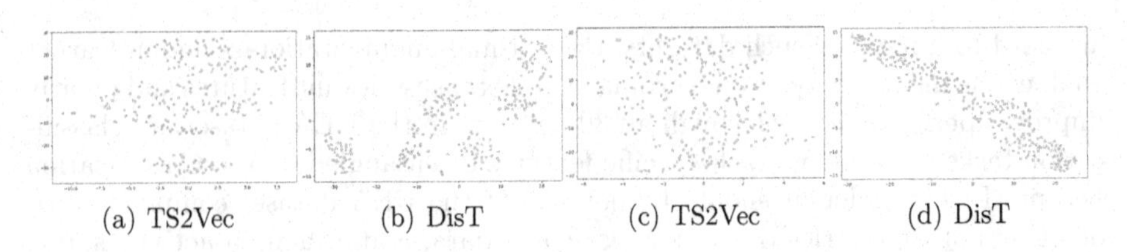

(a) TS2Vec (b) DisT (c) TS2Vec (d) DisT

Fig. 5. T-SNE visualization of learned features of on ERing and SelfRegulationSCP1 datasets of UEA, which contains 6 classes and 2 classes, respectively.

features related to seasonality and trends. Nonetheless, the proposed method still achieves SOTA level performance in forecasting when compare to unified representation learning methods. The relative improvements on MSE and MAE are 2.2% and 1.2% against the best representation learning method TS2Vec and are 27.5% and 18.1% against the best end-to-end models Informer.

4.5 Visualization Analysis

Figure 5 shows the visualization of the learned features using T-SNE [20] algorithm, and the visualized features in subfigs (b) (d) show that the learned representation of instances from the same class are more compactly clustered, demonstrating the effectiveness of the method.

5 Conclusion

In this paper, we have addressed the challenge of task-related feature learning in multivariate time series representation. By incorporating a feature decoupling module and adaptive data augmentation selection methods, DisT has successfully disentangled task-related features and achieved efficient feature learning. DisT prioritizes global features for time series classification tasks and captures trend and seasonal features for time series forecasting tasks. Our study highlights the importance of considering task-specific features and introduces a novel method to enhance representation learning in time series analysis.

References

1. Bagnall, A., et al.: The UEA multivariate time series classification archive. arXiv preprint arXiv:1811.00075 (2018)
2. Bai, S., Kolter, J.Z., Koltun, V.: An empirical evaluation of generic convolutional and recurrent networks for sequence modeling. arXiv preprint arXiv:1803.01271 (2018)
3. Bengio, Y., Courville, A., Vincent, P.: Representation learning: a review and new perspectives. IEEE Trans. Pattern Anal. Mach. Intell. **35**(8), 1798–1828 (2013)
4. Cleveland, R.B., Cleveland, W.S., McRae, J.E., Terpenning, I.: STL: a seasonal-trend decomposition. J. Off. Stat **6**(1), 3–73 (1990)

5. Demšar, J.: Statistical comparisons of classifiers over multiple data sets. J. Mach. Learn. Res. **7**, 1–30 (2006)
6. Eldele, E., et al.: Time-series representation learning via temporal and contextual contrasting (2021)
7. Franceschi, J.-Y., Dieuleveut, A., Jaggi, M.: Unsupervised scalable representation learning for multivariate time series. Adv. Neural Inf. Process. Syst. **32** (2019)
8. Harutyunyan, H., Khachatrian, H., Kale, D.C., Ver Steeg, G., Galstyan, A.: Multitask learning and benchmarking with clinical time series data. Scientific Data **6**(1), 96 (2019)
9. He, K., Fan, H., Wu, Y., Xie, S., Girshick, R.: Momentum contrast for unsupervised visual representation learning. In: Proceedings of the IEEE/CVF Conference on Computer Vision and Pattern Recognition, pp. 9729–9738 (2020)
10. Lai, G., Chang, W.-C., Yang, Y., Liu, H.: Modeling long-and short-term temporal patterns with deep neural networks. In: The 41st International ACM SIGIR Conference on Research and Development in Information Retrieval, pp. 95–104 (2018)
11. Le Guennec, A., Malinowski, S., Tavenard, R.: Data augmentation for time series classification using convolutional neural networks. In: ECML/PKDD Workshop on Advanced Analytics and Learning on Temporal Data (2016)
12. Li, S., et al.: Enhancing the locality and breaking the memory bottleneck of transformer on time series forecasting. Adv. Neural Inf. Process. Syst. **32** (2019)
13. Li, Y., Hu, G., Wang, Y., Hospedales, T., Robertson, N.M., Yang, Y.: Dada: differentiable automatic data augmentation. arXiv preprint arXiv:2003.03780 (2020)
14. Luo, D., et al.: Time series contrastive learning with information-aware augmentations. Proc. AAAI Conf. Artif. Intell. **37**, 4534–4542 (2023)
15. van den Oord, A., Li, Y., Vinyals, O.: Representation learning with contrastive predictive coding. arXiv preprint arXiv:1807.03748 (2018)
16. Theiler, J., Eubank, S., Longtin, A., Galdrikian, B., Farmer, J.D.: Testing for nonlinearity in time series: the method of surrogate data. Physica D: Nonl. Phenom. **58**(1–4), 77–94 (1992)
17. Tian, Y., Sun, C., Poole, B., Krishnan, D., Schmid, C., Isola, P.: What makes for good views for contrastive learning? Adv. Neural. Inf. Process. Syst. **33**, 6827–6839 (2020)
18. Tonekaboni, S., Eytan, D., Goldenberg, A.: Unsupervised representation learning for time series with temporal neighborhood coding. In: International Conference on Learning Representations
19. Tonekaboni, S., Eytan, D., Goldenberg, A.: Unsupervised representation learning for time series with temporal neighborhood coding. arXiv preprint arXiv:2106.00750 (2021)
20. Van der Maaten, L., Hinton, G.: Visualizing data using t-sne. J. Mach. Learn. Res. **9**(11) (2008)
21. Wang, H., Guo, X., Deng, Z.-H., Lu, Y.: Rethinking minimal sufficient representation in contrastive learning. In: Proceedings of the IEEE/CVF Conference on Computer Vision and Pattern Recognition, pp. 16041–16050 (2022)
22. Wang, Z., Xovee, X., Zhang, W., Trajcevski, G., Zhong, T., Zhou, F.: Learning latent seasonal-trend representations for time series forecasting. Adv. Neural. Inf. Process. Syst. **35**, 38775–38787 (2022)
23. Woo, G., Liu, C., Sahoo, D., Kumar, A., Hoi, S.: Cost: contrastive learning of disentangled seasonal-trend representations for time series forecasting. arXiv preprint arXiv:2202.01575 (2022)

24. Yu, H., Yang, H., Sano, A.: Leaves: learning views for time-series data in contrastive learning. arXiv preprint arXiv:2210.07340 (2022)
25. Yue, Z., et al.: Ts2vec: towards universal representation of time series. Proc. AAAI Conf. Artif. Intell. **36**, 8980–8987 (2022)
26. Zeng, A., Chen, M., Zhang, L., Qiang, X.: Are transformers effective for time series forecasting? Proc. AAAI Conf. Artif. Intell. **37**, 11121–11128 (2023)
27. Zerveas, G., Jayaraman, S., Patel, D., Bhamidipaty, A., Eickhoff, C.: A transformer-based framework for multivariate time series representation learning. In: Proceedings of the 27th ACM SIGKDD Conference on Knowledge Discovery and Data Mining, pp. 2114–2124 (2021)
28. Zhou, H., et al.: Informer: beyond efficient transformer for long sequence time-series forecasting. Proc. AAAI Conf. Artif. Intell. **35**, 11106–11115 (2021)

A Multi-view Feature Construction and Multi-Encoder-Decoder Transformer Architecture for Time Series Classification

Zihan Li[1(✉)], Wei Ding[1,3], Inal Mashukov[1], Scott Crouter[2], and Ping Chen[1]

[1] University of Massachusetts Boston, Boston, MA 02125, USA
zihan.li001@umb.edu
[2] University of Tennessee-Knoxville, Knoxville, TN 37996, USA
[3] Paul English Applied AI Institute, University of Massachusetts Boston, Boston, MA 02125, USA

Abstract. Time series data plays a significant role in many research fields since it can record and disclose the dynamic trends of a phenomenon with a sequence of ordered data points. Time series data is dynamic, of variable length, and often contains complex patterns, which makes its analysis challenging especially when the amount of data is limited. In this paper, we propose a multi-view feature construction approach that can generate multiple feature sets of different resolutions from a single dataset and produce a fixed-length representation of variable-length time series data. Furthermore, we propose a multi-encoder-decoder Transformer (MEDT) architecture to effectively analyze these multi-view representations. Through extensive experiments using multiple benchmarks and a real-world dataset, our method shows significant improvement over the state-of-the-art methods.

Keywords: Multivariate Time Series Classification · Multi-view Learning · Multi-Encoder-Decoder · Transformer · Variable-length Time Series

1 Introduction

Time series data provides a clear and dynamic way to record the evolution of different variables of a phenomenon over time. In dynamic scenarios, time becomes a crucial dimension for complete recording, and time series data can represent an ordered sequence of features based on time, efficiently captures the dynamic relationship within evolving phenomena. Currently, time series data is widely used in various fields, such as Biology, Physics, Meteorology and Sport Medicine, there are still a few significant challenges in time series data analysis. Although many studies providing strong solutions to deal with time series data, such as [2,4,7,9], there are still some limits and challenges in the field:

D.-N. Yang et al. (Eds.): PAKDD 2024, LNAI 14650, pp. 239–250, 2024.
https://doi.org/10.1007/978-981-97-2266-2_19

Time series analysis is very sensitive to the quality of data. Since the prediction is based on the dynamic trend of the relationship between features and time, it would have a significant impact to the model if erroneous data points existed. To solve this, we assume that any time series events consist of a set of time units with similar characteristics, we define them as atomic units of the data (more details in Sect. 4.1). These units are always repeated among the whole sequence of time series data since they are basic and key building blocks constituting the data. With a representation using a set of basic atomic units, noise and outliers can be reduced significantly, and only key characteristics of data can be kept in the new representation. The proposed method can extract and construct atomic units from the original sequences of time series data [4], which can significantly reduce the negative impacts caused by the data quality and make the prediction processing more robust compared to the current methods.

Time series data is often recorded in variable length. Considering the characteristics of time series data, it is hard to begin and finish the recording of phenomena concurrently for different individuals. For example, when we monitor people's activities during the day, we don't know how long each activity will last, and it is impossible to perform all actions in the same time period. In such cases, it is incompatible for use with traditional machine learning models due to its variable-length features. To deal with this issue, there are two common techniques used frequently: cutting off or padding the period with the same length for all records [9,10] or using fixed length sliding windows to represent the original data [23]. However, both of these methods will cause information loss due to the cut-off process. In contrast to these, the proposed method can keep maximum information by constructing a high-level summarization of the original variable-length time series data based on the extracted atomic units contained within the data. The atomic units serve as basic building blocks in data. The summarization of atomic units can provide a fix-length representation of the original variable-length time series data.

Time series data requires high computation cost. To make a time series dataset into a suitable input for a machine learning algorithm, the sliding windows technique is currently the most popular one, however, there is a significant limit of the sliding windows technique - the amount of data would increase rapidly depending on the width of a window and the length of a moving step, which means the cost of computation also would increase rapidly. Our proposed atomic based method can provide an adaptive choice of the size to fix the resource of computation. We are able to serve more flexibility by extracting fixed number of atomic units and building a new representation based on it, for example, we can reducing the number of features to a fixed number and decrease the complexity of computation significantly while producing competitive prediction results.

Time series data could be univariate or multivariate. While more features provide detailed insights, the challenge lies in managing collinearity to avoid overfitting during prediction processing. Our atomic based method can project variables to one or more spaces without high correlations by applying Gaussian

Mixture Model and generate multi-view representations according to different resolutions. Since these multiple views are extracted from single time series data, they can reveal as much as information under different granularity, which is hard to be extracted from the original time series data.

In summary, we propose an innovative feature construction approach that can generate multiple feature sets of different resolutions from a single dataset and produce a fixed-length representation of variable-length time series data. By applying the extracted atomic units from the original data, it allows the proposed model providing a competitive solution of noise, variable-length and computation cost. Our model, Multi-Encoder-Decoder Transformer (MEDT), attempts to encapsulate and utilize all global information about the time series data. These multiple feature sets provide multiple views on the same data and are fed into a multi-encoder-decoder Transformer architecture, which is inspired by multilingual neural translation. Our main contributions are as follows:

- We propose a novel multi-view feature construction approach to deal with variable-length time series data and noise. In order to keep as much as global information from the original time series data, our method can construct multiple sets of fixed length features representation based on atomic units of data. These multi-view representations capture information of different granularity from original data and produce more robust results.
- We develop a multi-encoder-decoder Transformer model to effectively analyse these multiple feature sets for time series classification since these multiple views describe the same underlying phenomenon, inspired by multilingual neural translation (e.g., different languages encode the same semantics).
- We provide more flexibility in the number of features, which can help reduce the complexity of computation and provide a more efficient method for time series prediction. By constructing selected number of new features summarizing original data, computation could decreases with a simple fixed length data input, while prediction quality improves.

2 Related Works

Numerous time series data analysis methods have been proposed. Recent methods include HIVE-COTE (Lines et al., 2018) [1], ROCKET (Dempster et al.,2020) [2], and TS-CHIEF(Shifaz et al., 2020) [3], which are considered to be the state-of-the-art when tackling time series classification problems. Other popular methods include CNN-based deep learning models such as REsNet (Fawaz et al., 2019b) [11] and InceptionTime (Fawaz et al., 2019a) [8]. However, these methods are computationally costly and complex and often fail to produce good results for datasets containing numerous samples of lengthy time series data. Dempster et al. (2021) [29] introduce a fast MiniRocket method which is 75 times faster than original Rocket. Gao et al. (2022) [30] provides a reinforcement learning framework for multivariate time series classification which can identify interpretable patterns without using neural network. Although these methods show improvements either on speed or accuracy. However, it is hard to find a method which can provide consistent and competitive accuracy versus other SOTA methods.

Currently, there are a lot of transformer based publications of time series analysis [27]. Transformer, although initially proposed for natural language translation and having demonstrated remarkable results in various NLP tasks, have found applications in time series tasks while also providing for efficient computation. In contrast to other popular sequential data classification methods, the classical Transformer model presented by Vaswani et al. (2017) [5] is based exclusively on the attention mechanism. The attention mechanism tends to global dependencies between input and output, while the architecture of the Transformer model allows for greater parallelization, resulting in a significantly more efficient and accurate classification [5]. Li et al. (2019) [13] and Wu et al. (2020) [14] have employed full encoder-decoder transformer architectures for univariate time series forecasting, outperforming traditional statistical methods and RNN-based models. Ma et al. (2019) applied transformers for the imputation of missing values in multivariate time series, showcasing their effectiveness in handling the data. Hao et al. (2020) [28] provides a 2-step attention-based CNN model which designs an attention mechanism to extract memories across all time steps and then applies another attention mechanism for variable selection. The work presented by Zerveas et al. (2021) [7] introduces a transformer-based framework for unsupervised representation learning of multivariate time series. This methodology leverages unlabeled data by pre-training a transformer model with an input denoising objective. Zhou et al. (2020) [12] introduced a Transformer-based architecture with two symmetric language-specific encoders. This Multi-Encoder-Decoder Transformer architecture effectively captures individual language attributes and employs a language-specific multi-head attention mechanism in the decoder module.

Our approach shows significant improvement over the state-of-the-art methods, as it performs feature extraction on variable-length time series data, learning to construct multiple robust fixed-length representations of the original information, with the different representations serving as inputs to the Multi-Encoder-Decorder Transformer model, leveraging the architecture's efficiency to capture maximal information from each feature set.

3 Problem Formulation

In this section, we formulate the problem solved in this paper in a mathematical way. Time series features record the data in ordered sequences of points over time, such as the gait force and the moving activities being recorded based on a sequence of time slots. We assume a time series sample X can be represented by $|T|$ ordered data points where T is the full-time period. In this case, each sample of records can be written as $X = \{x_{t_1}, x_{t_2}, \dots x_{t_i}; t_i \in T\}$, where x_{t_i} represents all features of X at time t_i, and a time series data set D with N features can be written as $D = \{X_1, X_2, \dots X_n; n \in N\}$. So the main problem can be formulated as follows [4]:

$$f(X_1, X_2 \ldots, X_n) : \mathcal{D} \to \mathcal{C} \tag{1}$$

where D is the input time series data and C is the class labels.

For the length of a sample $X_i \in D$, in most cases, it will vary individually, which means $|X_n| \neq |X_m|$ where $n, m \in N$. It is impossible to find a model f to handle inconsistent dimensional data.

Our solution is to extract a set of atomic units A by applying a data-driven summarized method $E(X) : D \to A$, where $A = \{a_1, a_2 \ldots a_k\}$ and k will be a fixed value. Given the set A, it allows us to construct a new fixed-dimensional summarized data D' based on the atomic units contained in it. The samples X' in the new data D' can be represented by the set of atomic units A. Now, it will be able to apply a clustering method f on the summarized data D'.

4 Methodology

In contrast to the existing methods, the proposed algorithm is applicable to both fixed and variable length time series data. The algorithm consists of the following phases:

- Feature Construction: Summarizing the variable-length time series data and constructing a fixed-length representation based on its atomic units.
- Multi-view Representation: Constructing multiple representations of the original dataset, each with varying number of features.
- Classification: Applying our Multi-Encoder-Decoder Transformer (MEDT) model to classify multivariate time series data instances.

4.1 Feature Construction

In time series data, there are always some repeated events that happen over time. The biggest challenge in the field is how to find patterns and relationships between these repeated events and time steps. In our study, we introduce the term of atomic units (Definition 1) which constitutes these events [4].

Definition 1. ***Atomic Unit***
Suppose a time series data is split into a sequence of small time periods, we assume that there are some repeated common characteristics among the sequence, this kind of common characteristics are named as atomic units.

Atomic units are the basic building block of a time series data. They appear in time series data repeatedly and play a significant role in representation. According to the different resolutions (number of features) required, we apply the Gaussian Mixture Model to extract a set of atomic units for each resolution. The new representations are ratio features built by each set of atomic units.

Gaussian Mixture Model (GMM) is a probability-based unsupervised clustering technique which is formed by several single Gaussian distributions [19].

With these individual Gaussian distributions, we can simply simulate the time series events by clustering the atomic units set A over time steps (T_k). Suppose we have time series data D and features X_n, the GMM clustering method is applied to find the best individual distribution of each atomic unit, then we are able to construct new ratio features based on the GMM clustering results.

Considering the time steps as the basic elements of a time series event, we can define a set of the distributions $p_k(x)$ over all time steps. And clustering each step into one of our designed atomic units following the distribution $p_k(x)$. We calculate the probability μ of each time step belongs to a atomic unit [4], where

$$\mu_k(x) = \frac{\pi_k N(x\|\mu_k, \sigma_k)}{\sigma_l \pi_l N(x\|\mu_l \sigma_l)} \tag{2}$$

The best matching atomic unit will be assigned based on the calculation of $argmax\mu_k(x)$ over all potential atomic units.

After pairing the time steps and atomic units by calculating $argmax\mu_k(x)$, we can construct a fix-length ratio feature to represent the original high dimensional time series events, where

$$r_k = \frac{|A_k|}{|A|} \tag{3}$$

The ratio can be calculated with the number of the atomic unit over the total number of atomic units. Then the time series D and be summarized as a vector $[\frac{|A_1|}{|A}], \frac{|A_2|}{|A|} \cdots \frac{|A_k|}{A}]$.

4.2 Multi-view Representation

Our feature construction approach can represent time series data with an arbitrary number of features. Since the created ratio features have a very low level of collinearity, we can generate multiple summarizations of the source data with different resolutions, which convert the source data into a set of multi-view representations based on atomic units.

Inspired by the idea of multilingual model [24], we create multiple representations with different numbers of features to describe single time series data. These representations are considered as multiple views of the data. We believe multi-view representations can help capture more information under different granularity [25,26]. To take as much as patterns in multi-view representations, we proposed a multi-encoder-decoder transformer model which uses multiple views as inputs.

4.3 Multi-Encoder-Decoder Transformer (MEDT) Classification

We propose a modified architecture of the model - one where each encoder-decoder pair will take as input two different representations of the same dataset. After constructing new features, we may choose to represent the original dataset with two new datasets of n and m features, passing each individually into either a single encoder or an additional decoder. For example, we may choose to construct two representations - one with 10 and another 20 with features. Thus, we would feed either the 10 or the 20 feature dataset into either the encoder or the decoder. If we decide to construct 8 different representations of the original dataset, we may choose to have 4 encoder-decoder pairs. The output of an encoder, along with the output of a decoder, passes through dense layers and a softmax layer, producing the final output of the Multi-Encoder-Decoder Transformer (MEDT).

The detailed architecture of the model is presented in Fig. 1.

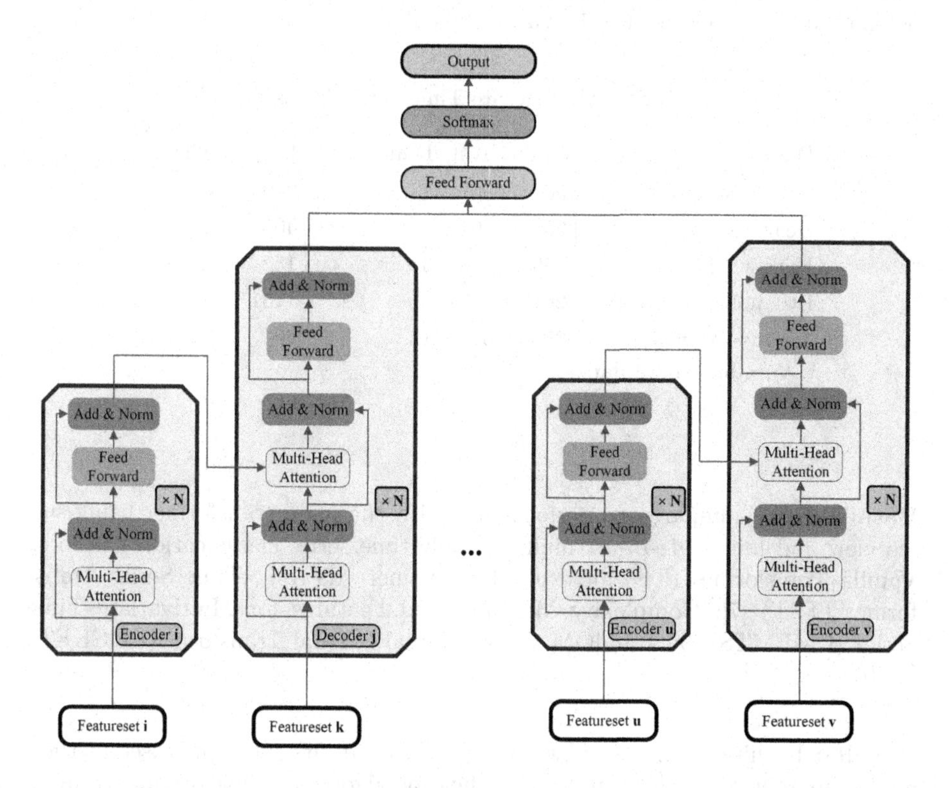

Fig. 1. Multi-Encoder-Decoder Transformer (MEDT) Architecture

5 Experiments

Our model is evaluated on the physical activities dataset [16] and 5 real-world datasets from the UEA multivariate time series classification (MTSC) archive [15]. We compare the performance of our model to the other four state-of-art methods.

5.1 Experiments Using Multivariate Time Series Data Benchmarks

We evaluate our method on 5 multivariate time series datasets from the University of East Anglia (UEA) Multivariate archive. We select the data from different domains, such as human activity, motion classification, and audio spectra classification, with various dimensions (from 3 to 1345), length (from 29 to 1751), and number of classes (from 2 to 26). We use 4 fixed-length datasets and 1 variable-length dataset in this study to show the robust and competitive performance of our method. The details are shown in Table 1.

Table 1. Multivariate Time Series Datasets

Dataset	Train	Test	Dimensions	Length	Classes
DuckDuckGeese	50	50	1345	270	5
Heartbeat	204	205	61	405	2
Handwriting	150	850	3	152	26
EthanolConcentration	261	263	3	1751	4
JapaneseVowels*	270	370	12	29	9

*: Variable-length dataset

Baseline: We compare our performance with the plain transformer using single view and four state-of-art multi-variable time series classification methods: Vanilla Transformer (TF) [5], Rule Transformer (RT) [6], Time Series Transformer (TST) [7], Complexity Measures and Features for Multivariate Time Series (CMFMTS) [9], RandOM Convolutional KErnal Transform (ROCKET) [2,10].

Results: In this section, we show the experiment results of our proposed method and compare it with other current classification algorithms. The original training and testing splits are used. Our method presents a competitive performance in most cases and in the first rank of average ranking, the details of results are shown in Table 2.

Our method (MEDT) shows the best performance in DuckDuckGeese and Heartbeat datasets. The DuckDuckGeese has the highest dimensions which is 1345, because of this, some competitors didn't provide results on this dataset, for instance, it is too large to run with the vanilla TF model. For other datasets, although our method stands at the second place among all models, it provides a very close performance to the best one. The JapaneseVowels dataset has variable-length time steps for each sample, and due to this issue, two of the competitors do not include it in their studies. Our proposed MEDT method presents the best overall average rank among all methods listed, which shows a solid good performance of it.

Table 2. Performance in Accuracy for Multivariate datasets

Dataset	TF	RT20%	TST(pre-train)	CMFMTS	ROCKET	MEDT
DuckDuckGeese	–	18.0%	–	51.0%	–	**51.99%**
Heartbeat	72.66%	73.17%	77.6%	76.8%	72.68%	**78.05%**
Handwriting	3.76%	26.24%	**35.9%**	27.4%	21.88%	28.5%
EthanolConcentration	25.81%	**41.44%**	32.6%	26%	27.38%	33.46%
JapaneseVowels*	93.4%	–	**99.7%**	83.7%	–	98.8%
Average Rank	5.25	3	1.75	3.4	4.67	**1.6**

*: Variable-length dataset

5.2 Experiment Using a Real-World Physical Activities Dataset

The physical activities dataset includes a variety of activities from 184 child participants between 8 years old and 15 years old. The original data was published in [16]. We collected the data in a normal and free-living environment which means we did not limit the type and length of participants' activities. Table 3 shows the detailed distribution for the related activities of each class.

The raw sensor recordings were cut into 12-second windows and generated model-based features with domain knowledge. Our method is feature-agnostic, so we use one of the most popular features, the percentile features, following the previous study in [17,18]. The 10th, 25th, 50th, 75th, and 90th percentiles are used in the summarization process to generate more robust features. The minimum and maximum are excluded to reduce potential outliers. Since the window length is 12 s and the resolution of the data is 1 s, in this case, the nearest points (2nd, 3rd, 6th, 9th, and 11th) are used for the features [4].

The dataset consists of time series signals of 36,670 window instances with 39 features after the data preparation step, representing the sensor readings. Summarizing global time-series data, we construct new features which represent the data in a fixed-length robust representation which only have 10 features. We show the classification results of both 10-feature data and the original 39-feature data as the baseline.

Instead of splitting the data at training, We pre-split the raw data with window instances (80% training and 20% testing) by the activity id, and then feed it to our summarization model. In this way, we are able to get a robust evaluation result excluding the repeated instances associated with a single activity. We also can avoid the bias of testing while predicting the type of physical activities with the benefit from the pre-split process.

Results: To compare the performance of our algorithm, we run the baseline model with identical hyperparameters, single view without summarization, same training and testing split. The results show that our summarized model completely outperforms the baseline classification model. Our model presents a good performance of overall accuracy at 93.24%, while, the baseline model has much

Table 3. The activity classes in each coarse categories and the number of 12-second windows recorded in the classes and categories.

Category and Class		Number of Windows	
Sedentary	Lying Rest	14755	16475
	Playing Computer Games	860	
	Reading	860	
Light Household and Games	Light Cleaning	840	2505
	Sweeping	865	
	Workout Video	800	
Moderate-Vigorous Household and Sports	Wall Ball	845	1570
	Playing Catch	725	
Walk	Brisk Track Walking	1210	3775
	Slow Track Walking	1000	
	Walking Course	1565	
Run	Track Running	485	485

lower accuracy at 86.30%. Typically, the baseline model requires much more time in the training. The training process takes 29372 s on the baseline model, but only half of that, 14981 s, on our proposed model with summarize mechanism. The confusion matrix of both models is presented in Fig. 2(a)and 2(b). In the assessment of sensitivity (Sens) and specificity (Spec), our MEDT model consistently outperforms across all sub-categories. Especially, in minor sub-categories such as LHH, MtV and Run, our method exhibits a superior level of sensitivity performance in comparison to the baseline model.

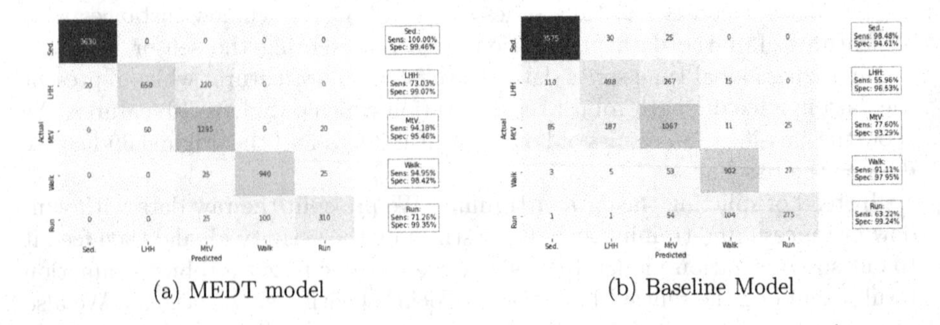

(a) MEDT model (b) Baseline Model

Fig. 2. The Confusion Matrix of Models

6 Conclusion

Although time series data offers insights into dynamic trends through collections of ordered data points, prevalent machine learning methods for time series are presenting challenges when dealing with noise, variable-length data and large data. We introduce a new approach utilizing GMM that provides fixed-length multi-view representations of variable-length time series data, allowing for compatibility with any classical machine learning methods. Leveraging this algorithm, we construct multiple representations from the original dataset, applying them to a Multi-Encoder-Decoder Transformer (MEDT) architecture for a comprehensive, multi-view classification approach. Through extensive experiments using multiple benchmarks and a real-world dataset, our method shows significant improvement compared to the state-of-the-art methods.

Acknowledgement. This material is based upon work partially supported by the National Institutes of Health under grant NIH 1R01DK129428-01A1 and National Science Foundation under NSF grants 2008202 and 2334665. Any opinion, findings, and conclusions or recommendations expressed in this material are those of the author(s) and do not necessarily reflect the views of the funding agencies.

References

1. Lines, J., Taylor, S., Bagnall, A.: Time series classification with HIVE-COTE: the hierarchical vote collective of transformation-based ensembles. ACM Trans. Knowl. Disc. Data (TKDD) **12**(5), 1–35 (2018)
2. Dempster, A., Petitjean, F., Webb, G.I.: ROCKET: exceptionally fast and accurate time series classification using random convolutional kernels. Data Min. Knowl. Disc. **34**(5), 1454–1495 (2020)
3. Shifaz, A., et al.: TS-CHIEF: a scalable and accurate forest algorithm for time series classification. Data Min. Knowl. Disc. **34**(3), 742–775 (2020)
4. Amaral, K., et al.: SummerTime: variable-length time series summarization with application to physical activity analysis. ACM Trans. Comput. Healthcare **3**(4), 1–15 (2022)
5. Vaswani, A., et al.: Attention is all you need. Adv. Neural Inf. Process. Syst. **30** (2017)
6. Bahri, O., Li, P., Boubrahimi, S.F., Hamdi, S.M.: Shapelet-based temporal association rule mining for multivariate time series classification. IEEE Xplore (2022). https://ieeexplore.ieee.org/stamp/stamp.jsp?tp=&arnumber=10020478. Accessed 02 Sept 2023
7. Zerveas, G., et al.: A transformer-based framework for multivariate time series representation learning. In: Proceedings of the 27th ACM SIGKDD Conference on Knowledge Discovery & Data Mining (2021)
8. Ismail Fawaz, H., et al.: Inceptiontime: finding alexnet for time series classification. Data Mining Knowl. Disc. **34**(6), 1936–1962 (2020)
9. Baldán, F.J., Benítez, J.M.: Multivariate times series classification through an interpretable representation. Inf. Sci. **569**, 596–614 (2021)
10. Bier, A., Jastrzębska, A., Olszewski, P.: Variable-length multivariate time series classification using ROCKET: a case study of incident detection. IEEE Access **10**, 95701–95715 (2022)
11. Ismail Fawaz, H., et al.: Deep learning for time series classification: a review. Data Mining Knowl. Disc. **33**(4), 917–963 (2019)
12. Zhou, X., et al.: Multi-encoder-decoder transformer for code-switching speech recognition. arXiv preprint arXiv:2006.10414 (2020)

13. Li, S., et al.: Enhancing the locality and breaking the memory bottleneck of transformer on time series forecasting. Adv. Neural Inf. Process. Syst. **32** (2019)
14. Wu, N., et al.: Deep transformer models for time series forecasting: the influenza prevalence case. arXiv preprint arXiv:2001.08317 (2020)
15. Bagnall, A., et al.: The UEA multivariate time series classification archive, 2018. arXiv preprint arXiv:1811.00075 (2018)
16. Crouter, S.E., Clowers, K.G., Bassett, D.R., Jr.: A novel method for using accelerometer data to predict energy expenditure. J. Appl. Physiol. **100**(4), 1324–1331 (2006)
17. Staudenmayer, J., et al.: An artificial neural network to estimate physical activity energy expenditure and identify physical activity type from an accelerometer. J. Applied Physiol. **107**(4), 1300–1307 (2009)
18. Trost, S.G., et al.: Artificial neural networks to predict activity type and energy expenditure in youth. Med. Sci. Sports Exerc. **44**(9), 1801 (2012)
19. Aitkin, M., Wilson, G.T.: Mixture models, outliers, and the EM algorithm. Technometrics **22**(3), 325–331 (1980)
20. Xu, C., Tao, D., Xu, C.: A survey on multi-view learning. arXiv preprint arXiv:1304.5634 (2013)
21. Dufter, P., Schmitt, M., Schütze, H.: Position information in transformers: an overview. Comput. Linguist. **48**(3), 733–763 (2022)
22. Costa-jussà, M.R., et al.: No language left behind: scaling human-centered machine translation. arXiv preprint arXiv:2207.04672 (2022)
23. Hota, H.S., Richa, H., Shrivas, A.K.: Time series data prediction using sliding window based RBF neural network. Int. J. Comput. Intell. Res. **13**(5), 1145–1156 (2017)
24. Devlin, J., et al.: Bert: pre-training of deep bidirectional transformers for language understanding. arXiv preprint arXiv:1810.04805 (2018)
25. Li, Y., Yang, M., Zhang, Z.: A survey of multi-view representation learning. IEEE Trans. Knowl. Data Eng. **31**(10), 1863–1883 (2018)
26. Xie, Z., et al.: Deep learning on multi-view sequential data: a survey. Artif. Intell. Rev. **56**(7), 6661–6704 (2023)
27. Li, H., et al.: MTS-LOF: medical time-series representation learning via occlusion-invariant features. arXiv preprint arXiv:2310.12451 (2023)
28. Hao, Y., Cao, H.: A new attention mechanism to classify multivariate time series. In: Proceedings of the Twenty-Ninth International Joint Conference on Artificial Intelligence (2020)
29. Dempster, A., Schmidt, D.F., Webb, G.I.: Minirocket: a very fast (almost) deterministic transform for time series classification. In: Proceedings of the 27th ACM SIGKDD Conference on Knowledge Discovery & Data Mining (2021)
30. Gao, G., et al.: A reinforcement learning-informed pattern mining framework for multivariate time series classification. In: The Proceeding of 31th International Joint Conference on Artificial Intelligence (IJCAI-2022) (2022)

Kernel Representation Learning with Dynamic Regime Discovery for Time Series Forecasting

Kunpeng Xu[1(✉)], Lifei Chen[2(✉)], Jean-Marc Patenaude[3], and Shengrui Wang[1(✉)]

[1] Université de Sherbrooke, Sherbrooke, QC, Canada
{kunpeng.xu,shengrui.wang}@usherbrooke.ca
[2] Fujian Normal University, Fuzhou, Fujian, China
clfei@fjnu.edu.cn
[3] Laplace Insights, Sherbrooke, QC, Canada
jeanmarc@laplaceinsights.com

Abstract. Correlations between variables in complex ecosystems such as weather and financial markets lead to a great amount of dynamic and co-evolving time series data, posing a significant challenge to the current forecast methods. Discovering dynamic patterns (aka regimes) is crucial to an accurate forecast, especially for the interpretability of the outcome. In this paper, we develop a kernel-based method to learn effective representations for capturing dynamically changing regimes. Each such representation accounts for the non-linear interactions among multiple time series, thereby facilitating more effective regime discovery. On the basis of regime information, we build a regression model to forecast all the variables simultaneously for the next multiple time points. The results on six real-life datasets demonstrate that our method can yield the most accurate forecast (with the lowest root mean square error) in comparison with seven predictive models.

Keywords: time series forecasting · kernel · self-representation learning

1 Introduction

Time-series forecasting is an important topic that continuously attracts a great deal of interest in a myriad of areas such as finance, medicine, meteorology, ecology, sociology, and many industrial sectors. In real applications, time series often comprise numerous short segments, each recurring within the series. These segments generally correspond to particular regimes/patterns in dynamically changing environments – *e.g.*, on the volatile financial market [6,13], stock prices might decline during wartime and subsequently rise with the onset of peace talks. Discovering and leveraging these underlying regimes has become an essential research topic for generating accurate and interpretable time-series forecasts.

To capture the dynamic behaviors of time series, several machine learning methods have been developed to explore the regimes for time series forecasting, *e.g.*, RSVAR [8] for health management, WCPD-RS [4] in financial,

D.-N. Yang et al. (Eds.): PAKDD 2024, LNAI 14650, pp. 251–263, 2024.
https://doi.org/10.1007/978-981-97-2266-2_20

and ObritMap [11] in IoT/sensor streams analysis. These models suggest that the presence of structural discontinuities in time series leads to a regime shift, wherein each regime represents distinct behaviors that reveal the underlying dynamics throughout time. In general, these models first analyze the overall regime shifts present in the time series data and subsequently employ the derived models to forecast future regimes. However, certain characteristics of time series can be hard to capture when time series exhibit nonlinearity, mixing, or noise. Furthermore, the requirement to predefine the number of regimes in many models, such as the Markov-based switching model [4], limits their flexibility in dynamically inferring and estimating regimes from data.

Another significant challenge arises from the complexity inherent in identifying regimes within multiple time series forecasting tasks [14]. This difficulty primarily stems from the interdependence and co-evolution of the time series – *e.g.*, to forecast the traffic for a particular road, it is necessary to consider the impact of traffic on adjacent roads; similarly, the fluctuating user engagement with music streaming services-evidenced by the decline in Pandora's click rates and the simultaneous surge in Spotify's from 2012 to 2022-suggests competitive dynamics, with Spotify seemingly attracting Pandora's user base. Exploring the interrelationships between series at different time intervals is crucial for regime identification and prediction.

To address these challenges, we propose a novel approach for multiple time series forecasting, emphasizing modeling of their evolving interactions and regime identification. Our method redefines regime identification from the perspective of self representation learning and transforms the challenge into a subspace clustering problem. This transformation allows for a more nuanced and granular analysis of multiple time series data, leading to a more precise and interpretable forecasting model. Our method has the following desirable properties:

(1) **Adaptive:** Automatically identify and handle regimes (patterns) exhibited by multiple time series, without prior knowledge about regimes.
(2) **Interpretability:** Convert heavy sets of time series into a lighter and meaningful structure through kernel representation, depicting the continuous regime shift mechanism over multiple time series in nonlinear space.
(3) **Effective:** Operate on multiple time series, explore the nonlinear interactions, and forecast the future values within an ecosystem consisting of multiple time series.

2 Related Work

Traditional time series models, such as general state-space models [3], including ARIMA and exponential smoothing, excel at modeling the complex dynamics of individual time series with sufficiently long histories. Although these methods are widely used in time series forecasting due to their simplicity and interpretability, they are local in the sense that one model is learned for each time series. Consequently, they cannot effectively extract information across multiple time series.

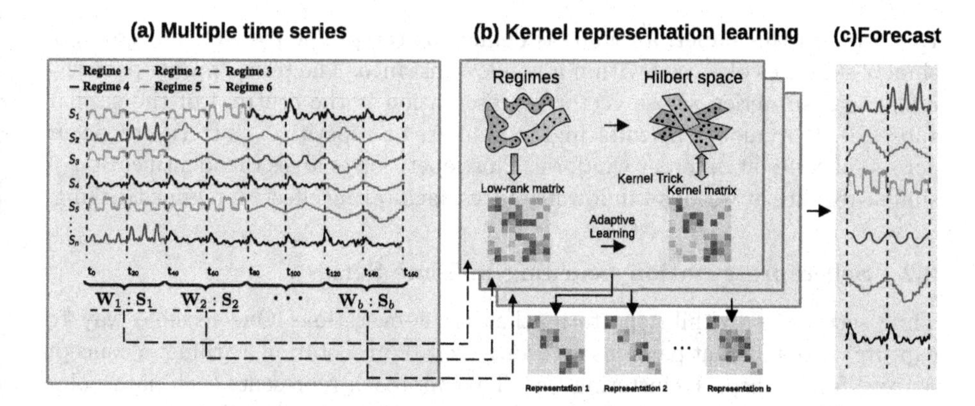

Fig. 1. Framework of our method. Given multiple co-evolving time series composed of various regimes in fixed-length windows, our method learns the kernel representation of these regimes for forecasting the next regimes.

Hochstein et al. [8] developed a multivariate smooth transition autoregression model to show how different time series are linearly dependent on each other. This model uses a vector autoregressive model for each regime. It is worth noting that this type of method attempts to capture the regime shift mechanism through a single transfer matrix, which unfortunately may be time-dependent for series that exhibit noncontiguous regimes. Matsubara et al. proposed the RegimeCast model [10], which learns the various patterns that may exist in a co-evolving environment at a given window and reports the pattern(s) most likely to be observed at a subsequent time. While the approach can report subsequent patterns, it does not capture possible dependencies between patterns. In their subsequent work [11], the authors introduced the deterministic OrbitMap model, designed to capture time-dependent transitions between exhibited regimes. However, their model relies on regimes that are labeled in advance. Recent research has demonstrated significant advancements in time series analysis through the use of deep neural networks [15,19–21]. However, the majority of these studies primarily focus on modeling and forecasting individual time series, often overlooking the interactions among multiple time series.

3 Preliminaries

3.1 Key Concepts

Time Series. A time series is a set of points ordered by a time index as follows: $S_i = \{(t_l, e_l^i)\}_{l=1}^m$, where t_l are regular time stamps, m the series length and e_l^i the series value at the specific time t_l. Here, the index i refers to the *i-th* time series in a set of N univariate time series $\mathbf{S} = \{S_i\}_{i=1}^N$.

Regime. In this paper, a regime is defined as the profile pattern of a group of similar subseries observed within a window instance. The term "profile pattern" refers to a subseries whose vector representation is the centroid of the similar subseries. Our model permits highly similar, or repetitive, patterns to occur across subseries at different windows. This repetition enables the identification of similar regimes at various window instances, facilitating effective regime tracking.

3.2 Self-representation Learning in Time Series

Time series often exhibit patterns that recur over time. One feasible way to capture these inherent patterns is through self-representation learning, a concept derived from subspace clustering [18]. This approach represents each data point in a series as a linear combination of others, formulated as $\mathbf{S} = \mathbf{SZ}$ or $S_i = \sum_j S_j Z_{ij}$, where $\mathbf{Z}$ is the self-representation coefficient matrix. In multiple time series, high Z_{ij} values indicate similar behaviors or regimes between S_i and S_j. The learning objective function is:

$$\min_{\mathbf{Z}} \frac{1}{2}||\mathbf{S} - \mathbf{SZ}||^2 + \Omega(\mathbf{Z}), \ \ s.t. \ \mathbf{Z} = \mathbf{Z}^{\mathrm{T}} \geq 0, \mathrm{diag}(\mathbf{Z}) = 0 \tag{1}$$

where $\Omega(\cdot)$ is a regularization term on $\mathbf{Z}$. The ideal representation $\mathbf{Z}$ should group data points with similar patterns, represented as block diagonals in $\mathbf{Z}$, each block signifying a specific regime. The number of blocks, k, corresponds to the distinct regimes.

The optimization of this problem can take various forms, influenced by the choice of $\Omega(\mathbf{Z})$. If $\Omega(\mathbf{Z}) = ||\mathbf{Z}||_1$, it results in classical Sparse Subspace Clustering [7]. Different norms for $\mathbf{Z}$ lead to various models like efficient dense subspace clustering (EDSC), the Frobenius norm in least-squares Regression (LSR) and the nuclear norm in Low-Rank Representation (LRR).

3.3 Kernel Trick for Modeling Time Series

Linear models in Euclidean space often struggle with capturing nonlinear relationships in multiple time series [16]. Kernelization techniques address this challenge by mapping data into higher, and in some cases, infinite-dimensional Hilbert spaces using suitable kernel functions [17]. This facilitates the identification of linear patterns within these transformed spaces. The process is facilitated by the "kernel trick", which employs a nonlinear feature mapping, $\Phi(\mathbf{S})$: $\mathcal{R}^d \rightarrow \mathcal{H}$, to project data $\mathbf{S}$ into a kernel Hilbert space $\mathcal{H}$. Direct knowledge of the transformation Φ is not required; instead, a kernel Gram matrix $\mathcal{K} = \Phi(\mathbf{S})^{\top}\Phi(\mathbf{S})$ is used. The Gaussian kernel, which results in an infinitely dimensional feature space $\mathcal{H}$, is notably prevalent in this context.

4 Proposed Method

For the sake of clarity, consider the set of time series depicted by Fig 1(a). We start by introducing our kernel representation learning. This process entails searching for homogeneous patterns via self-representation learning with a block diagonal regularizer in kernel space. With a given sliding window of size w, we can split time series into contiguous subseries of length w, *i.e.*, $\mathbf{S} = \bigcup_{p=1}^{b} \mathbf{S}_p$ and get the number of distinct regimes k by counting the number of distinct profile patterns across all window-stamps, $W_1, ..., W_b$. Based on the discovered regimes, we will be able to predict the regime switch and series values.

4.1 Kernel Representation Learning: Modeling Regime Behavior

In our approach, we circumvent the obstacle of discovering regimes for multiple time series, by solving a self-representation learning problem. This approach allows us to effectively cluster subseries, retrieved using a sliding window technique, into distinct regimes. We begin with the simplest case, where we treat the whole series as a single window.

Given a set of time series $\mathbf{S} = (S_1, \ldots, S_N) \in \mathcal{R}^{T \times N}$ as described in Eq. (1), its self-representation $\mathbf{Z}$ would make inner product $\mathbf{SZ}$ come close to $\mathbf{S}$ if we adopted the linear approach. However, the objective (1) falls short in capturing the nonlinear relationships between series. To address this issue, the time series can be mapped, by "kernel tricks", into a high-dimensional RKHS, where a linear pattern analysis will be performed. By integrating the kernel mapping, we present a new kernel representation learning strategy (as shown in Fig 1(b), with the following objective function:

$$\min_{\mathbf{Z}} ||\Phi(\mathbf{S}) - \Phi(\mathbf{S})\mathbf{Z}||^2, \ s.t. \ \mathbf{Z} = \mathbf{Z}^{\mathrm{T}} \geq 0, \mathrm{diag}(\mathbf{Z}) = 0 \quad (2)$$

Here, the mapping function $\Phi(\cdot)$ need not be explicitly identified and is typically replaced by a kernel $\mathcal{K}$ subject to $\mathcal{K} = \Phi(\cdot)^{\top}\Phi(\cdot)$.

Ideally, we hope to achieve the matrix $\mathbf{Z}$ having k block diagonals under some proper permutations if $\mathbf{S}$ contains k regimes. To this end, we introduce a regularization term to $\mathbf{Z}$ and transform Eq. (2) to:

$$\begin{aligned} \min_{\mathbf{Z}} ||\Phi(\mathbf{S}) - \Phi(\mathbf{S})\mathbf{Z}||^2 + \gamma \sum_{i=N-k+1}^{N} \lambda_i(\mathbf{L_Z}), \\ s.t. \ \mathbf{Z} = \mathbf{Z}^{\mathrm{T}} \geq 0, \mathrm{diag}(\mathbf{Z}) = 0 \end{aligned} \quad (3)$$

where $\gamma > 0$ defines the trade-off between the loss function and regularization terms, and $\lambda_i(\mathbf{L_Z})$ contains the eigenvalues of Laplacian matrix $\mathbf{L_Z}$ corresponding to $\mathbf{Z}$ in decreasing order. Here, the regularization term is equal to 0 if and only if $\mathbf{Z}$ is k-block diagonal (see Theorem 1 for details). Based on the learned high-quality matrix $\mathbf{Z}$ (containing the block diagonal structure), we can easily group the time series into k regimes using traditional spectral clustering technology [7].

Theorem 1. $\min \sum_{i=N-k+1}^{N} \lambda_i(\mathbf{L_Z})$ *is equivalent to* $\mathbf{Z}$ *is* k*-block diagonal.*

Proof. Due to the fact that $\mathbf{Z} = \mathbf{Z}^{\mathrm{T}} \geq 0$, the corresponding Laplacian matrix $\mathbf{L_Z}$ is positive semidefinite, and thus $\lambda_i(\mathbf{L_Z}) \geq 0$ for all i. The optimal solution of $\min \sum_{i=N-k+1}^{N} \lambda_i(\mathbf{L_Z})$ is that all elements of $\lambda_i(\mathbf{L_Z})$ are equal to 0, which means that the k smallest eigenvalues are 0. Combined with the Laplacian matrix property, the multiplicity k of the eigenvalue 0 of the corresponding Laplacian matrix $\mathbf{L_Z}$ equals the number of connected components (blocks) in $\mathbf{Z}$, and thus the soundness of Theorem 1 has been proved.

Optimization. The problem (3) can be solved by the ALM with Alternating Direction Minimization strategy. Normally, the representation matrix $\mathbf{Z}$ in Eq. (3) needs to be nonnegative and symmetric, which are necessary properties for defining the block diagonal regularizer. However, these restrictions on $\mathbf{Z}$ will limit its representation capability. For this reason, we propose a modified model by introducing an intermediate-term $\mathbf{C}$:

$$\begin{aligned} &\min_{\mathbf{Z},\mathbf{C}} \frac{1}{2}||\Phi(\mathbf{S}) - \Phi(\mathbf{S})\mathbf{C}||^2 + \frac{\beta}{2}||\mathbf{C} - \mathbf{Z}||^2 + \gamma \sum_{i=N-k+1}^{N} \lambda_i(\mathbf{L_Z}) \\ &\qquad s.t.\ \mathbf{Z} = \mathbf{Z}^{\mathrm{T}} \geq 0, \mathrm{diag}(\mathbf{Z}) = 0 \end{aligned} \tag{4}$$

The above two models (Eq. (3) and Eq. (4)) are equivalent when $\beta > 0$ is sufficiently large. As will be seen in optimization, a benefit of the relaxation term $||\mathbf{C} - \mathbf{Z}||^2$ is that it makes the objective function separable. More importantly, the subproblems for updating $\mathbf{Z}$ and $\mathbf{C}$ are strongly convex, leading to final solutions that are unique and stable.

Note that $\sum_{i=N-k+1}^{N} \lambda_i(\mathbf{L_Z})$ is a nonconvex term, and by introducing the property about the sum of eigenvalues, we reformulate it as $\min_{\mathbf{W}} < \mathbf{L_Z}, \mathbf{W} >$, where $0 \preceq \mathbf{W} \preceq \mathbf{I}, \mathrm{Tr}(\mathbf{W}) = k$. So Eq. (4) is equivalent to

$$\begin{aligned} &\min_{\mathbf{Z},\mathbf{C},\mathbf{W}} \frac{1}{2}||\Phi(\mathbf{S}) - \Phi(\mathbf{S})\mathbf{C}||^2 + \frac{\beta}{2}||\mathbf{C} - \mathbf{Z}||^2 + \gamma < \mathrm{Diag}(\mathbf{Z1}) - \mathbf{Z}, \mathbf{W} > \\ &s.t.\ \mathbf{Z} = \mathbf{Z}^{\mathrm{T}} \geq 0, \mathrm{diag}(\mathbf{Z}) = 0, 0 \preceq \mathbf{W} \preceq \mathbf{I}, Tr(\mathbf{W}) = k \end{aligned} \tag{5}$$

The optimization of Eq. (5) involves alternating updates of $\mathbf{W}$, $\mathbf{C}$, and $\mathbf{Z}$. Each subproblem is convex, allowing for closed-form solutions:

Updating W:

$$\mathbf{W}^{i+1} = \arg\min_{\mathbf{W}} < \mathrm{Diag}(\mathbf{Z1}) - \mathbf{Z}, \mathbf{W} >,\ \ s.t.\ 0 \preceq \mathbf{W} \preceq \mathbf{I}, \mathrm{Tr}(\mathbf{W}) = k \tag{6}$$

Updating C:

$$\mathbf{C}^{i+1} = (\mathcal{K} + \beta\mathbf{I})^{-1}(\mathcal{K} + \beta\mathbf{Z}) \tag{7}$$

Updating Z:

$$\mathbf{Z}^{i+1} = [(\hat{\mathbf{A}}+\hat{\mathbf{A}}^{\mathrm{T}})]_{+}, \text{where } \hat{\mathbf{A}} = \mathbf{A} - \mathrm{Diag}(\mathrm{diag}(\mathbf{A})), \mathbf{A} = \mathbf{C} - \frac{\gamma}{\beta}(\mathrm{diag}(\mathbf{W})\mathbf{1}^{\mathrm{T}} - \mathbf{W}) \tag{8}$$

4.2 Forecasting

We consider all windows to be known except the last one, for which we want to predict the series values. For b sliding windows $\{W_1, \ldots, W_b\}$, we obtain b kernel representations, each corresponding to a window. It is important to know that the regime R_i discovered from the subseries $\mathbf{S}_p$ within the $p^{th}(p \in [1, b])$ window might not be discovered (*i.e.*, there may be no series exhibiting this regime) in other subseries (*i.e.*, from other windows). This reveals the variety of regimes in time series and the demand for a dynamic representation. We predict the kernel representation for subseries in the next window W_{b+1} via a regression model λ, as follows:

$$\mathbf{Z}_{b+1} = \lambda(\mathbf{Z}_1, ..., \mathbf{Z}_b) + \mu_{b+1}, \tag{9}$$

where μ_{b+1} is the white Gaussian noise for reducing overfitting. Then, we forecast the value of the time series for the next window W_{b+1} based on the self-representation property:

$$\mathbf{S}_{b+1} = \hat{\mathbf{S}}_{b+1}\mathbf{Z}_{b+1} \tag{10}$$

Table 1. Data statistics

Data	# of series	Length of series
`Music`	20	219
`Electricity`	370	1,462
`Chlorine`	166	3,480
`Earthquake`	139	512
`Electrooculography`	362	1,250
`Rock`	50	2,844

In this case, we employ the same regression form to estimate $\hat{\mathbf{S}}_{b+1}$, *i.e.*, $\hat{\mathbf{S}}_{b+1} = \lambda(\mathbf{S}_1, ..., \mathbf{S}_b) + \mu_{b+1}$.

5 Experiments

5.1 Data

We collected six real-life datasets from various areas. The `Music` dataset from GoogleTrend event stream[1] contains 20 time series, each for the Google queries on a music-player spanning 219 months from 2004 to 2022. The `Electricity` dataset comprises 1462 daily electricity load diagrams for 370 clients, extracted from UCI[2]. From the UCR's public repository[3], we obtained four time-series datasets – *i.e.*, `Chlorine` concentration, `Earthquake`, `Electrooculography` signal, and `Rock`. Table 1 summarizes the statistics of the datasets.

[1] http://www.google.com/trends/.

[2] https://archive.ics.uci.edu/ml/datasets/.

[3] https://www.cs.ucr.edu/%7Eeamonn/time_series_data_2018.

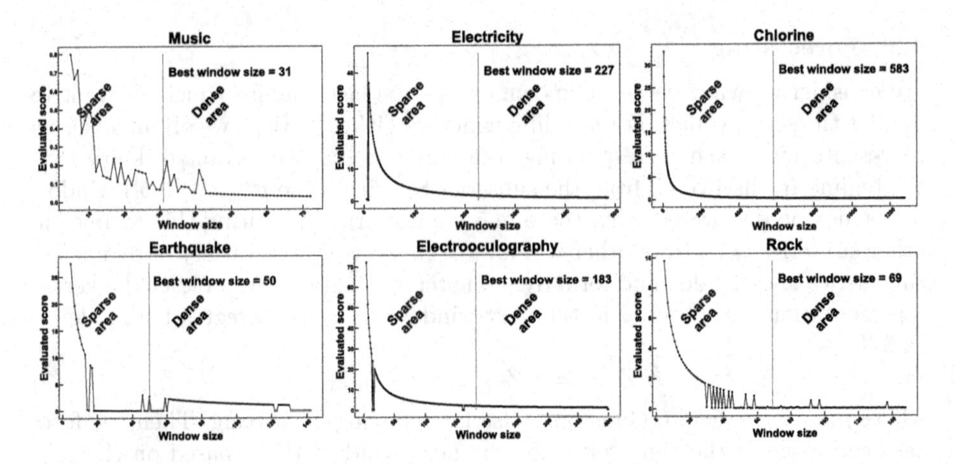

Fig. 2. The best window size (red line) for the six data sets (Color figure online)

5.2 Experimental Setup and Evaluation

In the kernel representation learning process, we used the Gaussian kernel of the form $\mathcal{K}(S_i, S_j) = exp(-||S_i - S_j||^2/d_{max}^2)$, where d_{max} is the maximal distance between series. Parameters γ in Eq. (3) is selected over [0.1,0.4,0.8,1,4,10] and set to be $\gamma = 0.8$ for the best performance.

We evaluate the forecasting performance of the proposed model against seven different models. Among them, four are forecasting models (ARIMA [3], KNNR [5], INFORMER [21], and a state-of-the-art ensemble model N-BEATS [12]), the other three are RS models (MSGARCH [1], SD-Markov [2] and OrBitMap [11]).

5.3 Regime Identification

In this subsection, we evaluate the capability of our model to identify regimes. During the learning process, a fixed window slides over all the series, generating subseries under different windows. We then learn a kernel representation for the subseries in each window. The quality of our kernel representation depends highly on how well the time series is split. Due to space limitations, our method for automatically estimating the optimal size of the sliding windows to obtain suitable regimes using the Minimum Description Length (MDL) technique can be found in Supplementary. Figure 2 exhibits the selected window sizes for the respective datasets: the length of 31 (resp., 227, 583, 50, 183, 69) window used for the `Music` (resp., `Electricity`, `Chlorine`, `Earthquake`, `Electrooculography`, `Rock`) data.

With these window sizes, we can plot the profile pattern to visualize the regime. In Fig. 3, each row displays the distinct regimes exhibited by co-evolving series in the six datasets, respectively. We discovered 4 different regimes in the `Music` time series, 3 in `Electricity`, 4 in `Chlorine`, 3 in `Earthquake`, 3 in `Electrooculography`, and 5 in `Rock`. For these real cases, we lack the ground

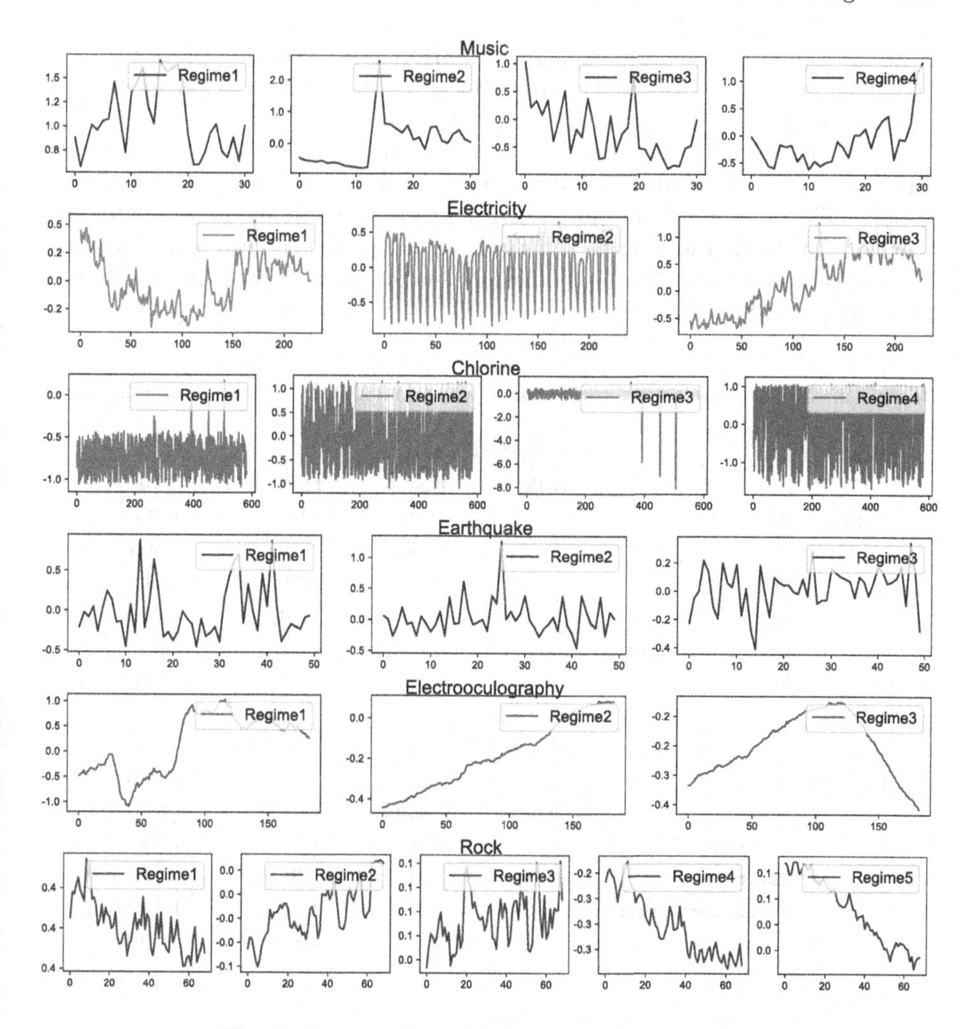

Fig. 3. Discovered regimes for the six real datasets

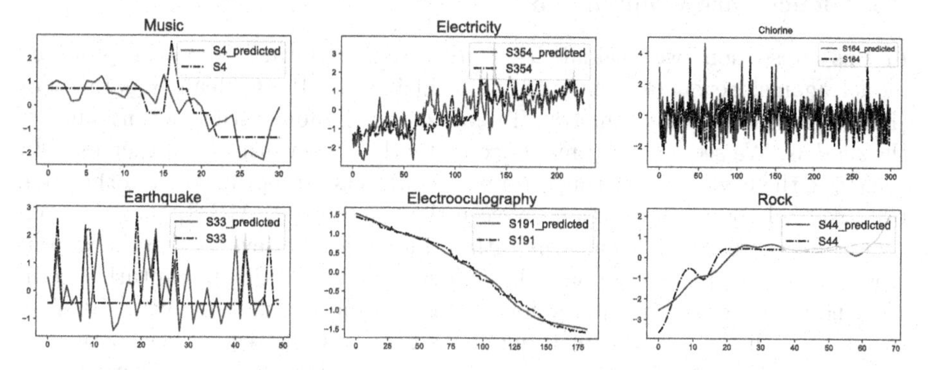

Fig. 4. True (black) and forecasted (blue) values for the six time series, each from a real dataset. (Color figure online)

truth for validating the obtained regimes. Fortunately, according to the forecasting which depends on the identified regimes, we will be able to better validate whether the identified regimes are the right ones.

Figure 4 illustrates the forecasted outcomes for six arbitrarily selected time series from the respective datasets, offering a demonstrative insight into the notable efficacy of our model in forecasting time series. It is important to note that this illustration is intended to showcase the proficient results achieved via our regime-based forecasting, a detailed evaluation of the forecasting ability will be presented in Sect. 5.4.

Table 2. Models' forecasting performance, in terms of RMSE, for the nine datasets

Models		Forecasting models			
		ARIMA	KNNR	INFORMER	N-BEATS
Datasets	`Music`	6.571	4.021	2.562	0.956
	`Electricity`	2.458	2.683	2.735	**1.593**
	`Chlorine`	8.361	6.831	3.746	1.692
	`Earthquake`	5.271	3.874	4.326	1.681
	`Electrooculography`	3.561	3.452	4.562	2.487
	`Rock`	6.836	6.043	5.682	2.854
Models		RS models			
		MSGARCH	SD-Markov	OrbitMap	Ours
Datasets	`Music`	2.641	3.234	1.244	**0.663**
	`Electricity`	2.425	2.439	1.835	1.644
	`Chlorine`	5.712	3.462	1.753	**1.387**
	`Earthquake`	4.213	3.573	**1.386**	1.392
	`Electrooculography`	3.566	3.571	3.251	**1.198**
	`Rock`	5.924	4.587	4.571	**1.699**

5.4 Benchmark Comparison

In this subsection, we evaluate the forecasting performance of our proposed model against seven different models, utilizing the Root Mean Square Error (RMSE) as an evaluative metric. Table 2 shows the forecasting performance of the models. We see that our model consistently outperforms the other models, achieving the lowest forecasting error on all datasets (except for the `Earthquake`, because of the weak correlation between the time series). ARIMA has the ability to capture seasonality patterns within time series; however, when the various seasonalities are noncontiguous, the models face difficulties in capturing complex, nonlinear dynamic interactions between time series. Notably, N-BEATS, the state-of-the-art deep network model, is generally the second-best performer owing to its ensemble-based strengths. However, it falls short in capturing complex regime transitions within multiple time series, revealing the limitations of

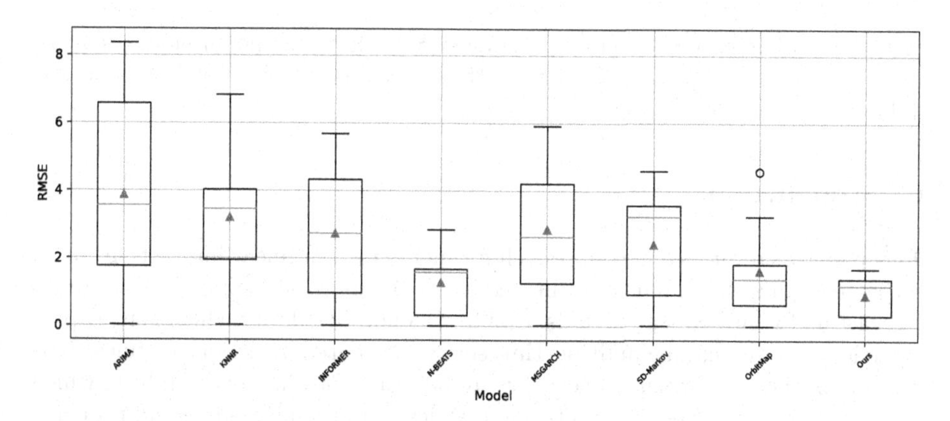

Fig. 5. Box Plot of RMSE Values for Each Model Across Datasets

a model geared solely for single time series forecasting. Meanwhile, OrbitMap, while also regime-aware, is hindered by its necessity for predefined regimes and struggles with handling multiple time series. Figure 5 illustrates the distribution of RMSE values for each model across all datasets, and it is evident that our model achieves the most favourable outcomes overall.

5.5 Ablation Study

We conduct ablation experiments to validate the efficacy of our model's kernel representation learning. We focus particularly on the regularization and kernelization techniques employed in (3). Our approach is compared against existing self-representation learning methods[9] - SSC, LSR, LRR, BDR, EDSC, SSQP, and SSCE, under varying values of γ. The comparative results are illustrated

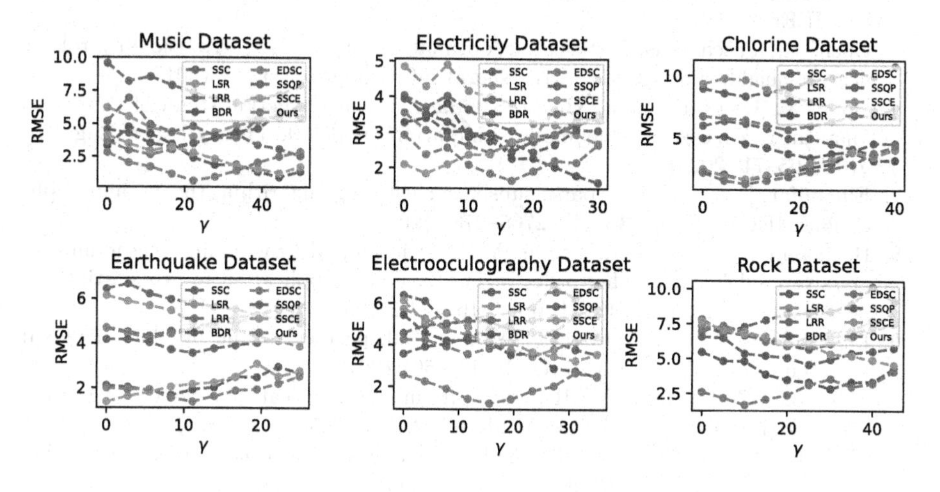

Fig. 6. RMSE across datasets for different regularizations at varying γ

in Fig. 6. This comparison clearly demonstrates that our model achieves superior performance, outperforming the other methods in terms of RMSE across different datasets for a range of γ values.

6 Conclusion

This paper introduces a new approach for modeling non-linear interactions in an ecosystem comprising multiple time series. This approach enhances time series forecasting for subsequent periods, thanks to a notable ability which is its capacity to identify and handle multiple time series dominated by various regimes. This is accomplished by devising a kernel representation learning method, from which the time-varying kernel representation matrices and the block-diagonal property are utilized to determine regime shifts. Furthermore, our model automatically uncovers various hidden regimes without requiring any prior knowledge about the series under investigation. Validation with real-world datasets has shown that our model surpasses existing models in terms of forecast accuracy.

References

1. Ardia, D., Bluteau, K., Boudt, K., Catania, L., Trottier, D.A.: Markov-switching garch models in r: the MSGARCH package. J. Stat. Softw. (2019)
2. Bazzi, M., Blasques, F., Koopman, S.J., Lucas, A.: Time-varying transition probabilities for Markov regime switching models. J. Time Series Anal. **38**, 458–478 (2017)
3. Box, G.: Box and jenkins: time series analysis, forecasting and control. In: A Very British Affair, pp. 161–215. Springer, Heidelberg (2013). https://doi.org/10.1057/9781137291264_6
4. Chatigny, P., Chen, R., Patenaude, J.M., Wang, S.: A variable-order regime switching model to identify significant patterns in financial markets. In: ICDM, pp. 887–892. IEEE (2018)
5. Chen, R., Paschalidis, I.: Selecting optimal decisions via distributionally robust nearest-neighbor regression. Adv. Neural Inf. Process. Syst. **32** (2019)
6. Dong, X., Dang, B., Zang, H., Li, S., Ma, D.: The prediction trend of enterprise financial risk based on machine learning arima model. J. Theory Pract. Eng. Sci. **4**(01), 65–71 (2024)
7. Elhamifar, E., Vidal, R.: Sparse subspace clustering: algorithm, theory, and applications. IEEE TPAMI **35**(11), 2765–2781 (2013)
8. Hochstein, A., Ahn, H.I., Leung, Y.T., Denesuk, M.: Switching vector autoregressive models with higher-order regime dynamics application to prognostics and health management. In: PHM, pp. 1–10. IEEE (2014)
9. Lu, C., Feng, J., Lin, Z., Mei, T., Yan, S.: Subspace clustering by block diagonal representation. IEEE TPAMI **41**(2), 487–501 (2018)
10. Matsubara, Y., Sakurai, Y.: Regime shifts in streams: real-time forecasting of co-evolving time sequences. In: ACM SIGKDD, pp. 1045–1054 (2016)
11. Matsubara, Y., Sakurai, Y.: Dynamic modeling and forecasting of time-evolving data streams. In: ACM SIGKDD, pp. 458–468 (2019)

12. Oreshkin, B.N., Carpov, D., Chapados, N., Bengio, Y.: N-beats: Neural basis expansion analysis for interpretable time series forecasting. arXiv preprint arXiv:1905.10437 (2019)
13. Qiao, Y., Jin, J., Ni, F., Yu, J., Chen, W.: Application of machine learning in financial risk early warning and regional prevention and control: a systematic analysis based on shap. World Trends Realities Accompany. Prob. Dev. **331** (2023)
14. Song, X., et al.: Zeroprompt: streaming acoustic encoders are zero-shot masked lms. arXiv preprint arXiv:2305.10649 (2023)
15. Xu, C., Yu, J., Chen, W., Xiong, J.: Deep learning in photovoltaic power generation forecasting: Cnn-lstm hybrid neural network exploration and research. In: The 3rd International scientific and practical conference "Technologies in education in schools and universities", Athens, Greece, 23–26 January 2024, vol. 363, p. 295. International Science Group (2024)
16. Xu, K., Chen, L., Wang, S.: Data-driven kernel subspace clustering with local manifold preservation. In: 2022 IEEE International Conference on Data Mining Workshops (ICDMW), pp. 876–884. IEEE (2022)
17. Xu, K., Chen, L., Wang, S.: A multi-view kernel clustering framework for categorical sequences. Expert Syst. Appl. 116637 (2022)
18. Xu, K., Chen, L., Wang, S., Wang, B.: A self-representation model for robust clustering of categorical sequences. In: U, L.H., Xie, H. (eds.) APWeb-WAIM 2018. LNCS, vol. 11268, pp. 13–23. Springer, Cham (2018). https://doi.org/10.1007/978-3-030-01298-4_2
19. Ye, J., et al.: Multiplexed oam beams classification via fourier optical convolutional neural network. In: 2023 IEEE Photonics Conference (IPC), pp. 1–2. IEEE (2023)
20. Ye, J., Solyanik, M., Hu, Z., Dalir, H., Nouri, B.M., Sorger, V.J.: Free-space optical multiplexed orbital angular momentum beam identification system using fourier optical convolutional layer based on 4f system. In: Complex Light and Optical Forces XVII, vol. 12436, pp. 70–80. SPIE (2023)
21. Zhou, H., et al.: Informer: beyond efficient transformer for long sequence time-series forecasting. In: Proceedings of the AAAI Conference on Artificial Intelligence, vol. 35, pp. 11106–11115 (2021)

Hyperparameter Tuning MLP's for Probabilistic Time Series Forecasting

Kiran Madhusudhanan[(✉)], Shayan Jawed, and Lars Schmidt-Thieme

Information Systems and Machine Learning Lab and VWFS Data Analytics Research Center, University Of Hildesheim, Hildesheim, Germany
{madhusudhanan,shayan,schmidt-thieme}@ismll.uni-hildesheim.de

Abstract. Time series forecasting attempts to predict future events by analyzing past trends and patterns. Although well researched, certain critical aspects pertaining to the use of deep learning in time series forecasting remain ambiguous. Our research primarily focuses on examining the impact of specific hyperparameters related to time series, such as context length and validation strategy, on the performance of the state-of-the-art MLP model in time series forecasting. We have conducted a comprehensive series of experiments involving 4800 configurations per dataset across 20 time series forecasting datasets, and our findings demonstrate the importance of tuning these parameters. Furthermore, in this work, we introduce the largest metadataset for time series forecasting to date, named TSBench, comprising 97200 evaluations, which is a twentyfold increase compared to previous works in the field. Finally, we demonstrate the utility of the created metadataset on multi-fidelity hyperparameter optimization tasks.

Keywords: Time Series Forecasting · Hyperparameter · Metadataset

1 Introduction

Time series forecasting is a machine learning technique that aims to capture historical patterns and use these patterns to predict the future values of the variables. Time series datasets are generated by various physical phenomena that change over time and can be described by different mathematical equations or functions. Using deep learning techniques to model the underlying distribution is a frequent approach. However, choosing the optimal hyperparameters for a learning algorithm is a challenging problem in deep learning, known as hyperparameter optimization (HPO). HPO methods have been widely researched and applied in domains such as computer vision [19], and tabular datasets [2], but they have received less attention within the time series domain.

For instance, the context length, a crucial parameter that determines the extent of immediate history available to the model for forecasting, is often

K. Madhusudhanan and S. Jawed—This authors contributed equally to this work.

D.-N. Yang et al. (Eds.): PAKDD 2024, LNAI 14650, pp. 264–275, 2024.
https://doi.org/10.1007/978-981-97-2266-2_21

Table 1. Summary statistics for evaluations considered in prior works.

Paper	Venue	# HPs	# Datasets	# Evaluations
[3]	ArXiv'22	107	44	4.7K
[5]	ECML/PKDD'23	200	24	4.8K
Our TSBench	–	4860	20	97 K

assigned a constant value across different datasets without careful tuning [14,22]. Despite evidence suggesting the importance of a context length as a hyperparameter [5,21], subsequent research appears to overlook this parameter's significance, maintaining a constant value across various datasets. While it could be argued that a longer context length is invariably beneficial, our experiments challenge this assumption. We demonstrate that the optimal context length is dependent on the dataset and varies according to the frequency and prediction horizon of the time series dataset.

Another point of contention within the time series forecasting community pertains to the use of validation splits. The validation split is typically chosen to best represent the test distribution. In most domains, a random sample of x% of the training dataset can be used for validation, as the test dataset is usually an unknown random subset of the dataset during the training process. However, this is not the case for time series forecasting where, the test split is definitively the last few samples when the samples are arranged in chronological order, i.e., the test split is a time-wise split. While current works employ a time-wise validation split, the community lacks consensus on whether the forecasting model should be retrained on the validation split using the same hyperparameters, as done in [15], or whether not to retrain on the validation split, as is common in other domains and in more recent time series forecasting papers such as [13,20,21]. This divergence underscores the need for further research and discussion within the community. In this paper, we rigorously benchmark across many configurations and datasets and analyze these specific validation split defining strategies that are applicable to time series forecasting.

Despite the general nature and extensive research in the field of time series forecasting, there is a noticeable lack of studies on hyperparameter optimization compared to other domains. In [5], the authors suggest the use of an AutoML framework that simultaneously optimizes the architecture and corresponding hyperparameters for a given dataset. However, this AutoML framework falls short in its ability to apply the knowledge gained from tuning one dataset to another. A study more closely related to our work is [3], where the authors compare various deep learning and classical models on 44 datasets and provide a metadataset consisting of evaluations and forecasts for all methods. However, this metadataset is limited to nearly five thousand runs, as reported in Table 1.

Our work diverges from the aforementioned studies based on the number of evaluations performed. We evaluate 4860 configurations of the state-of-the-art muli-layer perceptron (MLP) model, named NLinear [21] for 20 datasets, creating the largest metadataset of 97200 evaluations for time series forecasting. In addition to the evaluations and forecasts, our work differs in the fact that we adapt the NLinear to probabilistic outputs while logging also the gradient statistics as in [23], enabling the metadataset to be used for Learning Curve Forecasting techniques [8]. This comprehensive approach sets our work apart in the field of time series forecasting. Concretely, we summarize the contributions of the paper as follows.

1. We analyze the importance of time series specific hyperparameters like the validation strategy and context length for time series forecasting.
2. We introduce `TSBench`, a benchmark for multi-fidelity optimization that provides probabilistic time series forecasts for 97200 hyperparameter evaluations on 20 datasets from the Monash Forecasting Repository [7].
3. As a secondary task, we show the effectiveness of the TSBench dataset on multi-fidelity hyperparameter optimization.

2 Problem Statement

Given a time series $(x, y) \sim q$ drawn from an unknown random distribution q where $x = (x_1, \ldots, x_C)$ and $y = (x_{C+1}, \ldots, x_{C+\delta})$ represent the observed history and future values, respectively. Here, $x_i \in \mathbb{R}$ $(D \in \mathbb{N})$ denotes the observation made at relative time i. x and y are called univariate time series if $D = 1$ and multivariate time series if $D > 1$. Additionally, C and δ represents here the Context Length and Forecast Horizon, respectively.

Univariate time series forecasting task attempts to find a model $m : \mathbb{R}^C \rightarrow \mathbb{R}^\delta$ such that the expected loss $\ell : \mathbb{R}^\delta \times \mathbb{R}^\delta \rightarrow \mathbb{R}$ between the ground truth (y) and forecasted future values $(\hat{y} = m(x))$ is minimal:

$$\mathop{\mathbb{E}}_{(x,y)\sim q} \ell(y, m(x))$$

3 MLPs for Time Series Forecasting

3.1 Nlinear Model

In our research, we utilize a variant of linear MLP models, known as the NLinear [21]. This model employs an N-normalization technique to tackle the prevalent issue of distribution shift in time series datasets. The methodology involves subtracting the last value (x_C) of the sequence from the input (x), which is then passed through a linear layer to generate intermediate embedding (z). The subtracted values are subsequently added back to generate the forecasts $(\hat{y})$.

$$\begin{aligned} z &= x - x_C \\ z &= \text{Linear}(z) \\ \hat{y} &= z + x_C \end{aligned} \tag{1}$$

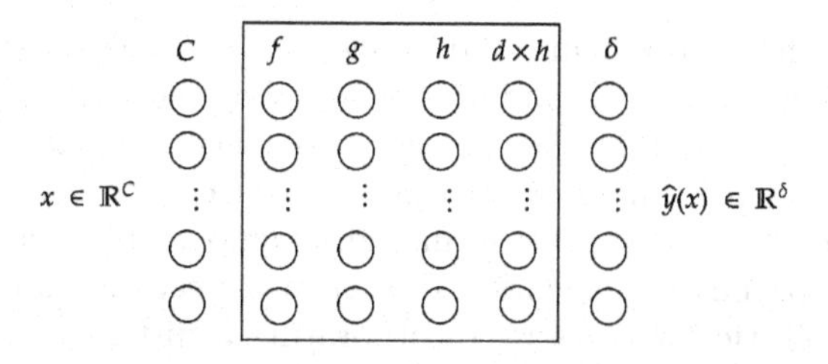

Fig. 1. Model Architecture. The parameters C and δ are indicative of the context length and forecast horizon, respectively. The hidden layers within the model are represented by f, g, and h, and are interspersed with ELU non-linearity. The parameter d signifies the distribution parameters that is learned per prediction time step.

The effectiveness of this approach has been demonstrated by [21] on seven benchmark datasets, thereby questioning the efficacy of current transformer-based models for time series forecasting. The study also claims that deeper models fail to enhance performance of the Linear model on the seven benchmark datasets. This raises an interesting question : Can linear models outperform their deeper non-linear counterparts in general?.

4 Hyperparameters

To address the previous question, we construct deeper model architecture variants in line with the methodology outlined by [9], by augmenting the NLinear model with three submodules, $(f \circ g \circ h)$, each implemented as a fully connected layer with ELU [4] activation function as in Fig. 1.

1. **Model Architectures**
   - `Base`: We choose the NLinear model as proposed in [21], a single layer network without any non-linearity as the base model. For instance, the structure could be represented as C-δ, where C denotes the Context length (Input layer) and δ (Output layer) signifies the forecast horizon.
   - `Diamond`: The depth of the NLinear model is augmented by incorporating hidden layers that feature an expanding layer at the center, separated by ELU nonlinearity, thereby adopting a diamond-like shape. The structure could C-f-g-h-δ be represented as C-32-64-32-δ.
   - `Contracting`: The hidden layers of the model are designed with decreasing number of neurons at each successive layer. e.g.: C-128-64-32-δ.
   - `Square`: The hidden layers have the same number of neurons at each hidden layer. e.g.: C-64-64-64-δ.
   - `Funnel`: The inverse of the `Diamond` architecture with a contracting layer at the middle of the network, e.g.: C-64-32-64-δ
   - `Expanding`: The hidden layers of the model are designed with decreasing number of neurons at each successive layer. e.g.: C-32-64-128-δ

The MLP architectures delineated previously have been adapted to leverage the probabilistic forecasting capabilities provided by the GluonTS framework [1]. This is achieved by predicting the parameters of distribution d at each predictive timestep. For instance, under the assumption of a Gaussian distribution, the model is tasked with learning the mean and variance as distribution parameters for each timestep. The aforementioned MLP architectures can be extended to generate probabilistic forecasts by incorporating a distribution layer atop the model, as depicted in Fig. 1. In our experimental setup, we utilize the Student-T distribution, which is the default distribution in GluonTS, and we experiment with varying the number of hidden parameters in the distribution layer.

2. **Distribution hidden layer** (d): The selection of the distribution hidden layer is made from a grid that includes the values 1, 2, and 10. Default size in GluonTS is 1, however the size is increased to 2 and 10 as increase in layer size have been observed to enhance the performance of the model.

4.1 Time Series Specific Configuration

The **context length**, also known as the lookback window, is a critical parameter in time series forecasting. It determines the amount of historical data used to predict future values of a time series. In their study [21], the authors examined the effects of context lengths on long-term forecasting. They found that the performance of most transformer-based methods deteriorates with an increase in context length. Similarly, [5] compared the importance of various parameters used in time series forecasting and found that context length is one of the most important hyperparameters to tune. Despite these findings, recent literature on time series forecasting often uses a constant context length across multiple forecasting tasks [14,20,22]. Our work aligns with these previous studies in analyzing the importance of context length. However, unlike [5], who consider context length as multiples of seasonality, we do not make any assumptions about the context length for a dataset. Our grid of context length ranges from very short to very long across all datasets.

3. **Context Length (C)**: In our study, the specific values we have selected for our grid are 2, 7, 24, 100, 300. These values represent a summary of the default context lengths provided in the Monash Forecasting Repository.

Validation Strategy for time series forecasting is not standardized in the literature. For instance, a well-known forecasting model NBEATS [15] uses a validation set to select the hyperparameters and then re-trains the model on the combined training and validation sets, while some recent forecasting models such as NLinear [21] skip the re-training step altogether. Therefore, we consider the validation strategy as another hyperparameter to be optimized.

4. **Validation Strategy**:
   - `Out-of-Sample` (OOS): The validation split is a time-wise split replicating the test split, however the model is not retrained.

Table 2. Summary of configurations used for generating the TSBench metadataset.

Configuration	Hyperparameter	Values
Time Series	Context Length	[2, 7, 24, 100, 300]
	Validation Strategy	[OOS, Re-OOS]
Model	Architecture	Base, Diamond, Contracting, Square, Funnel, Expanding]
	Distribution Hidden State	[1, 2, 10]
Training	Learning Rate	[0.01, 0.001, 0.0001]
	Weight Decay	[0, 0.1, 0.5]
	Seeds	[100, 101, 102]

- `Retrain-Out-Of-Sample` (Re-OOS): The validation split is a time-wise split replicating the test split, but the model is retrained on the validation split using the hyperparameters chosen from the validation split.
- `In-Sample` (IS): The validation split is randomly sampled from the dataset as in [17].

4.2 Training Specific Configurations

In our comprehensive evaluation, we take into account not only the specificities of time series and model configurations, but also the configurations of model training hyperparameters. To this extent we tune two training hyperparamters namely the learning rate and the weight decay. For the learning rate, we choose from the possible options of 0.01, 0.001 or 0.0001, while retaining the default value of 0.001 in the selection. To our understanding, previous studies have not considered regularized time series forecasting evaluations [3,5]. The significance of effective regularization for neural network models in the tabular domain has been emphasized in recent research [10]. Consequently, we have incorporated the option for our model to apply regularization through weight decay. Finally, we repeat each experiment 3 times for consistency and report the standard deviation. The configurations are summarized in Table 2.

4.3 TSBench-Metadataset

Metadatasets have shown to improve the performance of hyperparameter optimization [2,9] and often provide a qualitative basis to focus efforts in both manual algorithm design and automated hyperparameter optimization. In this paper, we follow the notable metadataset work by [18], where the authors record arguably the most important metafeatures required for the secondary tasks like learning curve forecasting [8] or transfer hyperparameter optimization [2]. We evaluate 4800 configurations per dataset for 20 datasets, each evaluated for 50 epochs, and log the results as our `TSBench` metadaset.

In our work, we collect metafeatures for each run at two levels of granularity. Firstly, we log metafeatures per configuration on a coarse scale, and secondly, on

Table 3. Summary of metrics logged for generating the `TSBench` metadataset.

Granularity	Metric	Value
Epoch	Losses	Negative log likelihood loss
	Metrics	MSE, MASE, MAPE, QuantileLoss at quantile interval of 10, RMSE, NRMSE, ND, MAE and weighted QuantileLoss
	Layer-wise Gradients	Max, Mean, Median, Std, and quantiles at intercal of 10
	Learning Rate	–
	Runtine	–
Configuration	Architecture	Activations, Model Architecture, Hyperparameters, Model Complexity,
	Dataset Features	Context Length, Prediction Length, Seasonality, Frequency

a fine-grained scale per epoch. At the coarse scale, we log the basic hyperparameter configurations mentioned in the previous sections and additionally include features such as the number of trainable parameters and dataset metafeatures like time series data frequency, seasonality, etc. At the fine-grained per epoch scale, we capture the train, validation, test losses and various metrics reported per epoch. This allows the user of the metadataset to perform hyperparameter optimizations based on metrics other than the train loss. GluonTS [1] offers numerous probabilistic and point-wise evaluation metrics for time series forecasting including the Quantile Loss, CRPS, seasonal error to name a few. In addition, learning curve forecasting methods like [8] could benefit from layer-wise gradient statistics such as the max, mean and quantiles as additional covariates to predict the trajectory of a particular hyperparameter run. `TSBench` also logs this information along with learning rate and runtime information. Table 3 provides an overview of all the logged metadataset features.

5 Experimental Setup

Evaluation: All models were trained using the negative log likelihood loss to generate probabilistic forecasts, and CRPS score [16] was reported as an uncertainty error metric in the supplementary material. However, in order to compare with the Monash Forecasting Repository results [7], we evaluate using MASE error metric.

$$\text{MASE} = \frac{1}{\delta}\sum_{j=0}^{\delta}\frac{|y_j - \hat{y_j}|}{|y_j - \hat{y_j}^{\text{Naive}}|} \tag{2}$$

Data: Monash Forecasting Repository [7], which is a collection of 50 datasets that are derived from 26 original real-world datasets by sampling time series data at different frequencies. We randomly select 20 datasets from this collection that do not have missing values and have varying characteristics, such as length, number of series, etc., to capture the diversity of real-world time series data.

Table 4. Comparison of `TSBench` results with the best performing model (TBATS) from Monash Benchmark reported on the MASE error. Best results are marked in bold. Standard deviations over multiple runs are indicated in brackets. We also provide the best overall result on the dataset across different models for reference.

Datasets	Train	Retrain	Monash - TBATS	Monash - Best Overall
Aus. Elecdemand	1.693 (0.133)	1.667 (0.199)	**1.174**	0.705
Bitcoin	7.370 (2.207)	8.351 (1.30)	**4.611**	2.664
FRED-MD	0.569 (0.019)	0.580 (0.029)	**0.502**	0.468
Hospital	0.787 (0.020)	0.794 (0.025)	**0.768**	0.761
KDD	1.172 (0.011)	**1.131 (0.016)**	1.394	1.185
M1 Monthly	1.565 (0.017)	1.577 (0.030)	**1.118**	1.074
M1 Quarterly	2.360 (0.239)	2.511 (0.283)	**1.694**	1.658
M1 Yearly	4.537 (0.028)	4.393 (0.018)	**3.499**	3.499
M3 Monthly	1.143 (0.006)	1.137 (0.011)	**0.861**	0.861
M3 Quarterly	1.327 (0.025)	1.313 (0.031)	**1.256**	1.117
M3 Yearly	4.168 (0.063)	35.640 (33.024)	**3.127**	2.774
M4 Hourly	1.421 (0.192)	**1.188 (0.114)**	2.663	1.662
M4 Weekly	0.438 (0.014)	**0.434 (0.002)**	0.504	0.453
NN5 Daily	**0.800 (0.004)**	0.803 (0.007)	0.858	0.858
NN5 Weekly	0.875 (0.063)	1.143 (0.050)	**0.872**	0.808
Tourism Monthly	**1.603 (0.002)**	1.628 (0.077)	1.751	1.409
Tourism Quarterly	1.731 (0.018)	**1.631 (0.038)**	1.835	1.475
Tourism Yearly	6.283 (1.816)	5.744 (2.22)	**3.685**	2.977
Traffic Hourly	0.802 (0.002)	**0.801 (0.013)**	2.482	0.821
Traffic Weekly	1.187 (0.006)	1.187 (0.010)	**1.148**	1.094

Framework: In our experiments, we set a batch size of 64 and a number of batches per epoch of 50 for all runs. The models were trained for a total of 50 epochs, and the model with the lowest validation loss was selected for evaluation on the test set. For the `Re-OOS` validation strategy, since the model was trained using an average loss instead of sum of losses, retraining the model with the validation split (`Re-OOS`) should have only a negligible impact on the chosen hyperparameters. `In-Sample` validation implementation was not straight forward in GluonTS, as the framework expects the splits to be in continuous time and limiting our study to `Re-OOS` and `OOS`. All experiments were conducted on Intel E5-2670v2 CPU cores using Pytorch 1.12, Pytorch-lightning 1.6.5, and MXNet 1.9. The training process took approximately one month on 40 nodes with the same CPU configuration. Code[1] is made public and can be easily reused to generate a larger metadataset with other state-of-the-art models from the GluonTS [1] framework, including NBEATS [15], DeepAR [17], among others. For HPO methods, we use the SMAC3 framework [12].

[1] https://github.com/18kiran12/TSBench.git.

6 Results

In Table 4, we present a comparison between the best performing NLinear and the best performing model TBATS from the Monash forecasting repository results. It is important to note that the TBATS model was able to outperform several powerful models, including NBEATS, DeepAR, and Transformer models across the different datasets, making it a really strong baseline to outperform.

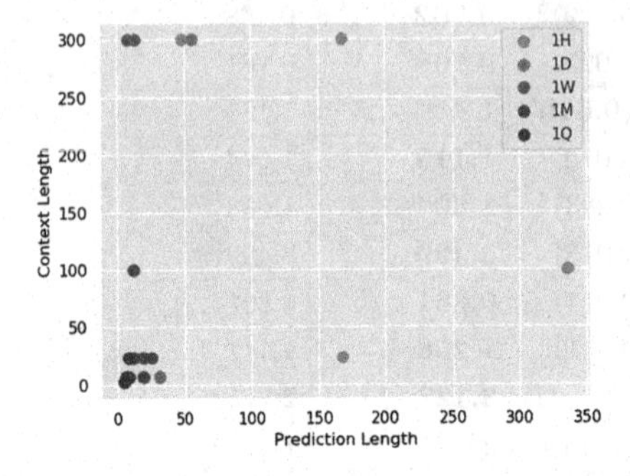

Fig. 2. Prediction length vs Context Length colored by Frequency of dataset. Longer Prediction length.

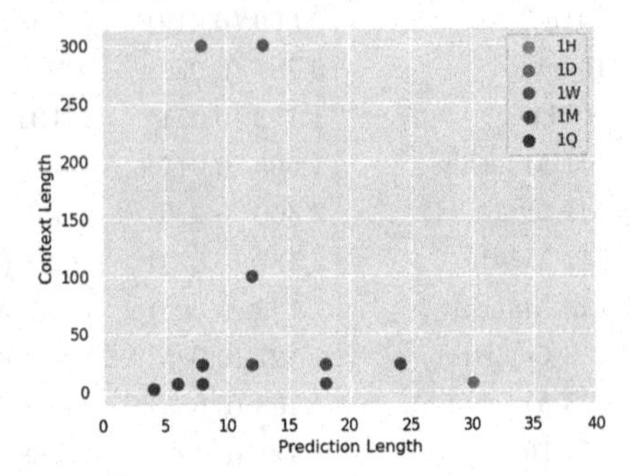

Fig. 3. Prediction length vs Context Length colored by Frequency of dataset. Shorter Prediction length.

RQ1 How Does NLinear Compare to Monash's Best Model? The NLinear model outperforms the best model TBATS from the Monash Forecasting repository on 7 out of 20 datasets of varying granularities. Additionally, when compared to the best overall results from the Monash, the NLinear model performs well on 5 out of 20 datasets. Given that a data-specific baseline is a challenging benchmark to surpass, the NLinear model performs on par with, if not better than, other deep learning baselines from the Monash forecasting results. This underscores the importance of considering carefully tuned linear models as a baseline for time series forecasting.

RQ2: Should We Retrain on the Validation Data? One of the contributions of this work is to address the uncertainty regarding whether deep learning models need to be retrained on validation data. Our results, presented in Table 4, indicate that `Re-OOS` offers only a slight advantage compared to `OOS`. This is consistent with current trends in time series forecasting, where retraining on validation data is ignored. We reason that this may be due to suboptimal hyperparameter fitting, as the addition of validation data may require changes to the hyperparameters selected on the training data. It is of significant importance to highlight that a considerable decline in the performance of the M3 yearly dataset was observed after the retraining process. This shows an extreme scenario where the retraining process adversely impacted the dataset's performance.

RQ3: What Is a Useful Context Length? The length of context required by a model is influenced by several factors. Our analysis suggests that the context length is a function of both the frequency of the dataset and the length of the forecast horizon. When given the option to select from a range of model complexities and context lengths, the model often chose longer context lengths for longer forecast horizons, as shown in Figs. 2 and 3. Additionally, our findings indicate that the frequency of the dataset also impacts the chosen context length. For instance, datasets with hourly (1H) and daily (1D) frequencies are more likely to have longer context lengths than those with yearly (1Y) or quarterly (1Q) frequencies. Figure 2 depicts the correlation between the length of the context and the prediction horizon, as well as the frequency of a specific time series dataset. Generally, a more extensive context is advantageous for the model when a larger forecast horizon is required. However, this pattern diminishes for monthly, yearly, and quarterly datasets. In these instances, the model performs satisfactorily with a shorter context length.

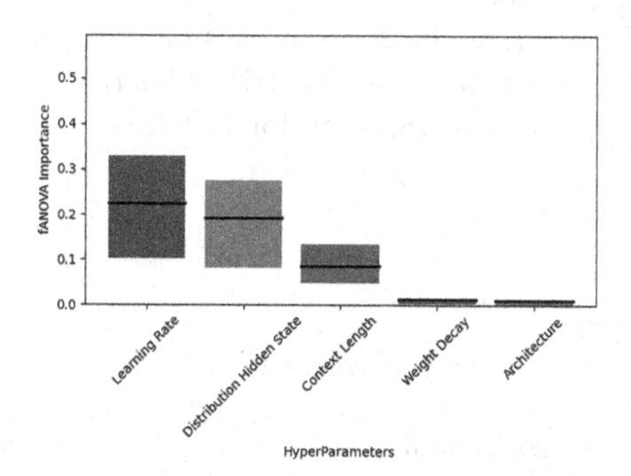

Fig. 4. Hyperparameter importance score

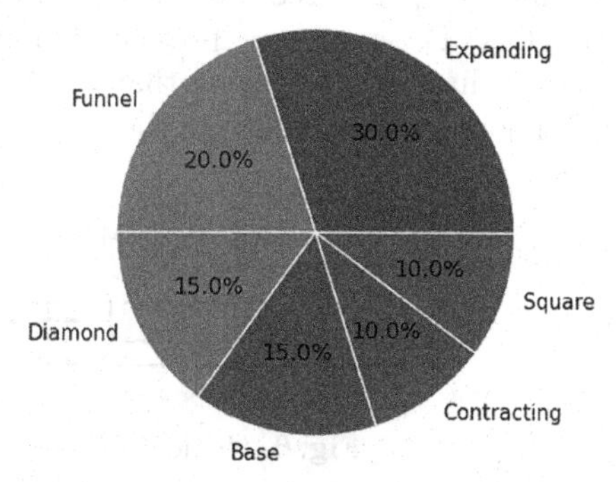

Fig. 5. Architecture selection globally across multiple datasets

RQ4: What Hyperparameters Are Important? To evaluate the significance of various hyperparameters in forecasting, we utilized an fANOVA test as outlined in [5]. This test employs a random forest model to capture the relationship between the hyperparameters and forecast accuracy, using the hyperparameters as input. A functional ANOVA is then applied to determine the importance of each hyperparameter. The results are depicted in Fig. 4. The initial learning rate selected appears to have a substantial impact on model performance, even when the model is configured with the Adam optimizer. Additionally, the hidden layer used to learn the probabilistic distribution of the forecast is also an important parameter. And most importantly, context length has a significant effect on model accuracy and needs to be carefully tuned per dataset.

RQ4: Can Linear Models Outperform Non-linear Models? In this study, we allowed the model to experiment with deeper architectures and ELU activations to determine whether a linear model consistently outperforms deeper models with various architectures, as described in Sect. 4.1. Our findings, presented in Fig. 5, indicate that deeper models can indeed be useful, however, considering the hyperparameter importance of architecture in Fig. 4, a Linear MLP is a strong baseline in most cases.

RQ5: Can TSBench Be Effectively Used for HPO? The utilization of HPO as a supplementary meta-task to demonstrate the efficacy of the constructed metadataset is a prevalent approach [2]. In this study, we applied four distinct HPO techniques to the TSBench dataset. We start with a rudimentary baseline that randomly selects a single hyperparameter from a pool of 50 trials. Secondly, we employed the HyperBand [11], a bandit-based HPO approach that conducts multiple successive halving operations to identify optimal configurations. Further, we utilized model-based algorithms such as SMAC [12] and BOHB, which utilize surrogate models to select the most promising hyperparameter evaluations. Specifically, SMAC employs a random forest model as its surrogate, while BOHB [6] adopts a Bayesian optimization algorithm. Each algorithm was permitted a total of 50 trials per dataset to select the best hyperparameters. The outcomes, presented in Fig. 6 across the 20 datasets in a critical difference diagram [2], reveal that SMAC, HyperBand and BOHB outperform the Random strategy showing the effectiveness of the `TSBench` for HPO.

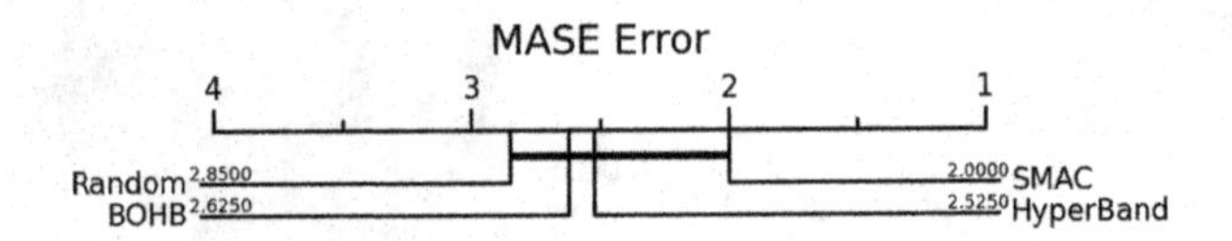

Fig. 6. Critical Difference Diagram Rank@50

7 Conclusion

Clarity regarding the significance of hyperparameters and the choice of validation strategy in time series forecasting literature is often lacking. This study aims to address these ambiguities by assessing the performance of the state-of-the-art MLP model on 20 univariate datasets. Our findings highlight the importance of tuning the context length for time series forecasting tasks and treating the validation strategy as a hyperparameter. Interestingly, while deeper MLP models may offer performance enhancements on certain datasets, our results affirm the robustness of a linear MLP model as a formidable baseline. Furthermore, we introduce TSBench, an extensive metadataset for time series forecasting to date, and demonstrate its efficacy in HPO tasks.

References

1. Alexandrov, A., et al.: Gluonts: probabilistic time series models in python. ArXiv (2019)
2. Arango, S.P., Jomaa, H.S., Wistuba, M., Grabocka, J.: Hpo-b: a large-scale reproducible benchmark for black-box hpo based on openml. In: NeurIPS Datasets and Benchmarks Track (2021)

3. Borchert, O., Salinas, D., Flunkert, V., Januschowski, T., Gunnemann, S.: Multi-objective model selection for time series forecasting. ArXiv (2022)
4. Clevert, D.A., Unterthiner, T., Hochreiter, S.: Fast and accurate deep network learning by exponential linear units (elus). In: ICLR (2015)
5. Deng, D., Karl, F., Hutter, F., Bischl, B., Lindauer, M.: Efficient automated deep learning for time series forecasting. In: ECML PKDD, pp. 664–680. Springer, Heidelberg (2023). https://doi.org/10.1007/978-3-031-26409-2_40
6. Falkner, S., Klein, A., Hutter, F.: Bohb: robust and efficient hyperparameter optimization at scale. In: ICML, pp. 1437–1446. PMLR (2018)
7. Godahewa, R., Bergmeir, C., Webb, G.I., Hyndman, R.J., Montero-Manso, P.: Monash time series forecasting archive. In: NeurIPS Datasets and Benchmarks (2021)
8. Jawed, S., Jomaa, H., Schmidt-Thieme, L., Grabocka, J.: Multi-task learning curve forecasting across hyperparameter configurations and datasets. In: ECML PKDD, pp. 485–501 (2021)
9. Jomaa, H.S., Schmidt-Thieme, L., Grabocka, J.: Dataset2vec: learning dataset meta-features. Data Min. Knowl. Disc. **35**, 964–985 (2021)
10. Kadra, A., Lindauer, M., Hutter, F., Grabocka, J.: Well-tuned simple nets excel on tabular datasets. In: NeurIPS, vol. 34, pp. 23928–23941 (2021)
11. Li, L., Jamieson, K., DeSalvo, G., Rostamizadeh, A., Talwalkar, A.: Hyperband: a novel bandit-based approach to hyperparameter optimization. JMLR **18**(1), 1–52 (2017)
12. Lindauer, M., et al.: Smac3: a versatile bayesian optimization package for hyperparameter optimization. JMLR **23**(54), 1–9 (2022)
13. Madhusudhanan, K., Burchert, J., Duong-Trung, N., Born, S., Schmidt-Thieme, L.: U-net inspired transformer architecture for far horizon time series forecasting. In: ECML/PKDD (2021)
14. Nie, Y., Nguyen, N.H., Sinthong, P., Kalagnanam, J.: A time series is worth 64 words: long-term forecasting with transformers. In: ICLR (2023)
15. Oreshkin, B.N., Carpov, D., Chapados, N., Bengio, Y.: N-BEATS: neural basis expansion analysis for interpretable time series forecasting. In: ICLR (2020)
16. Rasul, K., Sheikh, A.S., Schuster, I., Bergmann, U.M., Vollgraf, R.: Multivariate probabilistic time series forecasting via conditioned normalizing flows. In: ICLR (2021)
17. Salinas, D., Flunkert, V., Gasthaus, J., Januschowski, T.: Deepar: probabilistic forecasting with autoregressive recurrent networks. JMLR **36**(3), 1181–1191 (2020)
18. Shah, S.Y., et al.: Autoai-ts: autoai for time series forecasting. In: SIGMOD, pp. 2584–2596 (2021)
19. Ullah, I., et al.: Meta-album: multi-domain meta-dataset for few-shot image classification. In: NeurIPS, vol. 35, pp. 3232–3247 (2022)
20. Wu, H., Xu, J., Wang, J., Long, M.: Autoformer: decomposition transformers with auto-correlation for long-term series forecasting. In: NeurIPS, vol. 34, pp. 22419–22430 (2021)
21. Zeng, A., Chen, M., Zhang, L., Xu, Q.: Are transformers effective for time series forecasting? In: AAAI (2023)
22. Zhou, H., et al.: Informer: beyond efficient transformer for long sequence time-series forecasting. In: AAAI, vol. 35, pp. 11106–11115 (2021)
23. Zimmer, L., Lindauer, M., Hutter, F.: Auto-pytorch tabular: multi-fidelity metalearning for efficient and robust autodl. IEEE Trans. Pattern Anal. Mach. Intell. **43**(9), 3079–3090 (2021)

Efficient and Accurate Similarity-Aware Graph Neural Network for Semi-supervised Time Series Classification

Wenjie Xi[1](✉), Arnav Jain[2], Li Zhang[3], and Jessica Lin[1]

[1] George Mason University, Fairfax, VA, USA
{wxi,jessica}@gmu.edu
[2] The University of Texas at Austin, Austin, TX, USA
arnav@utexas.edu
[3] The University of Texas Rio Grande Valley, Edinburg, TX, USA
li.zhang@utrgv.edu

Abstract. Semi-supervised time series classification has become an increasingly popular task due to the limited availability of labeled data in practice. Recently, Similarity-aware Time Series Classification (SimTSC) has been proposed to address the label scarcity problem by using a graph neural network on the graph generated from pairwise Dynamic Time Warping (DTW) distance of batch data. While demonstrating superior accuracy compared to the state-of-the-art deep learning models, SimTSC relies on pairwise DTW distance computation and thus has limited usability in practice due to the quadratic complexity of DTW. To address this challenge, we propose a novel efficient semi-supervised time series classification technique with a new graph construction module. Instead of computing the full DTW distance matrix, we propose to approximate the dissimilarity between instances in linear time using a lower bound, while retaining the relative proximity relationships one would have obtained via DTW. The experiments conducted on the ten largest datasets from the UCR archive demonstrate that our model can be up to 104x faster than SimTSC when constructing the graph on large datasets without significantly decreasing classification accuracy.

Keywords: Time series classification · Semi-supervised learning · Graph neural network · Lower bound of DTW

1 Introduction

Time series classification is an important research topic and attracts a great amount of attention due to its widespread applications [2,3,19]. Despite its importance, the scarcity of labeled data in this field often poses significant challenges, as obtaining labeled data is difficult and expensive [13,17]. This limitation motivates researchers to investigate semi-supervised methods, which aim to

W. Xi and A. Jain—Equal Contribution.

D.-N. Yang et al. (Eds.): PAKDD 2024, LNAI 14650, pp. 276–287, 2024.
https://doi.org/10.1007/978-981-97-2266-2_22

mitigate the impact of limited labeled data availability [14]. Recently, Zha et al. [16] proposed Similarity-aware Time Series Classification (SimTSC), which integrates a batch Graph Convolution Network (GCN) [10] with Dynamic Time Warping (DTW) [12], a robust distance measure known for its effectiveness in allowing non-linear alignments in time. The superiority of DTW over other distance measures such as Euclidean Distance is evident in tasks involving time series classification and similarity search [3]. Specifically, SimTSC constructs a graph with each node corresponding to a time series instance to be classified; the edges are formed based on the DTW distances between the time series. SimTSC is versatile for three reasons. First, it only requires a batch dynamic subgraph and does not need the complete graph for the entire data to be stored in the GPU for training. Second, compared with the classic graph generation process, which simply uses the k-nearest neighbors (k-NN) to generate a static graph, SimTSC uses a graph via local k-NN generated from batch data, which allows the information in labeled data to be efficiently passed to the unlabeled data. Third, SimTSC is able to learn a warping-aware representation by inheriting the advantages of DTW [12].

Despite the advantages, SimTSC is not scalable for large datasets. The construction of its graph requires the complete pairwise computation of DTW distances. The complexity of calculating a single DTW distance between two time series instances is $O(L^2)$, where L represents the length of a time series instance. Therefore, the overall complexity for constructing the complete matrix scales significantly with the number of instances. For example, for the famous time series dataset, *StarLightCurves*, which contains 9,236 time series of length 1,024, DTW takes around two days on a computer equipped with an 12-core CPU and 32 GB of memory to calculate the distances between all pairs. Furthermore, since DTW computation uses dynamic programming, it is not parallelizable with GPU computations. While various techniques have been proposed to speed up DTW computation, they cannot be applied in this case since they work by pruning unnecessary computations for the similarity search problem, whereas SimTSC requires the full pairwise distance matrix for its training process.

To address this problem, we propose a graph-based time series classification technique with an efficient graph construction approach. Instead of computing the pairwise distance matrix using DTW, we propose to use a lower-bounding distance for DTW, which can significantly reduce the computational cost. In particular, we adapt LB_Keogh [8], a popular lower bound for DTW, with $O(L)$ time complexity. For the *StarlightCurves* dataset, it would reduce the computation time for all pairwise distances from two days to approximately 30 min. It is important to note that the idea of a lower bound was initially developed to accelerate the 1-NN similarity search problem by providing a reasonable and computationally efficient approximation of actual distances, thereby allowing pruning some candidates from consideration. Although not originally designed as a similarity measure by itself, our findings suggest that the LB_Keogh matrix can effectively replace DTW in guiding GCN, allowing it to learn a similar warping-aware representation at a fraction of the cost. We demonstrate through empirical evaluation that using the approximation matrix can achieve comparable performance to the original SimTSC model, but with a reduction in running time by two orders of magnitude.

In summary, our contributions are summarized as follows:

- We investigate an important and challenging problem of semi-supervised time series classification when only a few labeled samples are available.
- We introduce an efficient graph-based time series classification technique that employs a novel graph construction module with linear time complexity.
- Our experimental results demonstrate that our proposed method significantly speeds up graph construction for large datasets while achieving comparable performance to SimTSC by using an efficient lower bound distance-based graph to guide GCN, instead of using the more expensive DTW distance.

2 Related Work

2.1 Graph-Based Time Series Classification

Recently, the popularity of graph neural networks has led many researchers to study graph-based time series classification [4,6,11,16,18]. Most existing works either require actual graphs as an input or do not consider semi-supervised settings with only few labels in each class [1]. The most closely related work is SimTSC [16], which computes all pairwise distances between time series instances using the DTW distance measure, and constructs a graph from distance matrix, where each node represents a time series, and the connection between nodes represents their similarity. However, the computation of this matrix is highly time-consuming, especially for large datasets. Our proposed method takes an alternative approach to address the efficiency bottleneck by considering a cheaper distance computation that can still capture the relationships between instances.

2.2 Lower Bound of DTW

DTW [12] is an elastic distance measure that computes the dissimilarity between two time series by optimally aligning them in time. While DTW is a more robust measure compared to Euclidean distance, it is also much more computationally intensive. Driven by the need for efficient DTW-based similarity search, many lower-bounding distance measures for DTW have been proposed, including LB_Yi [15], LB_Kim [9], and LB_Keogh [8]. Since lower bounding distances are typically much cheaper to compute than the actual distances, they are often used to approximate the minimum distance between a pair of sequences. Such approximation enables the rapid elimination of certain unqualified candidates in similarity search without the need to calculate actual distances. Among them, LB_Keogh is simple to compute and remains one of the most competitive and tightest lower bounds for DTW-based tasks.

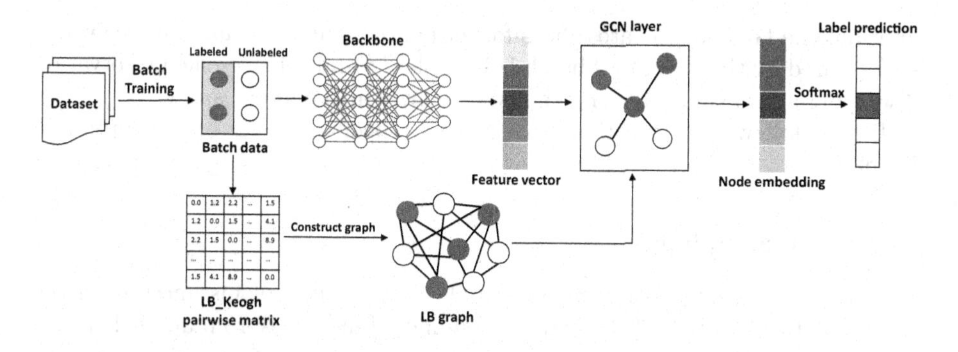

Fig. 1. Overall framework of our proposed model.

To the best of our knowledge, this is the first attempt to explore the potential of using a lower bounding measure for graph construction for graph-based semi-supervised time series classification models. This approach not only addresses the efficiency challenges of the existing method but also opens new pathways for research in efficient, scalable graph-based time series analysis.

3 Problem Formulation

In this paper, we study the semi-supervised time series classification problem with the transductive setting [16]. In this setting, the model is trained on the dataset $\mathcal{X}$ consists of both labeled and unlabeled data, including access to test data but not their labels. Specifically, the training data $\mathcal{X}^{train}$ consists of n time series instances $[X_1, X_2, \cdots, X_n]$ with their labels $Y^{train} = [y_1, y_2, \cdots, y_n]$ where $y_i \in \{1, \cdots, C\}$. The model also has access to some unlabeled data $\mathcal{X}^{unlabeled}$, and test data $\mathcal{X}^{test}$ without their labels.

4 Methodology

Figure 1 shows the overall framework of the proposed method. The framework consists of three steps: **1) Batch Sampling**. Given a dataset $(\mathcal{X}, Y^{train})$, we first form a batch consisting of an equal number of labeled and unlabeled data $\mathcal{X}^B$. **2) LB_Keogh Lower Bound Graph Construction**. We compute a pairwise lower bound distance matrix for $\mathcal{X}^B$ to generate a batch LB-graph G^B that describes the approximate warping-aware (dis)similarity between pairs of instances within each batch. The feature vector is obtained from the backbone network. **3) Graph Convolution and Classification** The feature vector is passed through Graph Convolution Network (GCN) [10] to aggregate the features across the generated G^B. Then the obtained node embedding is passed through a fully connected neural network and performs the classification task.

There are two main advantages of the framework: We can pass the label information from labeled instances to unlabeled instances through batched data. In addition, the GCN aggregates the information of similar instances under warping captured by the generated batch LB-graph G^B and achieves the preservation of warping-aware similarity on the instance level in the latent space.

In the following subsections, we provide specific details of our proposed model.

4.1 Batch Sampling

To form a batch $\mathcal{X}^B$ of size m, we sample $m/2$ number of instances from the training data $(\mathcal{X}^{train}, \mathcal{Y}^{train})$, and $m/2$ number of instances from unlabeled data $\mathcal{X}^{unlabeled}$ and $\mathcal{X}^{test}$. In this step, we ensure that each class has equal samples to avoid imbalance in the learning process. The batch data will be used to compute the embedding and the proposed batch LB-Graph.

4.2 LB_Keogh Graph Construction

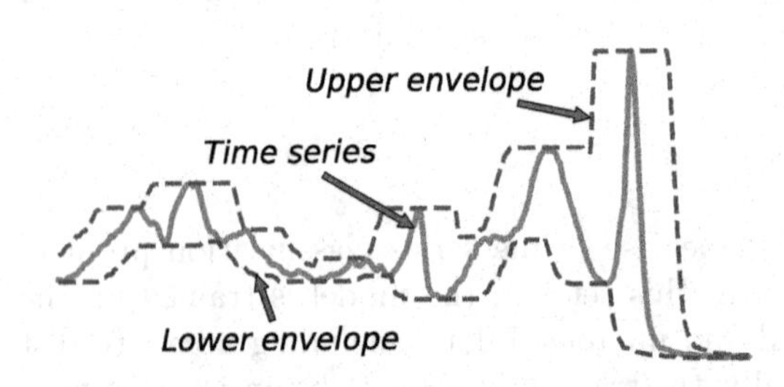

Fig. 2. A visual intuition of the lower envelope (blue dotted line) and upper envelope (green dotted line) of a sample time series (gray solid line) generated by LB_Keogh. (Color figure online)

We introduce our proposed LB_Keogh Graph (LB-Graph) construction in this subsection.

LB_Keogh [8] is originally proposed as a lower bound to prune unnecessary computations of DTW when performing similarity search. It is much cheaper to compute than the actual DTW distance, but still maintains the ability to allow non-linear alignment of points when comparing two time series.

Given two time series instances X_i and X_j from batch $\mathcal{X}^B$, LB_Keogh creates an envelope consisting of one upper bound and one lower bound time series (Fig. 2). As shown in Eq. 1, for each timestamp k:

$$\begin{aligned} u_{i,k} &= \max([X_{i,k-r}, X_{i,k-r+1}, \cdots, X_{i,k+r}]), \\ l_{i,k} &= \min([X_{i,k-r}, X_{i,k-r+1}, \cdots, X_{i,k+r}]), \end{aligned} \tag{1}$$

where r is the allowed range of warping. Then the LB_Keogh distance d^{LB} between instance X_i and X_j is calculated by Eq. 2:

$$d^{LB}(X_i, X_j) = \sqrt{\sum_{k=1}^{L} \begin{cases} (X_{j,k} - u_{i,k})^2, & if X_{j,k} > u_{i,k} \\ (X_{j,k} - l_{i,k})^2, & if X_{j,k} < l_{i,k} \\ 0, & \text{otherwise.} \end{cases}} \tag{2}$$

The pairwise LB_Keogh matrix D_{LB} given a batch $\mathcal{X}^B$ is hence defined as

$$D_{LB}(i,j) = d^{LB}(X_i, X_j), \quad X_i \in \mathcal{X}^B, X_j \in \mathcal{X}^B. \tag{3}$$

We next generate the graph $G^B(V,E)$ from D_{LB}. We perform a two-step approach to compute the edges.

First, we convert the LB_Keogh distance to similarity by:

$$A_{i,j} = \frac{1}{\exp(\alpha D_{i,j})}, \tag{4}$$

where α is the scaling parameter.

Then we generate a sparse graph based on A. For each row $A_i = [A_{i,1}, \ldots, A_{i,m}]$ in A, we select K candidates to form the sparse adjacency matrix. Specifically, to form a vector $\mathcal{Q}_i$, we select random K candidates with zero LB_Keogh distance, which indicates that the shapes of these instances fall within the enveloped area (i.e. they are likely to be similar to X_i). If there are fewer than K candidates with zero distance, then we pick K instances with the smallest LB_Keogh distances.

To incorporate this idea, we compute a sparse adjacency matrix $\bar{A}$ by:

$$\bar{A}_{i,j} = \begin{cases} 1/K, & j \sim \mathcal{Q}_i \quad \& \quad |\mathcal{Q}_i| \geq K; \\ A_{i,j}, & A_{i,j} \in \arg \text{TopK}(A_i) \quad \& \quad |\mathcal{Q}_i| < K; \\ 0, & \text{otherwise.} \end{cases} \tag{5}$$

where $\mathcal{Q}_i = \{k | D_{i,k} = 0\}$, and we randomly pick K samples from $\mathcal{Q}_i$.

Now we are ready to build the batch graph $G^B(V,E)$ where $|V| = m$, and the weights between node i and node j is defined as:

$$E_{i,j} = \frac{\bar{A}_{i,j}}{\sum_{b=1}^{m} \bar{A}_{i,b}}. \tag{6}$$

The overall LB graph construction algorithm is summarized in Algorithm 1.

4.3 Graph Convolution and Classification

Given a backbone network $f(.)$, we compute the feature embedding network $z = f(\mathcal{X}^B)$. Then the embedding z is passed through a series of graph convolutional network (GCN) layers and used to perform classification. Particularly,

$$z^l = GCN(G^B, z^{l-1}),$$
$$o = Softmax(z^l),$$

where l is the number of layers in this case. In our work, we use the same backbone network as SimTSC [16], which is a ResNet [7] with three residual blocks. Each block is connected by shortcuts and consists of three 1D convolutional layers followed by batch normalization and a ReLU activation layer. The kernel size of three 1D convolutional layers are 7, 5 and 3, respectively. The number of channels for all convolutional layers are 64. Finally, a global average pooling is applied to the output of the last residual block.

Algorithm 1. LB-Graph Construction

Require: $\mathcal{X}^B$: batch data, r: warping range, α: scaling factor, K: number of top neighbors
Ensure: G^B: the graph of the current batch
/* *Compute D_{LB} via Eqn. 3* */
1: $D \leftarrow$ ComputePairwiseLB($\mathcal{X}^B$)
/* *Compute A via Eqn. 4* */
2: **for** each $D_{i,j}$ in D **do**
3: $A_{i,j} \leftarrow 1/\exp(\alpha D_{i,j})$
4: **end for**
/* *Compute G^B via Eqn. 5 and Eqn. 6.* */
5: $G^B \leftarrow$ GenerateGraph(A, K)
6: **return** G^B

4.4 Advantages of Our Model

Compared with SimTSC [16], our proposed method has two advantages. First, we construct batch LB_Keogh graph, which can be seen as an alternative but more efficient solution to measure similarity. Compared to the original DTW distance computation, which is expensive via dynamic programming or recursion, LB_Keogh is simple and only requires $O(L)$ time complexity. Second, from Eq. 2, we can see that D_{LB} is highly parallelizable, thus more suitable to implement in GPU. Second, GCN does not require exact weights to perform aggregate information to propagate label information. Therefore, although LB_Keogh matrix D_{LB} containing zero values provides a less accurate warping measure than DTW, all of these instances fall into the envelop area and share some degree of warping similarity. Thus, a random subset of zero value in D_{LB} is sufficient to guide the training and propagate labels. Our experiments empirically demonstrate our model is statistically comparable with the performance of SimTSC.

5 Experimental Evaluation

In this section, we evaluate both the efficiency and accuracy of our proposed method.

Datasets. Our evaluation focuses on demonstrating the efficiency of our proposed model and the ability to achieve competitive performance with few labels available. We use *all* the large datasets from the well-known UCR time series classification archive [5], with the sequence length greater than 500, and the number of instances greater than 2000. We provide a detailed description of datasets in Table 1. The 10 qualified datasets lie in 5 major time series data types: sensor, ECG, motion, simulation, and image. We also provide the abbreviation names of some datasets that we used due to space limitation.

We follow the same dataset generation protocol as SimTSC [16], whereas we first split every dataset into the training set (80%) and the test set (20%).

Table 1. Description of ten large datasets

Datasets	Abbr.	Type	Class	Length	Instance
FordA	-	Sensor	2	500	4921
FordB	-	Sensor	2	500	4446
NonInvasiveFetalECGThorax1	NIFECGT1	ECG	42	750	3765
NonInvasiveFetalECGThorax2	NIFECGT2	ECG	42	750	3765
UWaveGestureLibraryAll	UWGLAll	Motion	8	945	4478
Phoneme	-	Sensor	39	1024	2110
Mallat	-	Simulated	8	1024	2400
MixedShapesSmallTrain	MSST	Image	5	1024	2525
MixedShapesRegularTrain	MSRT	Image	5	1024	2925
StarLightCurves	SLC	Sensor	3	1024	9236

We randomly sample β instances for each class from the training data to form $\mathcal{X}^{train}$ and Y^{train}. The remaining training data is used as $\mathcal{X}^{unlabeled}$. We test six different settings $\beta = \{5, 10, 15, 20, 25, 30\}$, and report the classification accuracy for each case. The experiments are conducted on a computer equipped with an AMD Ryzen 9 5900X 12-core processor, 32GB memory.

Hyperparameters. 1) 1NN-DTW. The warping window size is set to the minimum value between the length of the time series and 100. **2) SimTSC.** We keep all the settings the same as in the original paper. That is, we set the scaling factor α to 0.3 and the number of neighbors K to 3. The batch size is fixed to 128, and the number of epochs is fixed to 500. The number of GCN layers is set to 1. Adam with learning rate of 1×10^{-4} and weight decay of 4×10^{-3} is used as the optimizer. **3) Our model.** We fix the warping range r to be 5% of time series length and empirically set the scaling factor α to 11. All other settings are consistent with SimTSC.

5.1 Comparing with 1NN-DTW

We compare with 1NN-DTW, a well-known baseline in time series classification [3] and still performs well even with only few labeled data available [16].

Table 2 presents the classification accuracy of both 1NN-DTW and our proposed model across ten datasets. Our model wins 8 or 9 out of 10 in most cases, and ties with 1NN-DTW when $\beta = 5$. To further illustrate the superiority of the accuracy of our model, we perform the one-sided Wilcoxon signed-rank test. The p-values of the Wilcoxon test between two models on all β settings are 0.246,

Table 2. Accuracy scores of our model and 1NN-DTW on ten large datasets. The p-value is calculated using the Wilcoxon signed-rank test (one-sided).

Datasets	5 labels		10 labels		15 labels		20 labels		25 labels		30 labels	
	1NN-DTW	Ours	1NN-DTW	Ours	1NN-DTW	Ours	1NN-DTW	Ours	1NN-DTW	Ours	1NN-DTW	Ours
FordA	0.540	**0.793**	0.531	**0.826**	0.545	**0.816**	0.535	**0.825**	0.548	**0.862**	0.550	**0.864**
FordB	0.627	**0.812**	0.603	**0.806**	0.628	**0.807**	0.630	**0.820**	0.627	**0.836**	0.637	**0.847**
NIFECGT1	**0.665**	0.662	0.704	**0.743**	0.710	**0.830**	0.720	**0.855**	0.732	**0.867**	0.729	**0.860**
NIFECGT2	**0.757**	0.689	0.788	**0.798**	0.817	**0.856**	0.819	**0.856**	0.821	**0.874**	0.826	**0.879**
UWGLAll	**0.779**	0.447	**0.854**	0.549	**0.865**	0.622	**0.879**	0.628	**0.886**	0.682	**0.911**	0.677
Phoneme	0.185	**0.253**	0.204	**0.330**	0.209	**0.346**	0.258	**0.408**	0.277	**0.419**	0.282	**0.427**
Mallat	**0.944**	0.908	0.960	**0.963**	**0.971**	0.960	**0.971**	**0.971**	0.971	**0.972**	**0.977**	0.969
MSST	0.784	**0.871**	0.796	**0.883**	0.834	**0.908**	0.838	**0.914**	0.846	**0.931**	0.853	**0.925**
MSRT	**0.790**	0.775	0.819	**0.891**	0.843	**0.916**	0.855	**0.922**	0.851	**0.933**	0.853	**0.942**
SLC	0.559	**0.920**	0.683	**0.939**	0.763	**0.946**	0.798	**0.924**	0.807	**0.936**	0.780	**0.942**
wins	5	5	1	9	2	8	2	9	1	9	2	8
p-value	0.246	-	0.042	-	0.042	-	0.043	-	0.024	-	0.042	-

0.042, 0.042, 0.043, 0.043, 0.024, and 0.042, respectively. Notably, all p-values, except for $\beta = 5$, fall below 0.05, which corroborates that our model significantly outperforms 1NN-DTW in all tested settings except at 5 labels. This observation is consistent with results in the SimTSC paper, suggesting that classic 1-NN methods may perform well in extreme small-sample situations, while deep learning models may face challenges to perform well under similar conditions.

Figure 3 visually summarizes the accuracy comparison between our model and 1NN-DTW. In Fig. 3, the X-axis denotes the accuracy of our model, while the Y-axis represents the accuracy of 1NN-DTW. Each data point corresponds to one of the datasets evaluated. The point located in the lower triangle (shaded area) indicates the dataset where our model demonstrates superior performance over 1NN-DTW. Since the majority of data points are located in the lower triangle, we demonstrated our model outperforms 1NN-DTW.

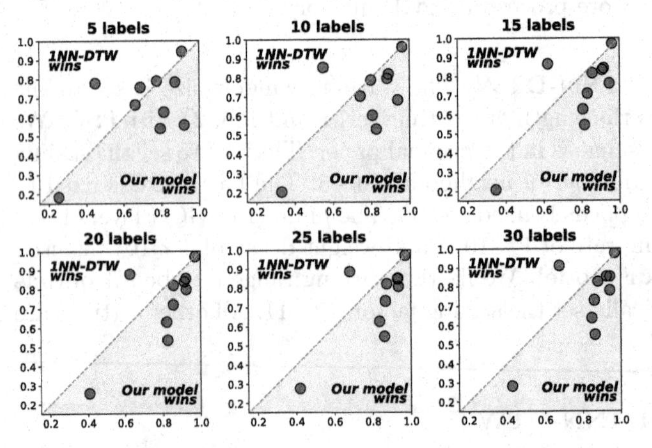

Fig. 3. Classification accuracy: our model (right bottom shaded) vs. 1NN-DTW (left top white).

5.2 Graph Construction Time Comparison with SimTSC

Table 3. Comparison of graph construction time

Datasets	Length	Instance	SimTSC (in hours)	Ours (in hours)	Speedup
FordA	500	4921	3.826	**0.127**	**30x**
FordB	500	4446	3.310	**0.099**	**33x**
NIFECGT1	750	3765	4.622	**0.082**	**56x**
NIFECGT2	750	3765	4.678	**0.081**	**58x**
UWGLAll	945	4478	10.385	**0.126**	**82x**
Phoneme	1024	2110	2.841	**0.029**	**96x**
Mallat	1024	2400	3.716	**0.036**	**104x**
MSST	1024	2525	4.014	**0.041**	**96x**
MSRT	1024	2925	5.045	**0.058**	**86x**
SLC	1024	9236	46.240	**0.566**	**82x**
Total			88.676	**1.247**	**71x**

SimTSC [16] is the most closely related work to our study. We compare computational time for graph construction, which is the major bottleneck in overall computation cost. For both methods, to ensure a fair comparison, we follow the same implementation process described in Zha et al. [16] via pre-computing the similarity matrix for the entire dataset first, then performing sampling for each batch. For our model, we use Python to implement the graph construction process. For SimTSC, since their original graph construction code in C is not executable on our computer, we use an equivalent C++ implementation to estimate their running time. We report the actual running time for both methods.

Table 4. Accuracy scores of our model and SimTSC on ten large datasets. The p-value is calculated using the Wilcoxon signed-rank test (two-sided).

Datasets	5 labels		10 labels		15 labels		20 labels		25 labels		30 labels	
	SimTSC	Ours	SimTSC	Ours	SimTSC	Ours	SimTSC	Ours	SimTSC	Ours	SimTSC	Ours
FordA	**0.795**	0.793	**0.832**	0.826	**0.827**	0.816	0.821	**0.825**	0.854	**0.862**	0.862	**0.864**
FordB	0.768	**0.812**	0.800	**0.806**	0.802	**0.807**	**0.824**	0.820	**0.842**	0.836	0.843	**0.847**
NIFECGT1	**0.692**	0.662	**0.790**	0.743	**0.861**	0.830	**0.887**	0.855	**0.906**	0.867	**0.908**	0.860
NIFECGT2	**0.733**	0.689	**0.823**	0.798	**0.866**	0.856	**0.896**	0.856	**0.912**	0.874	**0.917**	0.879
UWGLAll	**0.458**	0.447	**0.561**	0.549	0.601	**0.622**	**0.640**	0.628	0.671	**0.682**	0.650	**0.677**
Phoneme	**0.261**	0.253	0.313	**0.330**	**0.372**	0.346	0.405	**0.408**	0.414	**0.419**	**0.442**	0.427
Mallat	**0.943**	0.908	0.952	**0.963**	**0.968**	0.960	**0.974**	0.971	**0.976**	0.972	**0.975**	0.969
MSST	**0.880**	0.871	0.871	**0.883**	**0.914**	0.908	0.902	**0.914**	**0.933**	0.931	**0.940**	0.925
MSRT	0.732	**0.775**	**0.907**	0.891	0.912	**0.916**	0.919	**0.922**	0.909	**0.933**	0.925	**0.942**
SLC	0.914	**0.920**	**0.948**	0.939	0.931	**0.946**	**0.947**	0.924	**0.949**	0.936	**0.944**	0.942
wins	**7**	3	**6**	4	**6**	4	**6**	4	**6**	4	**6**	4
p-value	0.492	-	0.375	-	0.375	-	0.232	-	0.695	-	0.492	-

Table 3 shows the graph construction time in hours for SimTSC and our proposed model on the ten datasets. We can observe that even when we use the slower implementation (Python) for the graph construction module of our model and the much faster implementation (MATLAB) for SimTSC, our model is still 71x faster than SimTSC in total time, and up to 104x faster on one dataset. For the time required to complete all graph construction, our approach reduces it from 3 days 16 h to only 75 min. Note that an increase in time series length or number of instances makes the speed gap of our method more significant than that of SimTSC. The large time saving of our model makes our approach desirable for semi-supervised TSC on large data with few labels.

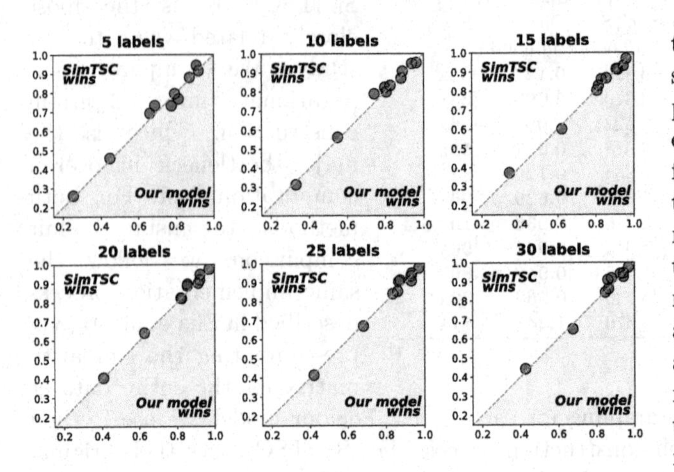

Fig. 4. Classification accuracy: our model (right bottom shaded) vs. SimTSC (left top white).

We also compare the accuracy to demonstrate that our proposed model still achieves competitive performance. To mitigate the impact of randomness in deep learning training, we run both methods three times and report the average accuracy of these runs. Table 4 shows the classification accuracy of SimTSC and our model. The number of wins of SimTSC is slightly more than ours in all settings. To show the difference in accuracy between the two models, we perform the two-sided Wilcoxon signed-rank test. The p-values between SimTSC and our model are 0.492, 0.375, 0.375, 0.232, 0.695 and 0.492, respectively. All the p-values are less than 0.05, meaning our model does not have a significant difference in accuracy from SimTSC.

In Fig. 4, the accuracy for our model is on the X-axis as well, and the accuracy for SimTSC is on the Y-axis. It can be observed that all points are near the midline, indicating that the two models do not have significant differences in accuracy.

6 Conclusion

In this paper, we study the semi-supervised time series classification problem under few labeled samples. We propose a new efficient graph-based time series classification with a novel efficient graph construction module that takes only linear time. Our proposed method resolves the scalability issue of constructing graphs from large datasets while maintaining the ability of graphs to capture warping similarity. Our work is the first to explore a graph construction with an efficient lower bounding based graph to guide GCN. The experimental results show that our model achieves two orders of magnitude of speedup on graph construction without significant loss of classification accuracy compared to SimTSC.

References

1. Alfke, D., Gondos, M., Peroche, L., Stoll, M.: An empirical study of graph-based approaches for semi-supervised time series classification. arXiv preprint arXiv:2104.08153 (2021)
2. Bagnall, A., et al.: The UEA multivariate time series classification archive, 2018. arXiv preprint arXiv:1811.00075 (2018)
3. Bagnall, A., Lines, J., Bostrom, A., Large, J., Keogh, E.: The great time series classification bake off: a review and experimental evaluation of recent algorithmic advances. Data Min. Knowl. Disc. **31**(3), 606–660 (2017)
4. Cheng, Z., et al.: Time2graph+: bridging time series and graph representation learning via multiple attentions. IEEE Trans. Knowl. Data Eng. **35**(2), 2078–2090 (2021)
5. Dau, H.A., et al.: The UCR time series archive. IEEE/CAA J. Automatica Sinica **6**(6), 1293–1305 (2019)
6. Duan, Z., et al.: Multivariate time-series classification with hierarchical variational graph pooling. Neural Netw. **154**, 481–490 (2022)
7. He, K., Zhang, X., Ren, S., Sun, J.: Deep residual learning for image recognition. In: Proceedings of the IEEE Conference on Computer Vision and Pattern Recognition, pp. 770–778 (2016)
8. Keogh, E., Ratanamahatana, C.A.: Exact indexing of dynamic time warping. Knowl. Inf. Syst. **7**(3), 358–386 (2005)
9. Kim, S.W., Park, S., Chu, W.W.: An index-based approach for similarity search supporting time warping in large sequence databases. In: Proceedings 17th International Conference on Data Engineering, pp. 607–614. IEEE (2001)
10. Kipf, T.N., Welling, M.: Semi-supervised classification with graph convolutional networks. arXiv preprint arXiv:1609.02907 (2016)
11. Liu, H., et al.: Todynet: temporal dynamic graph neural network for multivariate time series classification. arXiv preprint arXiv:2304.05078 (2023)
12. Sakoe, H.: Dynamic-programming approach to continuous speech recognition. In: 1971 Proceedings of the International Congress of Acoustics, Budapest (1971)
13. Tong, Y., et al.: Technology investigation on time series classification and prediction. PeerJ Comput. Sci. **8**, e982 (2022)
14. Wei, L., Keogh, E.: Semi-supervised time series classification. In: Proceedings of the 12th ACM SIGKDD International Conference on Knowledge Discovery and Data Mining, pp. 748–753 (2006)
15. Yi, B.K., Jagadish, H.V., Faloutsos, C.: Efficient retrieval of similar time sequences under time warping. In: Proceedings 14th International Conference on Data Engineering, pp. 201–208. IEEE (1998)
16. Zha, D., Lai, K.H., Zhou, K., Hu, X.: Towards similarity-aware time-series classification. In: Proceedings of the 2022 SIAM International Conference on Data Mining (SDM), pp. 199–207. SIAM (2022)
17. Zhang, L., Patel, N., Li, X., Lin, J.: Joint time series chain: Detecting unusual evolving trend across time series. In: Proceedings of the 2022 SIAM International Conference on Data Mining (SDM), pp. 208–216. SIAM (2022)
18. Zhang, X., Zeman, M., Tsiligkaridis, T., Zitnik, M.: Graph-guided network for irregularly sampled multivariate time series. arXiv preprint arXiv:2110.05357 (2021)
19. Zhang, X., Gao, Y., Lin, J., Lu, C.T.: Tapnet: multivariate time series classification with attentional prototypical network. In: Proceedings of the AAAI Conference on Artificial Intelligence, vol. 34, pp. 6845–6852 (2020)

STLGRU: Spatio-Temporal Lightweight Graph GRU for Traffic Flow Prediction

Kishor Kumar Bhaumik[1], Fahim Faisal Niloy[2], Saif Mahmud[3], and Simon S. Woo[1](✉)

[1] Sungkyunkwan University, Seoul, South Korea
{kishor25,swoo}@g.skku.edu
[2] University of California, Riverside, USA
fnilo001@ucr.edu
[3] Cornell University, Ithaca, USA
sm2446@cornell.edu

Abstract. Reliable forecasting of traffic flow requires efficient modeling of traffic data. Indeed, different correlations and influences arise in a dynamic traffic network, making modeling a complicated task. Existing literature has proposed many different methods to capture traffic networks' complex underlying spatial-temporal relations. However, given the heterogeneity of traffic data, consistently capturing both spatial and temporal dependencies presents a significant challenge. Also, as more and more sophisticated methods are being proposed, models are increasingly becoming memory-heavy and, thus, unsuitable for low-powered devices. To this end, we propose **S**patio-**T**emporal **L**ightweight Graph **GRU**, namely *STLGRU*, a novel traffic forecasting model for predicting traffic flow accurately. Specifically, our proposed *STLGRU* can effectively capture dynamic local and global spatial-temporal relations of traffic networks using memory-augmented attention and gating mechanisms in a continuously synchronized manner. Moreover, instead of employing separate temporal and spatial components, we show that our memory module and gated unit can successfully learn the spatial-temporal dependencies with reduced memory usage and fewer parameters. Extensive experimental results on three real-world public traffic datasets demonstrate that our method can not only achieve state-of-the-art performance but also exhibit competitive computational efficiency. Our code is available at https://github.com/Kishor-Bhaumik/STLGRU.

Keywords: Traffic Forecasting · Time Series · Graph Convolution

1 Introduction

A traffic network can be represented as a graph, with the locations of the sensors and the connections among them acting as the nodes and edges, respectively. In the same way, flow at a particular junction or node is defined as the total number of people or vehicles passing through that junction at a given time. Specifically, the goal of traffic flow prediction algorithms is to predict the flow of

D.-N. Yang et al. (Eds.): PAKDD 2024, LNAI 14650, pp. 288–299, 2024.
https://doi.org/10.1007/978-981-97-2266-2_23

future time steps by exploiting the complex spatialtemporal features of historical traffic data. Indeed, many cities are currently developing Intelligent Traffic Systems (ITS) [29] and predicting traffic flow is a key part of many of these systems' services. In particular, a large amount of collected traffic data have made urban data mining study much easier than ever before, such as traffic flow prediction [9], arrival time estimate [12], traffic speed analysis [2,4], and so on, thanks to the promising advancement of intelligent sensors. To be more specific, spatio-temporal traffic prediction aims to forecast future traffic trends by analyzing previous spatio-temporal features [30]. Furthermore, predicting traffic flow has become essential for several downstream applications, such as intelligent route planning [18], dynamic traffic management [27], and location-based services [16]. However, the efficiency and accuracy of traffic flow prediction algorithms are limited by the high variance in the spatial and temporal dimensions of traffic data. In addition, the observations made at different locations and time stamps are not independent, but they are rather dynamically correlated. Hence, traffic data has a nonlinear and complex spatial-temporal relationship, and its modeling is critical for designing effective prediction algorithms.

To address the aforementioned challenges, in this paper, we propose a novel traffic flow prediction model, called Spatio-Temporal Lightweight Graph GRU *(STLGRU)*. Our model takes advantage of graph convolution to model localized spatial relations. We then use an attention mechanism with a memory module to directly model the long-range local and non-local spatio-temporal dependencies. To update the memory, we use a gating mechanism, where our gating strategy records the key local and global spatio-temporal information and forgets the redundant ones when moving to the next time step. In addition, we carefully design our model to be lightweight, as the memory module uses fewer parameters than the existing baselines. Consequently, it can effectively learn long-range dependencies without the need to use multi-scale causal convolution or stacking past time step features. In summary, we make the following contributions:

- We propose *STLGRU*, a novel time series traffic flow prediction model. Our model captures the long-range global and local relationships of a traffic network more accurately by using memory-augmented attention module and gating mechanism.
- We carefully design our network to be lightweight by utilizing a memory module with minimal parameters, thus making it suitable for environments constrained by computational resources.
- We conduct extensive experiments on three popular traffic prediction benchmark datasets. Our results show that our model not only surpasses other baseline models in performance but also necessitates less memory usage in comparison.

2 Related Work

Spatio-Temporal Time Series Traffic Forecasting. Deep learning has been successfully applied to many tasks, such as image analysis [21,22], natural language processing [8], activity recognition [20] etc. Recently, such learning techniques have been quite extensively applied to traffic flow prediction task. Amongst these methods, STGCN [28] is the first pioneering work to model the traffic network with a fully convolutional structure. In this study, spatio-temporal relationships are effectively captured by including a graph convolution module inside temporal convolution modules. Moreover, DCRNN [17] introduces diffusion convolution to propagate information in the graph. PM-MemNet [15] learns to match input data to representative patterns with a key-value memory structure. Song et al. proposes STSGCN [23], which captures complex localized spatial-temporal correlation to find the heterogeneities in the spatial-temporal data. STSGCN [23] deals with spatial and temporal dimensions individually by utilizing various modules and calculates spatio-temporal attention within a restricted temporal frame.

In addition, Lin et al. [19] propose self-attention Conv-LSTM to capture long-range temporal dependencies for general spatio-temporal prediction task. A significant limitation of their approach is its reliance solely on convolution layers, confining their method to spatio-temporal prediction tasks representable by image grids. However, the traffic network has an inherent graph structure that needs to be exploited for reliable prediction. Yuzhou et al. [5] tackles this problem by enriching DL architectures with salient time-conditioned topological information of the traffic data. This study introduces the zig-zag persistence concept into time-aware graph convolutional networks.

However, most of the cutting-edge models fail to handle the challenge of being lightweight. RNN-based networks (including LSTM) are widely known to be difficult to train and computationally heavy [28]. For example, Mega-CRN [14] proposes Meta-Graph Convolutional Recurrent Network (MegaCRN) by plugging multiple Meta-Graph Learner powered by a MetaNode Bank into the encoder-decoder module. As a consequence, it becomes memory-heavy due to its large number of parameters. STSGCN [23] uses a certain length of time window to collect graph structure information and fuse the findings to forecast the following time steps. The computational cost is thus increased by employing repeated shots of graph aggregation. StemGNN [1] introduces a neural network that captures inter-series correlations and temporal dependencies in the spectral domain by aggregating numerous modules in separate blocks while disregarding the model's complexity. To solve the aforementioned issues, we present a simple but effective traffic forecasting model that is computationally cheap, lightweight, and capable of capturing both local and global long-range dependencies in a traffic network.

Attention Mechanism. Because of the high efficiency and versatility in modeling dependencies, attention mechanisms have been extensively used in a variety of domains [6, 24]. The basic principle behind attention mechanisms is to concentrate on the most relevant features of the input data [6]. Recently, researchers used attention processes to graph-structured data to model spatial correlations for graph classification [25]. We expand the attention method to synchronize spatial and temporal dependencies while sequentially predicting traffic data.

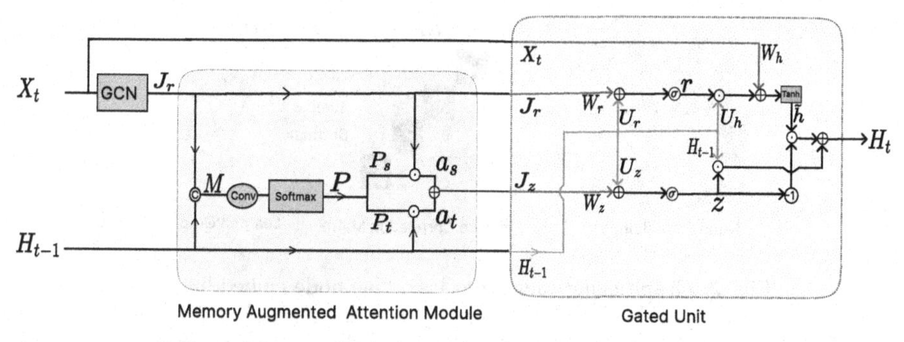

Fig. 1. Overall architecture of STLGRU designed for multivariate traffic forecasting. Our model consists of a memory-augmented attention module and a gated unit, which capture the long-range local and global dependencies. It takes input from a single time step with an initial hidden state and outputs a hidden state for the next time step.

3 Proposed Model

3.1 Preliminaries and Problem Definition

A graph $\mathcal{G} = (\mathcal{V}, \mathcal{E})$ models the traffic topological network, where $\mathcal{V}$ represents nodes and $\mathcal{E}$ signifies edges. An edge $e_{ij} \in \mathcal{E}$ connects nodes v_i and v_j, where each node has junctional features (e.g. inflow, outflow). Then, we define the spatio-temporal traffic data forecasting problem using a mapping function f_θ, where it takes the historical series $\langle X_{(t-T+1)}, X_{(t-T+2)}, \dots, X_t \rangle$. And, it predicts the future series $\langle X_{(t+1)}, X_{(t+2)}, \dots, X_{(t+T')} \rangle$, where T is the length of the historical series and T' is the length of the target forecast series, and, $X_i \in \mathbb{R}^{N \times C}$, where N is the number of nodes and C is the number of information channels (speed, flows, etc.). Thus, the time series forecasting model can be defined as follows:

$$\langle X_{(t-T+1)}, X_{(t-T+2)}, \dots, X_t \rangle \xrightarrow{f_\theta} \langle X_{(t+1)}, X_{(t+2)}, \dots, X_{(t+T')} \rangle$$

3.2 Graph Convolution

Here, we first define the graph convolution, where the initial input matrix is denoted as $X \in \mathbb{R}^{N \times T \times C}$. As our focus is solely on traffic flow, C is consequently set to 1, resulting $X \in \mathbb{R}^{N \times T \times 1}$. And, we take $X_{t'} \in \mathbb{R}^{N \times 1}$ as input from a single time step t, where $t \in T$, and pass it through a convolutional layer ξ_θ to transform the input feature into high-dimensional space C' to increase the representation power of the network as follows:

$$X_t = \xi_\theta \left(X_{t'} \right); \theta \in \mathbb{R}^{1 \times C'} \tag{1}$$

Then, $X_t \in R^{N \times C'}$ is used as an input to the original network at time step t.

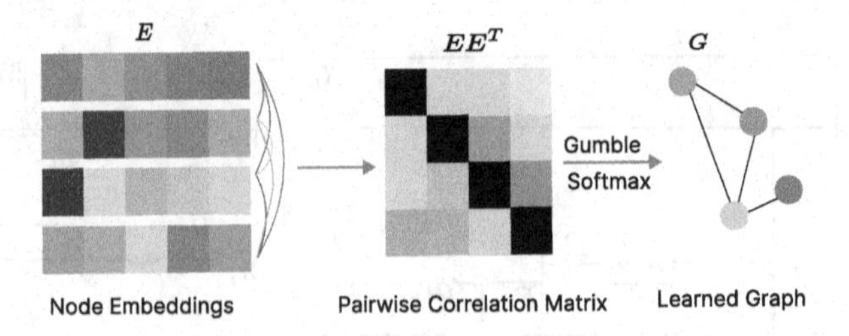

Fig. 2. Graph generation from learnable node embeddings E

As shown in Fig. 2, let $E \in \mathbb{R}^{N\times d}$ be the learned node embedding matrix, where d represents the embedding dimension. In addition, Ω represents the probability matrix, and each $\Omega_{ij} \in \Omega$ corresponds to the probability of preserving the edge between time series i and j, respectively. This relationship is formally expressed as follows:

$$\Omega = EE^T \tag{2}$$

In particular, we use the Gumbel softmax method [13] to obtain the final sparse adjacency matrix $A \in R^{N\times N}$ to effectively assure a sufficient amount of sparsity in the graph structure. And, let σ and τ be the activation function and the temperature variable, respectively. Then, we can define sparse adjacent matrix A as follows:

$$\begin{aligned} A = \sigma((log(\Omega_{ij}/(1-\Omega_{ij}) + (n^1_{ij} - n^2_{ij})/\tau) \\ s.t.\ n^1_{ij}, n^2_{ij} \sim Gumbel(0,1) \end{aligned} \tag{3}$$

Equation (3) implements Gumbel Softmax for our task, where $A_{i,j} = 1$ with probability $\Omega_{i,j}$ and $A_{i,j} = 0$ with the remaining probability. In particular, Gumbel Softmax maintains the same probability distribution as the normal Softmax, ensuring statistical consistency in generating the trainable probability matrix for the graph forecasting network. Next, let I be an identity matrix and D be a diagonal degree matrix satisfying $D_{ii} = \Sigma_j A_{ij}$. Then, the specific operation of graph convolution network (GCN) with the learnable weight $W \in R^{C'\times C'}$ can be expressed as follows:

$$GCN\,(X_t) = W(I + D^{-\frac{1}{2}}AD^{-\frac{1}{2}})X_t \in \mathbb{R}^{N\times C'} \tag{4}$$

3.3 Memory-Augmented Attention (MAA) Module

As discussed before, many state-of-the-art models struggle with maintaining a lightweight design. For instance, models proposed by Jiang et al. [1] and Yu et al. [2] have learned spatio-temporal relations by combining GCN and GRU modules and they further stack these fused modules multiple times. However, when stacking multiple layers to capture long-term dependencies in traffic data, they encounter a significant increase in memory usage during inference. To mitigate this issue, we introduce a memory-augmented attention (MAA) mechanism by continuously synchronizing relevant features in both spatial and temporal data in each timestep. In particular, Fig. 1 illustrates the MAA module's structure, where it combines the graph convolution output $J_r \in \mathbb{R}^{N\times C'}$ with the randomly initialized hidden input $H_{t-1} \in \mathbb{R}^{N\times C'}$ through concatenation, and pass it through a convolutional function as follows:

$$J_r = GCN\left(X_t\right), \tag{5}$$

$$M = J_r \oplus H_t, \tag{6}$$

$$P = \psi_w\left(M\right), \tag{7}$$

where ψ is a 1D convolutional function with parameter $w \in \mathbb{R}^{C'\times C'}$, and M and P both have the same dimension of $\mathbb{R}^{2N\times C'}$. We use the softmax specified by $P_{u,v}$ to calculate the attention score for both spatial characteristics from the GCN output and temporal features from the hidden input in the following way:

$$P_{u,v} = \frac{\exp P_{u,v}}{\sum_{v=1}^{N} \exp P_{u,v}}, u, v \in \{1, 2, \ldots, N\}. \tag{8}$$

After the above step, $P \in \mathbb{R}^{2N\times C'}$ is divided into P_s and P_t, which have the same size, $(P_s, P_t) \in \mathbb{R}^{N\times C'}$. Next, we element-wise multiply P_s and P_t with J_r and H_{t-1} respectively, as follows:

$$a_s = P_s \odot J_r, \tag{9}$$

$$a_t = P_t \odot H_{t-1}, \tag{10}$$

where $\odot$ represents the Hadamard product. Rather than exclusively representing spatial context, a_s also includes temporal information for a specific timestamp, while a_t serves a similar dual role, encompassing both spatial and temporal context. We then add these two context vectors, finally producing J_z as follows:

$$J_z = a_s + a_t; J_z \in \mathbb{R}^{N\times C'} \tag{11}$$

3.4 Memory Updating

Prior traffic forecasting models [10,11,28] often use graph and temporal convolution independently, overlooking the heterogeneities within spatial-temporal data. To tackle this problem, our approach involves a continuously synchronized gating mechanism to update the hidden state H_t, allowing MAA to capture long-range dependencies across both spatial and temporal domains effectively. The update process is defined as follows:

$$g = \sigma\left(W_z \cdot J_z + U_z \cdot H_{t-1}\right), \tag{12}$$

$$r = \sigma\left(W_r \cdot J_r + U_r \cdot H_{t-1}\right), \tag{13}$$

$$\tilde{h} = \tanh\left(W_h \cdot X_t + r * U_h \cdot H_{t-1}\right), \tag{14}$$

$$H_t = g * H_{(t-1)} + (1-g) * \tilde{h}, \tag{15}$$

where $(W, U) \in \mathbb{R}^{C' \times C'}$ are the learnable parameters and σ is the sigmoid function. Compared with the original memory cell in the GRU [7] that is updated only by current input X_t and previous hidden state H_{t-1}, Our proposed memory cell updates based on the original input X_t, graph convolution J_r, aggregated context vector J_z, and the previous hidden state H_{t-1}, which effectively captures both local and global spatio-temporal dependencies in real-time.

On the other hand, similar to the standard GRU mechanism, we use the final output at the last time step, denoted as $H_T \in R^{N \times C'}$, and process it through two fully connected layers for prediction as follows:

$$\hat{\mathcal{Y}} = \mathrm{Re}\,LU\left(H_T W_1 + b_1\right) \cdot W_2 + b_2, \tag{16}$$

where $\hat{\mathcal{Y}} \in \mathbb{R}^{N \times T'}$ denotes the prediction of the overall network, and $W_1 \in \mathbb{R}^{C' \times C'}, b_1 \in \mathbb{R}^{C'}, W_2 \in \mathbb{R}^{C' \times T'}$, and $b_2 \in \mathbb{R}^{T'}$ are learnable parameters. Finally, to train the model, we use the loss function as follows:

$$\mathcal{L}(\theta) = \left\| \widetilde{\mathcal{Y}} - \hat{\mathcal{Y}} \right\|_2^2 \tag{17}$$

where $\widetilde{\mathcal{Y}}$ denotes the ground truth and $\hat{\mathcal{Y}}$ denotes the prediction of the model, respectively.

4 Experimental Results and Analysis

Datasets. We perform experiments on three publicly available popular benchmark traffic datasets, which are PeMSD4, PeMSD7, and PeMSD8 from California Transportation Agencies [3]. In these datasets, each vertex on the graph represents a sensor node to collect the traffic flow data and the flow data is aggregated to 5 min. Thus, each hour has 12 data points in the flow data. We apply zero-mean normalization for preprocessing these datasets.

Baselines. We compare our proposed *STLGRU* against the following popular as well as SoTA baseline models on spatio-temporal prediction task: 1) Spatial-temporal synchronous modeling mechanism (STSGCN [23]), 2) Spectral Temporal Graph Neural Network for time series forecasting (StemGNN [1]), 3) Time Zigzags at Graph Convolutional Networks (Z-GCNETs [5]), 4) Graph-Wavenet (GW-Net [26]), 5) Pattern Matching Memory Networks (PM-MemNet [15]), and 6) Meta-Graph Convolutional Recurrent Network (Mega-CRN [14]). We use default settings for each baseline when performing comparisons.

Evaluation Metrics. We apply three widely used metrics to evaluate the performance of our model, (1) Mean Absolute Error (MAE), (2) Mean Absolute Percentage Error (MAPE), and (3) Root Mean Squared Error (RMSE).

Implementation Details. We divide all the datasets with a ratio 6:2:2 into training, testing, and validation sets, respectively. We use Adam optimizer with a learning rate of 0.001 and set 16 as the batch size. We conduct experiments

Table 1. The overall performance of STLGRU and baseline methods.

Datasets	Model	15 min			30 min			60 min		
		MAE	RMSE	MAPE	MAE	RMSE	MAPE	MAE	RMSE	MAPE
PeMSD4	STSGCN	19.41	30.69	14.82	21.83	31.33	15.54	23.19	33.65	16.90
	StemGNN	20.24	28.15	13.03	20.68	30.88	14.21	22.92	33.74	15.65
	Z-GCNETs	19.50	28.61	12.78	23.21	30.09	13.12	29.24	32.95	16.14
	GW-Net	18.15	25.24	13.27	22.12	30.62	16.28	21.85	33.70	17.29
	PM-MemNet	18.95	30.16	13.79	20.01	31.47	14.17	26.85	32.14	17.21
	Mega-CRN	19.25	24.88	12.72	19.60	25.96	13.84	22.82	26.33	14.87
	STLGRU(Ours)	**17.59**	**23.24**	**11.02**	**18.73**	**24.61**	**12.85**	**21.05**	**25.41**	**13.87**
PeMSD7	STSGCN	16.17	23.15	16.51	22.19	34.87	19.88	24.26	39.03	20.21
	StemGNN	15.77	22.68	13.97	22.38	33.69	18.99	24.54	34.41	19.45
	Z-GCNETs	15.64	25.19	15.47	23.78	33.64	19.05	26.12	34.78	23.47
	GW-Net	18.74	26.14	16.58	23.64	34.82	24.65	24.15	34.12	29.02
	PM-MemNet	15.25	24.14	15.17	21.12	34.41	19.97	25.39	33.50	21.29
	Mega-CRN	14.23	21.05	13.11	22.86	33.19	18.40	23.55	33.54	19.29
	STLGRU(Ours)	**13.79**	**19.12**	**12.31**	**20.89**	**31.45**	**15.56**	**23.06**	**32.19**	**19.12**
PeMSD8	STSGCN	15.97	23.14	14.79	16.45	24.78	18.47	19.13	29.80	18.96
	StemGNN	15.83	24.93	10.26	15.95	23.88	19.98	24.10	28.13	23.79
	Z-GCNETs	15.76	25.11	10.01	15.64	23.29	16.67	17.55	29.67	19.19
	GW-Net	14.95	24.92	12.79	15.92	24.99	18.97	17.69	28.92	22.67
	PM-MemNet	14.10	22.15	10.41	16.65	24.17	13.77	19.13	28.16	16.68
	Mega-CRN	14.07	22.53	9.54	16.10	22.42	17.97	18.12	27.29	21.05
	STLGRU(Ours)	**13.93**	**20.94**	**8.84**	**15.03**	**22.18**	**12.64**	**16.83**	**26.35**	**14.74**

with our model using non-overlapping time windows in the time series data. The entire experiments are run on a single GPU (Nvidia TITAN RTX). If the test scores of a baseline are unknown for a dataset, we run their publicly available code based on their suggested settings to obtain the results.

Results. Table 1 compares the performance of our model to the baseline models in 15, 30 and 60 min traffic forecasting, respectively. As shown in Table 1, our model outperforms all of the baseline models in both long and short-term forecasting. StemGNN, Z-GCNET, STSGCN, GW-Net, and PM-MemNet stack multiple layers of spatio-temporal modules by optimizing a probabilistic graph model. Our proposed method demonstrates improvements over the comparative models, achieving an average increase of 2.7%, 3.1%, and 2.3% in MAE, RMSE, and MAPE, respectively. Mega-CRN, which utilizes trainable adjacency matrices to understand node relationships and employs an encoder-decoder structure to manage traffic data heterogeneity, is also surpassed by our STLGRU model. Overall, STLGRU demonstrates superior performance, exhibiting average improvements of 2.9%, 3.1%, and 2.6% in MAE, RMSE, and MAPE, respectively.

Furthermore, Table 2 presents the maximum memory footprint, computational complexity, and the number of parameters of the baseline models on PeMSD4 dataset. Because we use the same model for each dataset, we present

Table 2. In-depth comparison of different model efficiency on PeMSD4. We show that STLGRU achieves high memory efficiency with less computational power and parameters in all three datasets. The second best is shown with underline (See Supp. for more results with additional datasets).

Model	Memory (MB)	FLOPs	Parameters
STSGCN	<u>1028</u>	282.24G	<u>550.48K</u>
GW-Net	1031	<u>189.16G</u>	610.25K
StemGNN	1220	378.98G	1.64M
Z-GCNETs	1473	389.49G	1.08M
Mega-CRN	1409	311.97G	669.14K
PM-MemNet	1052	421.49G	1.34M
STLGRU (Ours)	**990**	**77.93G**	**348.54K**

the experimental results with one dataset. Results with additional datasets are provided in Suppl. To compute a model's GPU memory usage during inference, we use the Linux command line "gpustat" with a minibatch size of 1 and with no gradients. We can observe that *STLGRU* requires the least memory during inference than baseline models. It also has the least computation complexity and number of parameters. In Table 2, we present the memory footprint, computational complexity, and the number of parameters used to train *STLGRU*. Our model stands out, as it demands the least memory, with fewer parameters, and ours exhibits reduced computational complexity. This efficiency positions our model as an ideal choice for integration into real-world, low-powered devices.

5 Ablation Study

We verify the effectiveness of *STLGRU* with additional ablation experiments. We dissect our model and focus on two main components: the Gumbel softmax and the memory augmented attention (MAA). As illustrated in Table 3, the absence of MAA leads to a remarkable decline in performance. The role of Gumbel softmax is pivotal in ensuring optimal sparsity within the graph. When we substitute Gumbel softmax with only the learnable embedding matrix, there is a noticeable decline in our model's performance. However, this is not surprising, given that irrelevant connections can reduce the model's ability to capture the dynamic interrelations between nodes accurately.

Afterwards, as demonstrated in Fig. 3, we compare our model against traditional spatio-temporal configurations. Specifically, we evaluate (1) Graph convolution for capturing spatial knowledge and 1D CNN to capture temporal dependencies, (2) Graph convolution for capturing spatial knowledge and LSTM to capture temporal dependencies, (3) Graph convolution for capturing spatial knowledge and vanilla GRU to capture temporal dependencies. From our observations in Fig. 3, it is evident that *STLGRU* consistently outperforms other methods significantly. We thus argue that memory-augmented attention can

Table 3. Ablation study for the effectiveness of the memory augmented attention (MAA) and gumble softmax module used in our method.

Gumble Softmax	MAA	Error Score (MAE)
×	×	23.12
×	✓	21.74
✓	×	19.83
✓	✓	**16.83**

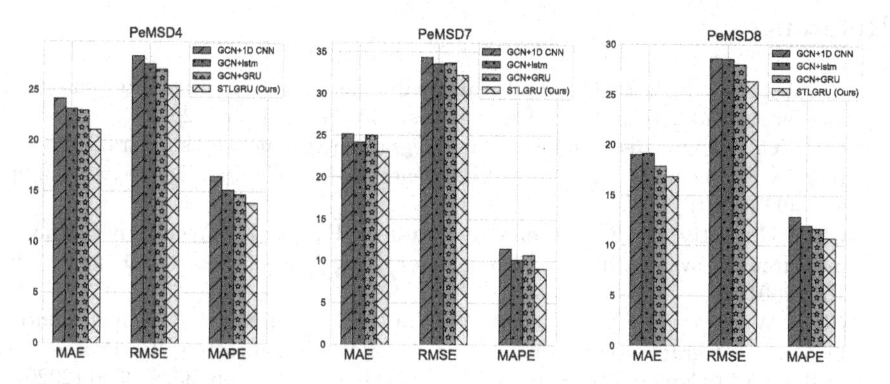

Fig. 3. Performance comparison of spatio-temporal models and *STLGRU* with different settings. MAE, RMSE and MAPE of 1-hour forecasting on three datasets are plotted.

capture more fine-grained spatio-temporal patterns and trace the crucial inter-dependencies among the road network.

6 Conclusion

In this work, we introduce *STLGRU*, a uniquely lightweight and efficient model for traffic flow prediction task. Our model incorporates a memory module enhanced with attention mechanism, capable of synchronizing spatial correlations within node networks and long-term temporal patterns in a continuous manner. Our experimental results showcase its superior performance across three benchmark traffic prediction datasets while maintaining a significantly reduced computational overhead compared to baseline models. For future work, we plan to adapt *STLGRU* for other spatial-temporal forecasting challenges, and explore how to model spatio-temporal dependencies when long-term data is scarce.

Acknowledgement. This work was partly supported by Institute for Information & communication Technology Planning & evaluation (IITP) grants funded by the Korean government MSIT: (No. 2022-0-01199, Graduate School of Convergence Security at Sungkyunkwan University), (No. 2022-0-01045, Self-directed Multi-Modal Intelligence for solving unknown, open domain problems), (No. 2022-0-00688, AI Platform to Fully Adapt and Reflect Privacy-Policy Changes), (No. 2021-0-02068, Artificial Intelligence Innovation Hub), (No. 2019-0-00421, AI Graduate School Support Program at Sungkyunkwan University), and (No. RS-2023-00230337, Advanced and Proactive AI Platform Research and Development Against Malicious Deepfakes). Lastly, this work was supported by Korea Internet & Security Agency (KISA) grant funded by the Korea government (PIPC) (No. RS-2023-00231200, Development of personal video information privacy protection technology capable of AI learning in an autonomous driving environment).

References

1. Cao, D., et al.: Spectral temporal graph neural network for multivariate time-series forecasting. Adv. Neural. Inf. Process. Syst. **33**, 17766–17778 (2020)
2. Chen, C., et al.: Gated residual recurrent graph neural networks for traffic prediction. In: Proceedings of the AAAI Conference on Artificial Intelligence, vol. 33, pp. 485–492 (2019)
3. Chen, C., Petty, K., Skabardonis, A., Varaiya, P., Jia, Z.: Freeway performance measurement system: mining loop detector data. Transp. Res. Rec. **1748**(1), 96–102 (2001)
4. Chen, W., Chen, L., Xie, Y., Cao, W., Gao, Y., Feng, X.: Multi-range attentive bicomponent graph convolutional network for traffic forecasting. In: Proceedings of the AAAI Conference on Artificial Intelligence, vol. 34, pp. 3529–3536 (2020)
5. Chen, Y., Segovia, I., Gel, Y.R.: Z-gcnets: time zigzags at graph convolutional networks for time series forecasting. In: International Conference on Machine Learning, pp. 1684–1694. PMLR (2021)
6. Cheng, W., Shen, Y., Zhu, Y., Huang, L.: A neural attention model for urban air quality inference: Learning the weights of monitoring stations. In: Proceedings of the AAAI Conference on Artificial Intelligence, vol. 32 (2018)
7. Chung, J., Gulcehre, C., Cho, K., Bengio, Y.: Empirical evaluation of gated recurrent neural networks on sequence modeling. arXiv preprint arXiv:1412.3555 (2014)
8. Deb, T., Sadmanee, A., Bhaumik, K.K., Ali, A.A., Amin, M.A., Rahman, A.: Variational stacked local attention networks for diverse video captioning. In: Proceedings of the IEEE/CVF Winter Conference on Applications of Computer Vision, pp. 4070–4079 (2022)
9. Diao, Z., et al.: A hybrid model for short-term traffic volume prediction in massive transportation systems. IEEE Trans. Intell. Transp. Syst. **20**(3), 935–946 (2018)
10. Fang, S., Zhang, Q., Meng, G., Xiang, S., Pan, C.: GSTNet: global spatial-temporal network for traffic flow prediction. In: IJCAI, pp. 2286–2293 (2019)
11. Guo, S., Lin, Y., Feng, N., Song, C., Wan, H.: Attention based spatial-temporal graph convolutional networks for traffic flow forecasting. In: Proceedings of the AAAI Conference on Artificial Intelligence, vol. 33, pp. 922–929 (2019)
12. He, P., Jiang, G., Lam, S.K., Tang, D.: Travel-time prediction of bus journey with multiple bus trips. IEEE Trans. Intell. Transp. Syst. **20**(11), 4192–4205 (2018)
13. Jang, E., Gu, S., Poole, B.: Categorical reparameterization with gumbel-softmax. arXiv preprint arXiv:1611.01144 (2016)
14. Jiang, R., et al.: Spatio-temporal meta-graph learning for traffic forecasting. In: Proceedings of the AAAI Conference on Artificial Intelligence, vol. 37, pp. 8078–8086 (2023)

15. Lee, H., Jin, S., Chu, H., Lim, H., Ko, S.: Learning to remember patterns: pattern matching memory networks for traffic forecasting. arXiv preprint arXiv:2110.10380 (2021)
16. Lee, W.H., Tseng, S.S., Shieh, J.L., Chen, H.H.: Discovering traffic bottlenecks in an urban network by spatiotemporal data mining on location-based services. IEEE Trans. Intell. Transp. Syst. **12**(4), 1047–1056 (2011)
17. Li, Y., Yu, R., Shahabi, C., Liu, Y.: Diffusion convolutional recurrent neural network: data-driven traffic forecasting. arXiv preprint arXiv:1707.01926 (2017)
18. Liebig, T., Piatkowski, N., Bockermann, C., Morik, K.: Dynamic route planning with real-time traffic predictions. Inf. Syst. **64**, 258–265 (2017)
19. Lin, Z., Li, M., Zheng, Z., Cheng, Y., Yuan, C.: Self-attention convlstm for spatiotemporal prediction. In: Proceedings of the AAAI Conference on Artificial Intelligence, vol. 34, pp. 11531–11538 (2020)
20. Mahmud, S., et al.: Human activity recognition from wearable sensor data using self-attention. arXiv preprint arXiv:2003.09018 (2020)
21. Niloy, F.F., Amin, M.A., Ali, A.A., Rahman, A.M.: Attention toward neighbors: a context aware framework for high resolution image segmentation. In: 2021 IEEE International Conference on Image Processing (ICIP), pp. 2279–2283. IEEE (2021)
22. Niloy, F.F., Bhaumik, K.K., Woo, S.S.: CFL-net: image forgery localization using contrastive learning. arXiv preprint arXiv:2210.02182 (2022)
23. Song, C., Lin, Y., Guo, S., Wan, H.: Spatial-temporal synchronous graph convolutional networks: a new framework for spatial-temporal network data forecasting. In: Proceedings of the AAAI Conference on Artificial Intelligence, vol. 34, pp. 914–921 (2020)
24. Vaswani, A., et al.: Attention is all you need. In: Advances in Neural Information Processing Systems, vol. 30 (2017)
25. Veličković, P., Cucurull, G., Casanova, A., Romero, A., Lio, P., Bengio, Y.: Graph attention networks. arXiv preprint arXiv:1710.10903 (2017)
26. Wu, Z., Pan, S., Long, G., Jiang, J., Zhang, C.: Graph wavenet for deep spatial-temporal graph modeling. arXiv preprint arXiv:1906.00121 (2019)
27. Yang, Q., Koutsopoulos, H.N., Ben-Akiva, M.E.: Simulation laboratory for evaluating dynamic traffic management systems. Transp. Res. Rec. **1710**(1), 122–130 (2000)
28. Yu, B., Yin, H., Zhu, Z.: Spatio-temporal graph convolutional networks: a deep learning framework for traffic forecasting. arXiv preprint arXiv:1709.04875 (2017)
29. Zhang, J., Wang, F.Y., Wang, K., Lin, W.H., Xu, X., Chen, C.: Data-driven intelligent transportation systems: a survey. IEEE Trans. Intell. Transp. Syst. **12**(4), 1624–1639 (2011)
30. Zhao, X., Fan, W., Liu, H., Tang, J.: Multi-type urban crime prediction. In: Proceedings of the AAAI Conference on Artificial Intelligence, vol. 36, pp. 4388–4396 (2022)

[illegible] S. [illegible] [illegible] [illegible] (2016).

[illegible] S. [illegible] [illegible] [illegible] 40 [illegible] 11 [illegible]

[illegible] [illegible] (2019).

[illegible] [illegible] [illegible] (2019).

[illegible] [illegible] (2020).

[illegible] [illegible] 2021 [illegible] (2022).

[illegible]

[illegible]

[illegible] (2017).

[illegible] (2019).

[illegible] [illegible] 2016.

[illegible] [illegible] (2020).

Author Index

D.-N. Yang et al. (Eds.): PAKDD 2024, LNAI 14650, pp. 301–302, 2024.
https://doi.org/10.1007/978-981-97-2266-2

Printed in the United States
by Baker & Taylor Publisher Services